Getting It Right in Science and Medicine

Hans R. Kricheldorf

Getting It Right in Science and Medicine

Can Science Progress through Errors? Fallacies and Facts

Hans R. Kricheldorf
Institute for Technical and Macromolecular Chemistry
Universität Hamburg
Hamburg, Germany

ISBN 978-3-319-30386-4 ISBN 978-3-319-30388-8 (eBook)
DOI 10.1007/978-3-319-30388-8

Library of Congress Control Number: 2016934591

Printed on acid-free paper

This Springer imprint is published by Springer Nature
The registered company is Springer International Publishing AG Switzerland

For my mother Orlaug Christie v. Aadna

Preface

This book may be understood as a second revised and augmented edition of the German version *Erkenntnisse und Irrtümer in Medizin und Naturwissenschaft* published by Springer Spektrum in 2014.

Hamburg, Germany — Hans R. Kricheldorf

Acknowledgments

The author wishes to thank Prof. Dr. Saber Chatti (INSA Institute, Lyon, France) for drawing the formulas and Dr. Norbert Czerwinski (TU Karlsruhe, Germany) for digitalization of all figures. The author feels particularly indebted to Dr. Tobias Wassermann (Springer, Heidelberg) for careful revision of the entire manuscript.

Contents

Part I
Insights and Definitions

Chapter 1
Introduction

> The Ten Commandments of God are so clear because their formulation was not influenced by a commission of experts.
> (Charles de Gaulle)

Comments on and critiques of science have come and continue to come from various groups of authors. For example, journalists and philosophers have developed an increasing tendency during the past 30 years to complain about scientific progress and its technical utilization, mainly considering negative consequences such as the rapid increase in allergies, environmental pollution, climate change, and consumption of landscape. These critics ignore the enormous benefits resulting from scientific discoveries and inventions, for example, with regard to food production and progress in medicine (see Sects. 3.2 and 5.3, and Chap. 6). Their attitude is also hypocritical, because their professional activities rely on scientific and technical innovations. For example, none of the vehicles used for transportation, such as bicycles, cars, trains, buses, steamships, and airplanes, grow in nature, and paper, ballpoints, computers, and printing machines are also the result of scientific discoveries and technical inventions.

Another group of critics, with a tradition of almost 500 years, are theologians, regardless of whether the Bible or the Koran forms their theoretical background. A detailed discussion of the numerous critical comments and arguments written by theologians over a period of 500 years is, of course, beyond the scope of this book. However, one important but rarely discussed point should be mentioned. All monotheistic religions have in common that God/Allah is considered to be the only creator of the universe and of all living organisms, including humankind. If so, scientists studying the properties of nature directly deal with God's/Allah's own work. In contrast, theologians exclusively study manuscripts and printed texts written by men or women, because not a single sentence exists that was written by God/Allah. Most frequently, theologians extract and interpret secondary and tertiary literature. Therefore, carefully interpreted scientific findings offer a closer and more trustworthy look at God's/Allah's work and intention. Certainly,

H.R. Kricheldorf, *Getting It Right in Science and Medicine*,
DOI 10.1007/978-3-319-30388-8_1

individual scientists may lack self-criticism and, thus, may be responsible for overinterpretations of scientific results, but such misbehavior occurs in all areas of human activity, including theology.

This book mainly focuses on discussing the skeptical comments and critiques of science contributed by two other groups of authors, namely historians or philosophers of science and theoretical scientists or experimental scientists. Historians, philosophers, and theoretical scientists like to discuss fundamental structures, limits of cognition, reliability, sense, and justification of scientific research. In numerous contributions to this field, the philosophy and theory of science are discussed without clear definition or description of what the term "science" really means. Does science, for instance, include sociology and anthropology, or even the humanities? Therefore, to avoid misunderstandings the meaning of science, as it is understood in this book, is defined in the first section of Chap. 2.

Typical consequences of insufficient differentiation or lack of definition are comments and conclusions that sound strange when applied to the natural sciences. For example, the British philosopher Stephen Toulmin (1922–2009) wrote in his book *Foresight and Understanding – an Inquiry into the Aim of Science* (p. 62): "Just as the question 'Is this music good of its kind?' is distinct from the question 'Is this good music?' so we find scientists asking both 'Is this event a natural and self-explanatory one of its kind?' and also 'Is this an example of the most natural and self-explanatory sort?'" On pp. 15 and 16, he wrote: "It is, in fact, doubtful whether a final account could ever be given of the aims of science: especially one which was both exhaustive and brief. … Science has not one aim but many, and its development has passed through many contrasted stages. … There is no universal recipe for all science and all scientists any more than there is for all cakes and all cooks. There is much in science which cannot be created according to set rules and methods at all. … Science as a whole – the activity, the aims, its methods and ideas – evolve by variation and selection." The response of the author of this book is given in Sects. 2.1–2.5.

Another example of a funny description of science can be found in the work of the French historian Jacques Barzun (1907–2012). One of his books is entitled *Science: The Glorious Entertainment* and on p. 77 he says about scientists: "What science is in their view amounts to an earthly translation of the kingdom of heaven where fitness and perfection rule and nothing is other than it seems." This and other statements by Barzun are discussed in Sect. 4.1.

One more example of a work that suffers from a lack of definitions and differentiation is the famous book *Against Method* by the Austrian philosopher Paul Feyerabend (1924–1992). Going through the text, the reader must learn step by step that arguments and conclusions have various roots, ranging from astronomy and physics to sociology and historical research. His stance is discussed in Sect. 4.1. Examples extracted from the fourth edition (pp. ixxx) are given below:

> Science is an essentially anarchic enterprise: theoretical anarchism is more humanitarian and more likely to encourage progress than its law-and-order alternative.

> There is no idea, however ancient and absurd, that is not capable of improving our knowledge. The whole history of thought is absorbed into science and is used for improving every single theory. Nor is political interference rejected. It may be needed to overcome the chauvinism of science that resists alternatives to the status quo.

> This is shown both by examination of historical episodes and by an abstract analysis of the relationship between idea and reaction. The only principle which does not inhibit progress is: Anything Goes!.

Over the past 200 years, numerous sociologists, but also biologists and philosophers, have attacked reductionist tendencies in science without condemning science as a whole. Depending on the working field, reductionism is a philosophical or scientific position that claims that a complex system is nothing but the sum of its components and that an account of it can be reduced to accounts of its individual constituents. A fundamental criticism of science relies on the assumption that science is per se and automatically reductionist, which is certainly an exaggeration. In Sect. 4.2, the advantages and disadvantages of reductionist concepts in science are commented on in more detail.

Fundamental criticism of the trustworthiness of empirical research was presented by the philosophers Ludwig Wittgenstein (1889–1951) and, above all, Karl Popper (1902–1994). Those theorists, their coworkers, and followers believed that inductive conclusions are not reliable and that any hypothesis or theory may be falsified . Hence, the philosophical approach of this group of skeptic theorists was called falsificationism. Popper also described science as a theory of theories. Sections 2.4 and 3.3 are devoted to the falsification of their critique.

Yet, fundamental criticism concerning the reliability of empirical research and scientific knowledge also comes from experimental scientists, mainly from physicists. The following statement by Max Born (1882–1970), awarded the Nobel Prize for Physics in 1954, is characteristic: “Ideas such as absolute certainty, absolute accuracy, final truth, and so forth are inventions of the human imagination and should be avoided in science.”

However, strange comments on and fundamental criticisms of science were not only uttered by physicists and theorists infected by physics, but also by biologists. For example, the American professor of biology Robert Shapiro says in his book *Origins* (dealing with the origin of life on earth) in a chapter entitled “Science, Realm of Doubt” (p. 33):

> I have chosen this title to make the strongest possible contrast between the common view of science described above and its essence. Science is not a given set of answers, but a system for obtaining answers. The method by which the search is conducted is more important than the nature of solution. Questions need not be answered at all, or answers may be provided and then changed. It does not matter how often or profoundly our view of the universe alters, as long as these changes take place in a way appropriate to science. For the practice of science, like the game of baseball, is covered by definite rules.

This characterization of science needs opposition for two reasons. First, if asking and answering, including repetitive modification of answers (insights), does not have any final target, then science is nothing more than a scholarly but pseudoscientific social game and a gigantic waste of tax revenues. The primary aim is certainly a reliable and as precise as possible description and analysis of natural phenomena. The second aim is the utilization of discoveries and inventions to improve the welfare and prosperity of humankind. Furthermore, Shapiro ignores the roots of science. Since emancipation from apes, humankind has striven to learn more and more about regular processes in nature and to extrapolate experience into the future (see Chap. 3). This capability of the human brain, existing at a lower level in the brains of other mammals, was not designed by evolution to play games, but as a strategy to enable survival for at least a few million years in a world dominated by microbes, arthropods (insects), and natural catastrophes.

Finally, the latest book by the biologist Rupert Sheldrake should be mentioned, which appeared in 2013 under the title *The Science Delusion*. This book contains numerous critical questions, but almost no suggestions of better alternatives. It contains the following statement (p. 6): "In this book I argue that science is being held back by centuries-old assumptions that have hardened into dogmas. The sciences would be better off without them: freer, more interesting, and more fun. The biggest scientific delusion of all is that science already knows the answers. The details still need working out, but, in principle, the fundamental questions are settled."

A similar statement was presented decades ago by the philosopher Ludwig Wittgenstein (1889–1951) in his *Tractatus* (6, 371–372):

> The whole modern conception of the world is founded on the illusion that the so-called laws of nature are the explanations of natural phenomena. Thus, people today stop at the laws of nature, treating them as something inviolable just as God and Fate were treated in past ages.
>
> In fact, both are right and both wrong: though the view of the ancients is clearer insofar as they have a clear and acknowledged terminus, while the modern system tries to make it look as if everything were explained.

These statements have to be qualified as untruth. Wittgenstein's comment is perhaps a reflex on the scientific worldview of the physicists at the end of the nineteenth century (see below and Sect. 2.2). However, every modern scientist endowed with at least a minimum of self-critique knows that the sea of unknown and unexplored facts and theories is many orders of magnitude larger than the nutshell of knowledge in which he moves forward. Complementary to this, the Austrian scientist Adolf Pichler (1817–1900) merits the citation: "Scientific research is always on the move and will never come to an end."

On page 8 of his book *The Science Delusion* Sheldrake also states that medicine and science support the following dogma: "Mechanistic medicine is the only kind that really works." The author of the present book has never met any physician or surgeon who adhered to this dogma. Furthermore, Sheldrake ignores and defames self-understanding and the intentions of psychoanalysis, psychotherapy, and psychosomatic medicine (see Chap. 6).

The numerous partially strange, partially misleading, partially defaming, and partially incorrect comments on science delivered by various groups of authors have stimulated this author to revise and reshape the picture of science from the viewpoint of an experimental chemist. The author's view is based on 50 years of experience in experimental research and it is the view of a non-physicist. This second point deserves explanation.

In 1873 James C. Maxwell published in his textbook *A Treatise on Electricity and Magnetism* mathematical equations explaining the phenomena of electricity and magnetism, including a better understanding of the nature of light. Together with previously achieved results in the areas of mechanics, optics, and the nature of elements the physicists believed at the end of the nineteenth century that they could explain almost all the fundamental principles and properties of the world. They believed that physics is the leading branch of the natural sciences, and they felt called upon to explain to all other scientists and interested laics what the world looks like. The discoveries and calculations of Max Planck and Albert Einstein after the turn of the century required considerable revision of the existing picture of the world, but this revision also stimulated new important discoveries and insights in the fields of astronomy, cosmology, and physics. Hence, physicists kept the tendency to consider physics as the leading science.

Yet, as exemplarily demonstrated in Chaps. 2, 3, 4, and 5 and in Sect. 4.1, conclusions and interpretations based on physics and its history are not necessarily representative for all natural sciences and may even be misleading. Furthermore, the numerous revisions of seemingly established theories, which were necessary in the history of astronomy, cosmology, and physics, made a significant contribution to the fact that philosophers, theoreticians, and physicists themselves became skeptical about the reliability of insights and knowledge elaborated in all scientific disciplines.

From the viewpoint of the author, modern science provides worldwide a steady flow of data every day, which is accompanied on a numerically much, much lower level by a flow of minor and major errors, mistakes, and fallacies. Errors and mistakes are unavoidable, because scientists are not perfect robots. However, it is also characteristic of modern science that it involves a self-healing process. This means that the permanent flow of results and mistakes is accompanied by a flow of revisions of previous errors and mistakes. This automatism arises from the fact that any step into a new field is based on knowledge, methods, and materials acquired by previous research activities. In this way, all previous results are sooner or later reexamined, and errors and mistakes are revised. In other words, ongoing research has a "Janus character" looking into the future and checking the results of the past. As demonstrated in Part II, any revision of a big mistake automatically entails a big step forward.

In summary, this book serves two purposes:

First, it deals with the questions of what is meant by science and whether science can provide trustworthy information and knowledge despite numerous mistakes, errors, and fallacies (Part I).

Second, it illustrates with important examples selected from medicine and various natural sciences, how mistakes and errors were made and revised, thereby also shedding light on the history of science (Part II).

Bibliography

Barzun J (1964) Science: the glorious entertainment. Harper & Row, New York

Feyerabend P (2010) Against method. Verso, London, NY (first published by New Left Books, 1975)

Popper K (1959) The logic of scientific discovery, 1st edn. Hutchinson, London

Shapiro R (1987) Origins - a sceptic's guide to the creation of life on earth. Bantham Books, Toronto, NY

Sheldrake R (2013) The science delusion. Hodder & Stoughton, London

Toulmin S (1961) Foresight and understanding. Hutchinson, London

Wittgenstein (1921) Logisch-Philosophische Abhandlung. Annalen der Naturphilosophie 14 (translated by C. K. Ogden in 1922 under the title *Tractatus logico-philosophicus*)

Chapter 2
What Is the Meaning of Science?

2.1 How Can We Define Science?

> The strongest arguments prove nothing so long as the conclusions are not verified by experience. Experimental science is the queen of sciences and the goal of speculation.
> (Roger Bacon)

In his book *Asimov's New Guide to Science* the Russian author and science fiction expert Isaac Asimov (1920–1992) offered a plausible explanation for the origin of science, a shortened version of which is cited here (1987 edition, pp. 3–5):

> Almost in the beginning was curiosity: Early in the scheme of life, however, independent motion was developed by some organism. It meant a tremendous advance in the control of the environment. A moving organism no longer had to wait in stolid rigidity for food to come its way, but went out after it. Thus adventure entered the world—and curiosity. The individual that hesitated in the competitive hunt for food, that was overly conservative in its investigation, starved. Early on, curiosity concerning the environment was enforced as the price of survival. As organisms grew more intricate, their sense organs multiplied and became both more complex and more delicate. More messages of greater variety were received from and about the external environment. At the same time, there developed (whether as cause or effect we cannot tell) an increasing complexity of the nervous system, the living instrument that interprets and stores the data collected by the sense organs. . . . There comes a point, where the capacity to receive, store, and interpret messages from the outside world may outrun sheer necessity. An organism may be sated with food, and there may, at the moment, be no danger in sight. What does it do then? . . . If curiosity can, like any other human drive, be put to ignoble use—the prying invasion of privacy that has given the word its cheap and unpleasant connotation—it nevertheless remains one of the noblest properties of the human mind. For its simplest definition is 'the desire to know.' . . . Thus, the desire to know leads in successive realms of greater etherealization and more efficient occupation of the mind—from knowledge of accomplishing the useful, to knowledge of accomplishing the esthetic, to 'pure' knowledge.

Unfortunately, Asimov did not provide a compact definition of science. From this point of view, his book shares the character of many other books, essays, and articles dealing with science, as already mentioned above. The reluctance of many

H.R. Kricheldorf, *Getting It Right in Science and Medicine*,
DOI 10.1007/978-3-319-30388-8_2

authors to define science may have three reasons. First, several authors apparently do not understand the purpose, usefulness, and character of definitions correctly. A typical example is the following comment by Toulmin (on p. 18 of *Foresight and Understanding*): "Definitions are like belts. The shorter they are, the more elastic they have to be. A short belt reveals nothing about its wearer: by stretching, it can be made fit almost anybody. A short definition applied to a heterogeneous set of examples has to be expanded and contacted, qualified and reinterpreted, before it will fit every case." These words may be enjoyed by philosophers, but from the viewpoint of scientists a definition has neither to fit anybody, nor to accommodate anybody, nor to fit a heterogeneous set of examples. A definition has to provide a description that is as precise as possible of a term or phenomenon, avoiding misunderstandings and avoiding overlapping with terms that may look similar at first glance.

The second reason that certain authors might avoid a definition of science is because they believe that the meaning of science is clear and all readers have the same meaning and definition in mind. However, such a conviction stands in sharp contrast to the numerous different comments published about the nature and aims of science, as exemplarily demonstrated in the introduction to this book (Chap.1). Third, other authors perhaps do not dare to provide a definition because they are afraid of attracting criticism. Yet, if an author is a scientist and not willing to make clear statements, he should perhaps change his profession and turn to politics. The author of this book presents his understanding and definition of science here at the beginning of the text, not because of the assumption that he has found the optimum definition for all time, but to provide a precise basis for consistent discussions.

The term "science" as it is used throughout this book means the sum of all natural sciences, such as astronomy, biology, chemistry, geology, pharmacy, and physics. Because of characteristic differences in their methods of inquiry relative to those of the natural sciences, psychology, sociology, anthropology, and the humanities are not included in the term science as it is used in this book. After this primary definition, a secondary definition may follow:

Science means observation and description of all natural phenomena (including experiments in laboratories) and explanation of these phenomena on the basis of the laws of nature and their interactions.

Fundamental research is, in turn, defined as the search for laws of nature and for a better understanding of their consequences and interactions.

Other descriptions or definitions of science are discussed in Sects. 2.3, 2.4, 4.1, 5.1, and 5.2. To avoid misunderstandings, it should be emphasized at this point that the above definitions are not meant as justification for reductionalism as the sole intellectual strategy in scientific research. Aristotle's antireductionist conclusion "The whole is more than the sum of its parts" may be a law of nature limited to living organisms (see Sect. 4.2). Furthermore, the human consciousness may formulate questions concerning the entire universe or individual humans (e.g., what is the purpose of life?) that cannot be answered by scientific methods.

Because the above definition of science emphasizes the interaction of laws of nature or, in other words, the simultaneous influence of several laws on one phenomenon, this definition also includes a new branch of science, systems science. Systems science may be understood as a kind of "metascience" of all traditional natural sciences. The American theoretician George J. Klir wrote in the first chapter of his textbook *Facets of Systems Science*: "Systems science is that field of inquiry whose object of study are systems. ... To be made operational this definition requires that some broad and generally accepted characterization of a concept of a system is established. ... However, when separated from its specific connotations and uses the term system is almost never explicitly defined. ... To begin our search for a meaningful definition of the term system from a broad perspective let us consult a standard dictionary. We are likely to find that a system is a set or arrangement of things so related or connected as to form a unity or organic whole (*Webster's The New World Dictionary*), although different dictionaries may contain stylistic variations of this particular formulation. It follows from this commonsense definition that the term system stands in general for a set of some things and a relation among the things. Formally, we have:

$$\mathrm{S} = (\mathrm{T}, \mathrm{R})$$

whereas S, T, and R denote, respectively, a system of things distinguished within S and a relation (or possibly a set of relations) defined on T. Clearly the thinghood and systemhood properties of S reside in T and R respectively. ... For example, a collection of books is not a system, only a set. However, when we organize the books in some way, the collection becomes a system. ... From the standpoint of classical science, systems science is clearly cross-disciplinary. ... Classical science and systems science may be viewed as complementary dimensions of modern science."

Characteristic of the basic methodology of scientific research is the search for observations, measurements, and experiments that are reproducible regardless of their location, regardless of time, and regardless of the properties of the researcher (a restriction of this statement is discussed in Sect. 2.2). In this regard, natural sciences differ from all other sciences and research activities.

Although the individual branches of traditional science, such as biology and physics, apply the same fundamental methodology (as defined above) they differ in certain formal aspects, and for a proper understanding of Part I of this work it is useful to keep these differences in mind.

Importance for the Scientific View of Life and the Universe

As a result of their different working fields, the various branches of science made and continue to make considerably different contributions to the scientific worldview and to the self-understanding of humankind. Nowadays and in the near future, the most important contributions come from theoretical physics in combination with nuclear physics, from astronomy, from the theory of evolution in combination with genetics, and from cerebral research. Since the redefinition of the elements and the elimination of vitalism in the middle of the nineteenth century, chemistry has not made an important contribution (if molecular genetics is attributed to biology). However, in the future chemistry has a chance to deliver an extraordinarily important contribution, namely if it can prove or disprove that life on earth can spontaneously emerge from dead matter.

Importance for Everyday Life
Concerning the level of modern civilization and any progress made in medicine, chemistry has provided more important contributions than any other branch of science. For instance, with the exception of raw wood and stone, virtually all other materials are produced by chemical processes. More than 90 % of the food supply of the western civilization depends on the availability of fertilizers, antibiotics, and agrochemicals such as insecticides and fungicides. Furthermore, more than 90 % of all remedies and medicaments are produced by pharmaceutical companies. Moreover, the availability of electricity and all vehicles, from bicycles to airplanes, is based on the production of metals and polymers (e.g., plastics and elastomers; see Sects. 9.4 and 9.5) by chemical companies (see also Sect. 3.2 and Chap. 6).

Degree of Abstraction
The degree of abstraction is highest for physics, in general, and for theoretical physics, in particular. Working fields that are concerned with the description of natural phenomena, such as landscapes, sediments, habitats, and herds of animals, represent the lowest level of abstraction. This differentiation does not involve any value judgment. All branches of science began with observations and descriptions of natural phenomena, and astronomy demonstrates that observation and description are still an important kind of research activity in modern science.

Extent of Experimental Research in Laboratories Chemistry and physics form together one pole, because more than 95 %, perhaps even more than 99 %, of all empirical data result from experimental work in laboratories. Those working fields of biology, geology, and meteorology concerned with description of natural phenomena represent the opposite pole.

Frequency of Experiments
In this dimension, certain working fields of physics represent one pole and chemistry the opposite pole. This classification deserves an explanation. Physicists working with particle accelerators can typically perform 10–50 new experiments per year. A small group of physicists working on fundamental research with laser light can often only perform between two and ten new experiments per year, when a new apparatus or method is developed. As an example of a medium-sized working group of chemists at a university, the author presents his own group. This group usually comprises between 12 and 17 coworkers consisting of master students, Ph. D. students, postdocs, one or two technicians, and one assistant professor. Almost all experiments are conducted in standard glassware or simple reactors and most chemicals are available from chemical companies. On this basis, 500–800 experiments are performed every year, and each experiment entails at least two

measurements with the purpose of elucidating whether the experiment is successful or not.

These numbers are by no means extreme; this means that a research group of chemists can usually conduct 20–100 times more experiments per year than physicists working with particle accelerators or developing new, complex instruments and methods. These numbers do not imply any judgment about the value and importance of the working groups or experiments. However, a scientist who can perform or supervise several hundreds of experiments per year has two advantages. First, it is easier and usually much cheaper to check the reproducibility of important experiments. Second, the scientist is in a better position to observe routine aspects of scientific research. This means that it is easier to observe how and why errors and mistakes arise again and again, and it is easier to observe the self-healing mechanism as a consequence of ongoing research.

Finally, a much shorter differentiation between biology, chemistry, and physics, as found in the "Handy Guide to Modern Science" (see Internet), should be mentioned:

1. If it's green or wiggles, it's biology.
2. If it stinks, it's chemistry.
3. If it doesn't work, it's physics.

2.2 What Is a Law of Nature?

Laws have two sources, humans themselves and nature. Leaders of clans, kings and emperors, or democratic institutions such as parliaments enact laws to regulate the social life of people living together in small or large societies. Nature (in this book meaning the universe or the entire creation) presents all its structures and activities in the form of laws to those willing to find correlations between, and explanations for, natural phenomena and experiments in laboratories. Definitions of the term "law of nature" are usually absent from textbooks of biology, chemistry, and physics. *Encyclopedia Americana* offers the following definition:

> A scientific law is a general statement that purports to describe some general fact or regularity of the universe. For example, Newton's law of gravitation asserts that every pair of bodies exerts a mutual attraction directly proportional to the product of their masses and inversely proportional to the square of the distance between them. Any such regularity may be termed law of nature.

Although this definition is in principle correct and useful, it has two weak points. First, it suggests a confusion of regularity or rule, on the one hand, and law, on the other hand, a point discussed in more detail in Sect. 3.4. Second, it is focused on the universe and gives an example concerning a physical property of the universe. Yet, it is not clear to what extent the properties of living organisms are included. In this book, a law of nature is understood as a property of nature, including any kind of living or non-living object. Laws of nature are responsible for the reproducibility of phenomena, which within a certain frame of validity (see Sect. 2.3) are independent of time, location, and the properties of the researcher.

However, the German philosopher Emanuel Kant (1724–1994) and later other philosophers held that laws of nature are a property of the human brain and not of nature itself. Yet, this view is neither progressive nor helpful, because it ignores the fact that the human brain is itself part of nature. Philosophers have the tendency to believe that their brain came from somewhere outside the universe. However, the human brain is the result of a long evolution of the central nervous system, which exists in all higher animals. There are no facts indicating that evolution had the goal of producing philosophers. Evolution of the central nervous system had the purpose of supporting the survival of new species in their struggle for life and broadening the diversification of species. With the modern human brain, evolution has surprisingly developed an organ which allows nature to reflect itself. Therefore, the author prefers to say that laws are the language that nature uses for rational communication with the human brain. A quite similar view has already been formulated by the physicist and Nobel Prize laureate Werner Heisenberg (1901–1976): "Natural science does not simply describe and explain nature, it is part of the interplay between nature and ourselves."

Problems with the proper understanding of the term "law of nature" arose and still arise from the fact that the physicists of the nineteenth century overloaded this term with attributes that in the aftermath were all found to be incorrect or misleading. The physicist Erwin Schrödinger (1887–1967, Nobel Prize 1933) began his frequently cited and published inaugural speech at the Technical University of Zürich in 1922 with the following statement (p. 10 of the German edition translared by the author): "Laws of nature are obviously nothing more than a sufficiently confirmed regularity of phenomena. ... Physical research has unambiguously demonstrated over the past four or five decades that, at least in an overwhelming number of phenomena, the regularity and constancy of which have founded the postulate of causality, the common roots of their strict regularity are accidental events. ... Each physical phenomenon, for which strict regularity is observed, is based on the actions and motions of many thousands, mostly billions, of atoms and molecules. ... The simplest and most transparent example for a statistical understanding of laws of nature is the properties of gases, the discovery of which also marks the historic origin of statistical laws in science." In the subsequent text (pp. 11 and 12) Schrödinger discusses the kinetic theory of gases in detail and continues on p. 13: "I would be able to contribute and explain still a much larger number of experimentally and theoretically exactly studied phenomena, for example, that the uniform blue color of the sky results from random variation of the air density. Another example is the strictly regular decay of radioactive substances, which results from irregular decay of radioactive atoms, whereby it seems a matter of chance, which atom will decay soon, or tomorrow, or within 1 year."

Schrödinger certainly delivered a correct description of most, if not of all physical laws However, a law of nature is not a property of physicists. Schrödinger, like other physicists (see below), did not take into account that conclusions and interpretations elaborated in physics are not automatically valid in all branches of science. For instance, the many thousands of biochemical and physiological reactions underlying the biological functions of all living organisms are not based on random motions or statistical reactions of molecules. The contrary is true. The generation of a biological function or signal (e.g., synthesis of an enzyme within seconds) requires trillions and quadrillions of almost parallel reactions, whereby molecules of identical structure perform exactly the same reaction. Furthermore, all the different molecules that contribute to a single process generating a biological function or physiological signal react in a cooperative mode and never at random. The transformation of photons into signals of the optical nerve, partially described in Sect. 10.5, is a typical example of such a chain of coordinated reactions. At this point, but also with respect to the following text, the professor of evolutionary biology, Ernst Mayr (1904–2005) needs to be quoted: "Biology is not a second physics."

In the second half of the twentieth century and in the twenty-first century, the term "law of nature" has attracted much criticism, partially from philosophers (see Sects. 2.3 and 2.4), partially from theorists or historians (see Chap. 5), and partially from physicists. Most of this criticism is stimulated by the fact that the physicists of the nineteenth century overloaded this term with attributes typical for their physical world view. This historical scenario has two main roots. First, physics, above all astronomy, may be considered to be the oldest branch of modern science, and as a result of the influence of mathematics, it soon reached a high level of abstraction. Second, the physicists believed at the end of the nineteenth century that almost all fundamental and important laws of nature were known. This scientific world view is illustrated by the answer of Professor Jolly, physicist at the University of München, when the young Max Planck asked him in 1874 whether it makes sense to study physics (Max Planck was an excellent musician and considered studying classical music). Jolly answered no, "because in this branch of science almost all important aspects are explored and only a few minor problems are still open." In other words, the physicists, but not only physicists, at that time considered physics to be the leading and representative branch of science (see Sect. 4.1). This mentality entailed at least three important fallacies, which are discussed next.

First, the physicists considered that only physical laws were fundamental laws of nature. The numerous laws found by biologist, chemists, geologists, and other scientists were called biological, chemical, or geological laws. They were at best third-rate laws of nature. From Hermann Helmholtz (1821–1894) the following statement is known: "The final aim of all kinds of science is to find mechanical explanations." Even after 1900, Sir Ernest Rutherford (see Sect. 10.2) remarked that "All science is either physics or stamp collection."

Presumably, not all physicists shared this short-sighted and arrogant view. Nonetheless, these comments are certainly representative of the mentality of physicists at the end of the nineteenth century. This narrow-minded view of science prevented physicists from becoming aware of the following issues.

Second, for physicists of the nineteenth century, correct understanding of the laws of nature included necessarily a mathematical formula. Forerunners of this mentality were Thales of Miletus (624–546 B.C.) and Galileo Galilei (1564–1642). Galilei published in 1623 in his book *Il Saggiatore* the following statement (equivalent translation): "Mathematics (he was focused on geometry) is the alphabet God used to write the book of the universe." In 1786, the philosopher Emanuel Kant (1724–1804) wrote in his work *Metaphysische Anfangsgründe der Naturwissenschaft* (Metaphysical Foundations of Natural Science) the following insight: "I declare any philosophy or theory of science contains only that much science as it contains mathematics."

In his latest book *Gottes Würfel* (The Dice of God) the German professor of physics Helmut Satz says on p. 205: "It is frequently said that mathematics is the language of physics when God wants to talk to man. This may be, although God is certainly polyglot and capable of sending a message via music or poetry. Nevertheless, it is hard to ignore that he finally returns to mathematics again and again. Otherwise it is hard to understand, why the arrangement of blossoms on all flowers follows a series of numbers, first elaborated by the mathematician Leonardo da Pisa, better known as Fibonacci, to describe the growing of a colony of rabbits."

This statement, although not quite correct (what is true for sunflowers is not true for all flowers) is certainly much more pleasing and flexible than Kant's view of science.

Nonetheless, even Satz's comment is too one-sided. The physicists ignore for instance, that chemistry has developed its own formula language, and in this language both sides of an equation are connected by two reaction arrows and not by a sign of equality. The chemical formula language was developed in the nineteenth century and completed by a publication of the Dutch chemist Hendrik van t'Hoff (1852–1908) in 1874. For this achievement and other discoveries he was awarded the first Nobel Prize in Chemistry in 1901. It is also worth noting here that the formula language of chemistry was not only decisive for progress in modern chemistry and pharmacy, it also supported and supports progress in all other natural sciences and in medicine, because disciplines exist in all branches of science where the structure and reactivity of molecules play a significant role.

Furthermore, physicists and other scientists tend to ignore the fact that fundamental laws of nature can be formulated using words or tables without any need for mathematical or chemical equations. Examples from chemistry are Mendeleev's Periodic Table (see Sect. 9.1) and the law of neutralization. This law says that mixing of equivalent amounts of acid and base (more precisely, equal numbers of acidic protons and hydroxide ions) yields water and salts. A fundamental insight of biology says that the individuals of all vertebrate species must die. A fundamental principle geologists have to learn is the finding that the spatial arrangement of sediments or layers of rocks is directly correlated with the timely sequence of the events that produced those layers (see Sect. 10.6).

It is of course indisputable that mathematics has provided and continues to provide the most efficient mental tools for the progress of modern science and the mental basis for all technical applications of scientific discoveries. However, Kant's extreme stance considering mathematics as an indispensable and decisive constituent of the definition of science is inacceptable and must be rejected. If Kant was right, the immensely important discoveries and insights of Louis Pasteur, Eduard Buchner, Charles Darwin, Alfred R. Wallace, and Barbara McClintock (see Chap. 8) were all outside science—an absurd vision. Finally, a prominent critique of too much mathematics in science should be cited. Albert Einstein (1879–1955) confessed: “Since the mathematicians have invaded my theory of relativity, I don’t understand it myself anymore.”

Third, the physicists of the nineteenth century assumed that their laws of nature were effective and valid at any time and everywhere in the universe. This characterization has rightly been attacked by physicists of the twentieth and twenty-first century, but without considering that the target of their critique is the mental heritage of their scientific ancestors and was not held by all scientists. Biologists, chemists, and geologists were already conscious in the nineteenth century that the validity of their laws was confined by a narrow frame of temperatures. In the second half of the nineteenth century, chemists had begun to explore the thermal stability of organic chemicals and to study the chemical structures of degradation products. They found that all the organic molecules under investigation decomposed at temperatures in the range of 250–350 °C. Nowadays, it is known that the upper limit for the survival of complex organic compounds is around 500 °C. Inorganic materials such as salts and minerals may show, in rare cases, a short-term stability up to 3500 °C. At higher temperatures chemical bonds cannot exist anymore. In contrast, almost all biologically active molecules decompose in the temperature range of 150–250 °C. Furthermore, all chemical reactions have a low temperature limit. When all translational and vibrational motions are frozen, chemical reactions can no longer take place.

The recent critique of physicists concerning the term law of nature, as defined by the physicists of the nineteenth century, only repeats what other scientists had learned from the laws of other natural sciences more than a century before.

The validity of all laws of nature has limits. Any law of nature is only effective and detectable within a certain frame of validity. For instance, most laws of nature that are effective in the biosphere of earth are not effective in the center of a black hole. Whether, and to what extent, the frame of validity of a certain law of nature is explored depends on the interest, money, time, and human resources available for this purpose in the international community of scientists.

Finally, it is worth mentioning that the terms “law of nature” and “natural law,” which sound similar at first glance, have quite differing meanings. Natural law means a kind of philosophy that certain rights or values are inherent in human nature. According to the *Internet Encyclopedia of Philosophy* the term natural law is ambiguous and refers to a type of moral theory as well as to a type of legal theory. The core claims of the two kinds of theory are logically independent, but both theories intersect. Certain writers still contribute to a confusion of both terms in the twenty-first century. For instance, the American historian Perez Zagorin (1920–2009) published in 2009 a book entitled *Hobbes and the Law of Nature*. Three sections or chapters have the following titles: “Law of Nature,” “Enter the Law of Nature,” and “The Sovereign and the Law of Nature.” However, the entire content of this book is devoted to history and the meaning of natural law and has nothing to do with science.

A complementary example is an article by the English philosopher William Kneale (1906–1990) entitled *Natural Laws and Contrary-to-Fact Conditionals*, which exclusively deals with laws of nature. This problem is characteristic of the English language and of the Anglo-Saxon understanding of laws.

2.3 More About Laws of Nature

After World War II the paradigm changes in physics initiated by Max Planck, Albert Einstein, Louis de Broglie, Werner Heisenberg, and others stimulated (mainly among Anglo-Saxon philosophers of science) a vivid discussion about the meaning and role of laws of nature. Books or articles by the American philosophers Peter Achinstein, John W. Carroll, Frederick Dretske, Gilbert Harman, and Nelson Goodman; the Australian philosophers David Armstrong, Brian Ellis, and John L. Mackie; the British philosophers A. J. Ayer, Helen Beebee, Alexander Bird, William Kneale, and Stephen Mumford; the Canadian philosopher Norman Schwartz; and the German philosopher Carl G. Hempel should be mentioned as representative examples (see Bibliography). The problems and topics discussed by those philosophers concern metaphysical aspects and not real science. For example, Armstrong's introduction to his book *What is a Law of Nature?* begins with the sentence: "The question 'what is a law of nature?' is a central question for the philosophy of science. But its importance goes beyond this relatively restricted context to embrace general epistemology and metaphysics." Ellis says in the beginning of a review article that "Stephen Mumford's book *Laws in Nature* is an important contribution to metaphysics." Nonetheless, a few characteristic topics that merit a short comment should be mentioned here.

Philosophers of science have developed their own terminology, which has little in common with the terminology of scientists. For example, Armstrong distinguishes between spatio-temporally limited laws, infinitely qualified laws, instantiated and uninstantiated laws, disjunctive laws, bridge laws, functional laws, probabilistic laws, and iron and oaken laws. Schwartz presents a further classification in the preface of his book *The Concept of Physical Law*: "In the following pages you will find nothing about the taxonomy of laws, for example the distinction between causal laws, laws of concomitance, laws of dynamics, and functional laws. Similarly, you will find nothing about empirical laws and theoretical laws and nothing about the difference between low-level and high-level laws, nothing about basic laws and derived laws. You will find nothing about the difference between those laws whose non-logical and non-mathematical terms refer only to observables and those laws that contain some descriptive terms that refer to unobservable (or theoretical) entities."

Schwartz also explains: "I prefer the term 'physical law' to either 'law of nature' or 'natural law' so as to avoid any seeming connection with the doctrine of 'natural law' in all." This substitution obscures the fact that laws of nature are properties of nature as defined in Sects. 2.1 and 2.2, and it is not clear whether the term physical law includes biological and chemical laws. On p. 4 he declares: "Scientific laws are conceptually distinct from physical laws. Only the barest handful of scientific laws are physical laws." Further confusion arises from the following statements (p. 14): "More specifically, physical laws occur among only two classes of general propositions: universal propositions and statistical propositions" and "Physical laws, however, are not empirical generalizations, i.e., propositions inductively generated from empirical data" (an opposing view by scientists is discussed in Sect. 2.4).

The debate about the metaphysical meaning of laws of nature mainly concerns two competing theories. The "regularity theory" holds that laws of nature are statements of the uniformities or regularities in the world. They are mere descriptions of natural phenomena and events. The "necessitarian theory" holds that laws of nature are the principles that govern the natural phenomena of the world. That is, the natural world obeys the laws of nature or, in the words of Beebee: "Laws are relations of necessity between universals." For example Armstrong, Dretske, and Tooley are counted among the advocates of the necessitarian theory. They argue that a physical (also called "nomic" or "nomological") necessity is inherent in the laws of nature. This physical necessity is a property of nature and, thus, inherent in matter and in the structure of the universe.

Beebee, Bird, and Ellis are representative of philosophers favoring the regularity theory. Beebee entitled an article explaining her concept "The Non-Governing Conceptions of Laws of Nature," Bird entitled one of his articles "The Ultimate Argument against Armstrong's Contingent Necessitation View of Laws," and Ellis declared in his article "Looking for Laws" (p. 338): "The things that exist are thus supposed to determine what the laws of nature are, rather than the laws determine how things should behave." On p. 439, he states that "laws of nature are descriptive of reality not prescriptive." For reasons discussed below and in Sects. 2.2, and 2.4, this debate looks artificial and obsolete to scientists. It is partly based on a misleading analogy of laws of nature and the laws existing in a human society. Some philosophers seem to understand the role of the human brain in analogy to that of a dictator or parliament that can modify or change laws at will, and members of the society have to obey them.

A topic that is intensively discussed in this context is the difference between "accidental truth" and a law of nature. In his book *Laws of Nature* Carroll outlines the problem as follows: "The perplexing nature of the problem of laws emerges upon realizing there are laws and accidental properties (phenomena) that appear to be very similar indeed. Suppose that the fastest that any raven has ever traveled is exactly 30 m/s. Then consider this generalization:

1. All ravens have speeds <31 m/s.
2. All signals have speeds <300,000,001 m/s.

Generalization (1) is true, but not a law. What about (2)? Suppose that this special aspect of reality is correct; it is a law that no signals travel faster than light and the speed of light is slightly less than 300,000,001 m/s. So it is plausible to think that (2) is both true and a law. Now it is extremely difficult to describe precisely the significant difference between (1) and (2)."

For scientists, Carroll's example is not a theoretical problem but may involve experimental difficulties, because any differentiation between laws and accidental truths or other speculative regularities and hypotheses requires experimental results. However, Carroll's argumentation about ravens is not well suited for scientific analysis. A more realistic scenario looks like this: A population of ravens in Kent (UK) and another in Cornwall (UK) were studied and no raven travelling at a speed above 30 m/s was found. Certainly, no scientist would conclude that all ravens worldwide and all ravens of all times fly at speeds below 31 m/s. The experimental evidence does not suffice for this conclusion. That all ravens in southern England fly at speeds below 31 m/s is clearly an accidental truth. Now, let the speed limit be raised to 3001 m/s. With such a high limit the statement "all ravens have speeds less than 3001 m/s" may become a law of nature. It is possible to demonstrate that:

1. The lungs (and eyes) of ravens and all other birds cannot resist the air pressure at a speed of 3001 m/s
2. The power of their muscles can never suffice to reach such a high speed

Even easier to prove and to understand is a similar "exclusion law" that purports that no ravens ever fly at temperatures above 500 °C. Regardless of which property or function of a living or non-living object is considered, decisive for identification of a law of nature is the experimental evidence. However, terms such as experiment, measurement, and reproducibility are usually absent from the texts of philosophers.

Another strange kind of argumentation by philosophers of science is a comparison with "another world." For example, Bird presents in his aforementioned article about Armstrong's necessitarian theory the following comment (p. 147): "For example and in particular, properties [of materials] do not essentially or necessarily have or confer any dispositional character or power. Being made of rubber confers elasticity on an object, but it does not do so necessarily. Being negatively charged confers on objects the power to repel other negatively charged objects, but not necessarily. In other possible worlds rubber objects are not elastic, negatively charged objects attract rather than repel one another." The meaning of "other possible worlds" is not defined. If other inhabitable planets are meant, this argumentation is wrong because it is well known from space research that all the chemical and physical laws discovered on Earth are also valid on other planets. In a certain temperature range, rubber has the same elasticity on Mars as it has on Earth. If a world outside our universe is meant, the above argumentation is sheer non-sense because nobody knows anything about a world outside our universe, and human logic and all scientific methods are confined to the universe we live in.

In summary, the relevance of what philosophers debate about laws of nature is for scientists best expressed in an aphorism of the physicist Richard Feynman (1918–1988, Nobel Prize 1965): "Theory of science is as useful to scientists as ornithology is to birds"! The ravens from the example above will certainly agree.

However, three closely related questions that are of outmost importance for both scientists and theoreticians should be discussed here in more detail:

– What does the term "frame of validity" (introduced in Sect. 2.2) really mean?

- Does a law of nature exist at a certain time, a certain location, and certain temperature when its consequences are not detectable under these conditions?
- Were all laws of nature "born" at the first moment of the Big Bang?

The following, similarly important, questions are discussed in other subchapters:

- Do inductive inferences based on reproducible experimental facts yield trustworthy laws of nature as foundation of the natural sciences? Discussion of this question is given in Sect. 2.4.
- Does scientific research yield absolute truths (in the terminology of philosophers)? This question is discussed in Sect. 2.5.

What does "frame of validity" mean? As outlined in Sect. 2.2, all laws of nature have limits with regard to time, location, and temperature or with regard to two or all three parameters. The frame of validity is therefore a set of parameters (or reaction conditions) that allows the detection of reproducible phenomena obeying the law under consideration. Whether this frame has sharp borderlines or (as usual) consists of a gradual transition to other laws of nature is not relevant for this discussion. It is another logical consequence of this consideration that when the conditions of a scientific inquiry that do not allow for the discovery of a law of nature are shifted in such a way that they overlap with the frame of validity, the existence and effectivity of the law immediately becomes detectable.

A first experimental argument in favor of the above consideration results from the discovery of black holes. The physicists say that most laws of nature identified on the surface of the earth during the past 500 years are at not detectable in a black hole. The experimental verification of this statement is difficult to achieve and thus a fictitious experiment which, in principle, allows for verification on earth should be discussed. Consider a chemical laboratory that can repeatedly be heated to temperatures above 500 °C. In the beginning, a chemist studies the synthesis of penicillin from various starting materials at 20 °C. Afterwards the laboratory is heated to 500 °C. The penicillin and all organic substances the chemist has used are destroyed by thermal degradation, yielding charcoal and various gases. Under these conditions, organic chemistry is no longer feasible and the chemical laws that govern the synthesis and structural analysis of penicillin are not detectable. Now, the laboratory is cooled down to 20 °C. The chemist can enter the laboratory again and repeat the syntheses of penicillin. This process can be repeated again and again provided the laboratory is cleaned after each heating cycle. In other words, when a chemist equipped with a cooled, fire-proof protective suit enters a laboratory at 500 °C and asks nature, whether the molecules A, B, or C are suitable for the synthesis of penicillin, nature does not answer. However, in a second identical laboratory, the temperature of which is fixed at 20 °C, the synthesis of penicillin can be studied at any time and without interruption. Any experiment or study of natural phenomena is a question to nature and it is a fundamental requirement of any scientific inquiry to find the set of parameters (reaction conditions) that allow nature to answer.

The same line of argumentation can be considered from another point of view. At the time when all planets existing in the universe had surface temperatures above

1000 °C, none of the millions of biological and chemical laws that have been discovered on earth or that are foreseeable today (due to the existence of many millions of species of microbes, plants, and animals) were observable and detectable in the universe at that time.

The only consistent interpretation of these arguments is the assumption that all laws of nature are permanently present in the universe, but their detectability is limited by a certain set of parameters, their frame of validity. If, as defined above (Sects. 2.1 and 2.2), laws are fundamental and characteristic properties of nature, they were born together with all other fundamental aspects of our world at the very first moment of the Big Bang. In other words, the universe, our world, is characterized by the sum of all laws of nature plus space and time. As a result of the existence of a biosphere on earth (and perhaps on other planets), the current state of the universe represents a preliminary maximum of laws of nature that are detectable. However, the number of laws of nature inherent in the universe is potentially endless and, regardless of how long humankind survives, the limited capacity of the human brain (even when augmented by computers) will never be able to explore all the laws and their complex interactions. Therefore, it is wrong or at least premature to call this concept deterministic. Such a classification would overrate existing scientific knowledge and underestimate the richness and complexity of nature. In the words of R. Feynman: "The imagination of nature is far, far greater than the imagination of men." However, this concept is consistent with the initially given definition of laws as the language that nature uses for the rational communication with the human brain, which is itself part of this nature. The last words should be left to Max Planck: "Science cannot solve the ultimate mystery of nature and that is because in the last analysis we ourselves are part of nature, and therefore, part of the mystery that we are trying to solve."

2.4 Critique of Empirical Research as Source of the Laws of Nature

When, after 1933, the Nazis gained increasing influence on the political and social life in Austria, numerous artists, scientists, and philosophers with a Jewish background immigrated to foreign countries, notably to England and the USA. Among them was a small group of philosophers who were later nicknamed the "skeptic philosophers." The most prominent figures were Ludwig Wittgenstein (1889–1951) and Karl R. Popper (1902–1994). The most severe and most widely known fundamental critique of empirical research originates from Popper and, therefore, this chapter is primarily devoted to an analysis of his arguments. However, concerning his critique of inductive inferences as source of reliable knowledge, he had two forerunners, namely the British philosopher and politician John S. Mills (1806–1873) and Wittgenstein (see Sect. 3.3).

In 1934 Popper published the book *Logik der Forschung* (The Logic of Scientific Discovery), which was translated into English in 1959. The following citations and page numbers concern the 6th edition (1972). At the beginning of the first chapter Popper asks (p. 27): "But what are these methods of empirical research? And what do we call empirical science?" He gives the answers in the subsequent section with a critique entitled "The Problem of Induction" and this critique begins with the following statements:

> According to the widely accepted view—to be opposed in this book—the empirical sciences can be characterized by the fact that they use inductive methods, as they are called. According to this view, the logic of scientific discovery would be identical with inductive logic, i.e., with the logical analyses of these inductive methods. It is usual to call an inference 'inductive' if it passes from singular statements (sometimes also called particular statements), such as accounts of results of observations or experiments, to universal statements, such as hypotheses or theories.
>
> Now it is far from obvious, from a logical point of view, that we are justified in inferring universal statements from singular ones, no matter how numerous; for any conclusion drawn in this way may always turn out to be false: no matter how many instances of white swans we may have observed, this does not justify the conclusion that all swans are white.
>
> The question whether inductive inferences are justified, or under what conditions, is known as the problem of induction.

In the subsequent sections entitled "Deductive Testing of Theories," "The Problem of Demarcation," "Experience as Method," Scientific Objectivity and Subjective Conviction," and "Elimination of Psychologism" Popper never comes back to concrete examples illustrating or proving his inferences and statements. Therefore, a discussion of his frequently cited "swan argument" is of particular importance. Popper, like other philosophers, distinguishes between two types of statements: the singular (particular) statement and the universal (general) statement. The singular statement concerns a single observation, such as "the swan in front of me is white." When all swans an observer comes across are white he achieves a series of identical singular statements and may finally formulate the universal conclusion: all swans are white. Popper points out that this inductive inference is falsified as soon as a colored swan is observed. Popper draws the final conclusion that a series of identical singular statements (observations) never yield reliable universal statements and, therefore, laws of nature can never be reliably determined from reproducible experimental results followed by an inductive inference.

However, the swan argument combines a correct with an incorrect inference, because Popper does not differentiate between the properties of a living or non-living object, which are variable without affecting the definition of the object, and properties that are constituents of their definition. The color of swans is not relevant for the definition of swans (*cygninae*). In the case of swans and many other animals, races exist that display a variation in color as an evolutionary consequence of their adaption to different environments. Therefore, it is true that even the largest number of white swans does not guarantee that all swans observed in the past, today, or in the future will be white (dark colored swans were indeed discovered in Australia).

For comparison, the following universal statement should be examined. All swans (like other birds) have wings equipped with feathers and their nerve signals propagate via a sodium (Na^+)–potassium (K^+) exchange across the nerve membranes. Where the nerves are interrupted by synapses, the nerve signals propagate via neurotransmitter molecules. Imagine that a kind of bird is discovered that at a distance looks like a swan, but closer inspection shows that it has wings covered by hair or the propagation of nerve signal involves other chemical mechanisms. The discovery of such a bird does not falsify the above universal statement, because such a bird does not obey the definition of swans.

An analogous example is from the domain of plants: All oak trees (*Quercus rubra* and *Quercus petraea*) produce acorns that after falling to earth serve as seeds for new oak trees. Their green leaves produce glucose via photosynthesis and the glucose serves as starting material for the synthesis of cellulose. Suppose that trees are found that look like oaks but do not produce acorns and their leaves do not produce glucose via photosynthesis. The discovery of such trees does not falsify the above universal statement, because they are outside the definition of oak trees.

Finally, a third example from the domain of non-living objects: All rubies are red gemstones consisting of aluminum oxide (Al_2O_3) colored by traces of the element chromium. Other red minerals having a different chemical composition, or aluminum oxide crystal having another color (e.g., corundum or sapphire) do not obey the definition of ruby.

Considering that millions of plant and animal species exist on earth, many millions of universal statements could be formulated that represent at least partial definitions of living organisms, and many more such statements can be made about non-living objects. Together with the structural formulas of chemical compounds (see Sect. 2.4) all these descriptions and definitions represent the largest body of reliable knowledge in science. In other words, the swan argument is wrong whenever it is used as a critique of inductive inferences and empirical research.

Popper also wrote on p. 28: "Thus to ask whether there are natural laws known to be true appears to be only another way of asking whether inductive inferences are logically justified." The answer is: Yes, they are justified and laws of nature may represent an absolute truth, as demonstrated in Chap. 3.

Based on his misinterpretation of the swan argument and its underlying logic, Popper continues on p. 40: "Now in my view there is no such thing as induction. Thus inference of theories from singular statements which are verified by experience (whatever that means) is logically inadmissible." On p. 28, he states: "Now this principle of induction cannot be a purely logical truth like a tautology of an analytical statement. Indeed, if there were such a thing as a logical principle of induction, there would be no problem of induction; for in this case all inductive inferences would have to be regarded as purely logical or tautological transformations like inferences in deductive logic. Thus the principle of induction must be a synthetic statement, that is, a statement whose negation is not self-contradicting but logically possible."

These statements provoke the following questions: What does the negation of the law of the lever look like and what is its logic justification? Furthermore, any baby must start breathing immediately after the navel cord is cut. What does the negation of this natural law look like and what is the logical justification in this case?

On p. 59 Popper says: "The experimental sciences are systems of theories. The logic of scientific knowledge can, therefore, be described as a theory of theories. Scientific theories are universal statements. Like all linguistic representations they are systems of signs and symbols." This description of empirical sciences is not agreeable. First of all, empirical science begins with accumulation of observations or experimental data. If, for certain observations or data, reproducibility irrespective of time and location can be attested, the observer can continue on two tracks:

I. The observer recognizes in his consciousness or subconsciousness a law of nature and extrapolates the reproducibility into the future to profit from reliable predictions. This process does not require the design of hypotheses.
II. The observer develops a hypothesis that combines two aspects: (A) an explanation based on comparisons with other observations (experimental data) and/or with existing theories; and (B) prediction of new experiments to test the hypothesis, which after confirmation achieves the status of an established theory.

Hence, the development of a theory and its comparison with other theories is the last step of research activities. Proceeding according to (I) is typical for children and for early humankind (Neolithic and before). Yet, even the brains of higher animals behave in this way, as illustrated below, and other mammals certainly do not formulate hypotheses nor use signs and symbols.

However, even in modern science such behavior may make sense. Consider a chemist who has developed a new cheap synthesis for an antibiotic. Prior to technical production, optimization of the yield is required to reduce the costs and to improve the purity of the product. The optimization is usually achieved by systematic variation of those parameters that have a significant influence on the yield, such as reaction time, temperature, and concentration. The dependence of the yield on each parameter represents a law of nature. The reproducibility of the optimized procedure serves, in turn, as basis for the technical production (see Sect. 3.2). The successful technical production confirms every day the reproducibility and predictability of the empirical research underlying the production.

This entire process from the first synthesis to the technical production does not require a hypothesis or theory about the reaction mechanism of the involved molecules, although such a hypothesis might be helpful and desirable. In summary, Popper's definition of empirical science is one-sided and misleading and his wrong understanding of science culminates in the statement "science is a theory of theories". This statement is characteristic for a physic-infectef theoretician with little knowledge of the scientific fundament of the human civilization (Chapter 3.2), of chemistry in general (Chapter 9) and of the progress of medicine (Chapters 6 and 7).

Popper's belief that inductive inferences do not yield reliable universal statements and insights in laws of nature has led him to construct a theory of science focused on the aspect of falsifiability of hypotheses and theories. A statement is called falsifiable if it is possible to find an observation or logical argument that proves the statement to be false. Popper used the falsifiability of hypotheses and theories for the demarcation between science and non-science. His concept was labeled "falsificationism" and he and his followers were nicknamed falsificationalists. He explained his concept and its deviation from positivism as follows (p. 40): "The criterion of demarcation inherent in inductive logic—that is, the positive dogma of meaning—is equivalent to the requirement that all statements of empirical science (or all meaningful statements) must be capable of being finally decided with respect to their truth and falsity; we shall say that they must be conclusively decidable. This means that their form must be such that to verify them and to falsify them must both be logically possible. Thus [the Austrian philosopher] Schick says, 'a genuine statement must be capable of conclusive verification' and [the Austrian physicist] Waismann says still more clearly, 'If there is no possible way to determine whether a statement is true, then that statement has no meaning whatsoever. For the meaning of a statement is the method of its verification.' Now in my view there is no such thing as induction. Thus inference to theories from singular statements which are verified by experience (whatever that means) is logically inadmissible. Theories are, therefore, never empirically verifiable. ... These considerations suggest that not the verifiability but the falsifiability of a system is to be taken as criterion of demarcation."

Finally he says: "My proposal is based upon an asymmetry between verifiability and falsifiability, an asymmetry which results from the logical form of universal statements. For these are never derivable from singular statements but can be contradicted by singular statements."

This latter statement underlines Popper's misunderstanding of inductive conclusions and is here confronted with non-falsifiable universal statements from biology, physiology, chemistry, and astronomy:

- All men must die. This law is a special case of a more universal law saying that the individuals of all species, whose reproduction is exclusively based on sexual acts, must eventually die. For a falsification of this law, an individual (human or other mammalian) would need to be found that lives for eternity.
- All humans (*Homo sapiens*) produce enzymes for the digestion of starch (rice, corn, wheat, potatoes), but not for the digestion of cellulose (cotton, wood), although both types of biopolymers consist of glucose units. Any individual whose digestion works differently is outside the definition of *Homo sapiens*.
- The human brainstem regulates breathing so that the pH in the cerebrospinal liquid does not exceed the limits of 7.35 and 7.45. At higher or lower pH values the nerves are rapidly damaged, with lethal consequences. A living organism working at a significantly higher or lower pH does not have a human brain.
- The sun and other stars of similar mass have a well-defined *vita*. Their birth begins with the contraction of a cloud of hydrogen atoms, they pass a maximum of radiation in the form of a red giant, a maximum of temperature in the form of a white dwarf, and they finally cool down asymptotically approaching the temperature of the environment. To falsify this law, one would need to find a star that maintains a significantly higher level of radiation or temperature than its environment for all eternity.

Another type of critique of Popper's falsification theory was contributed by the theoretician and historian Thomas Kuhn. In his famous book *The Structure of Scientific Revolution* (for review see Chap. 5) he wrote (p. 146/147): "A very different approach to his whole network of problems [is verification of a theory possible?] has been developed by Karl Popper who denies the existence of any verification procedure at all. Instead he emphasizes the importance of falsification, i.e., of the test that, because its outcome is negative, necessitates the rejection of an established theory. ... But falsification though it surely occurs, does not happen with, or simply because of, the emergence of anomaly or falsifying instance. Instead it is a subsequent and separate process that might equally well be called verification since it consists in the triumph of a new paradigm over the old one."

Unfortunately, Popper cannot answer the question: What does science mean if a reliable determination of natural laws is not feasible? If science is merely an enterprise yielding falsifiable theories, it is nothing more than an intellectual game. Further arguments in favor of an empiricist view of science are presented in Sects. 3.1, 3.2, and 3.3 and below.

2.5 Is There an Absolute Truth in Science?

Truth is a term that is very much liked by poets, theologians, and philosophers, including philosophers of science, whereas scientists usually prefer to talk or write about reliable knowledge, consistent interpretation, correct information, unequivocal evidence, definite proof, etc. Regardless of this difference, the question of whether scientific research can provide reliable knowledge/absolute truth or not needs to be answered. In this connection, Max Born's statement, cited in Chap. 1, should be recalled: "Ideas such as absolute certainty, absolute accuracy and final truth are inventions of the human imagination and should be avoided in science." A similar statement was given by the physicist and Nobel Prize laureate Werner Heisenberg (1901–1976): "Therefore, it will never be possible by pure reason to arrive at some absolute truth." These statements are not acceptable for at least two reasons. First, they are a kind of oxymoron, because they negate themselves. If absolute truth does not exist, this conclusion necessarily implies that it is itself not an absolute truth and Born's and Heisenberg's statements are a pseudo-insight; this is not helpful for a correct understanding of natural sciences.

Over the past 100 years, philosophers of science have extensively discussed whether scientific theories may represent an absolute truth, or a gradual approach to truth, or a truth-likely interpretation of scientific results. Almost all these discussions suffer from the following deficits:

- Almost all philosophers of science lack any personal experience of experimental research.
- They rarely cite concrete examples illustrating and proving their concepts and conclusions. If they cite scientific results and theories they allude to the history

of astronomy and physics, assuming that this history is representative of any kind of natural science and of any kind of research.
- Their texts suggest that they do not have any knowledge of chemistry, geology, physiology, etc.
- Two terms that are of outmost importance for a proper understanding of science, namely "law of nature" and "reproducibility" (of experiments or observations) are not usually found in their texts.

Hence, experimental scientists have the impression that those philosophers discuss a Fata Morgana of science but not real science. In the words of the theoretical physicist R. Feynman in *The Pleasure of Finding Things Out*, published in 1999 (p. 173): "And so what science is, is not what the philosophers have said it is and certainly not what the teacher editions say it is."

However, exceptions, such as the Canadian philosopher Ian Hacking, are exempt from this critique. In 1983, Hacking wrote in his book *Representing and Intervening* (p. 149): "Philosophers of science constantly discuss theories and representations of reality but say almost nothing about experiment, technology, or the use of knowledge to alter the world." Furthermore, the American philosopher Moti Mizrahi in a recent article (2013) entitled "What is Scientific Progress? Lessons from Scientific Practice" demonstrated by means of two discoveries in the field of physiology that the understanding of science as it is presented by most philosophers, is too narrow and one-sided. The historic examples cited by Mizrahi are the discovery of blood groups by Karl Landsteiner (1836–1943, Nobel Prize in 1930) and the work of Ivan P. Pavlov (1849–1936) on the physiology of digestion (Nobel Prize in 1904).

The titles of articles and books cited below should illustrate the problems and questions frequently discussed in philosophical treatises (the meaning of truth in science is, of course, also debated in other publications):

- *Likeliness of Truth*, Oddie G (1986), D. Reidel, Dordrecht
- *Truthlikeness*, Niiniluoto I (1987), D. Reidel, Dordrecht
- *Realism and Truth*, Devitt M (1991), Blackwell, Oxford
- "Approximate Truth and Scientific Realism," Weston T (1992) Philosophy of Science 59:52
- *How Science Tracks Truth*, Psillos S (1994), Routledge, London
- *Science, Truth and Democracy*, Kitcher P (2001), Oxford University Press
- "Revising Beliefs Towards Truth," Niiniluoto I (2011), Erkenntnis 75:165
- "Progress as Approximation to Truth: A Defense of the Verisimilitudinarian Approach," Cevolani G, Tambolo L (2013), Erkenntnis 78:921

According to a citation in the *Stanford Encyclopedia of Philosophy*, the American philosopher Larry Laudan (born 1941) reached the conclusion that "knowing the truth is an utopian task." This statement is almost identical to those of Born and Heisenberg and it deserves the same critique. However, any use or discussion of the term "truth in science" has to consider that truth frequently has a religious or other transcendent connotation. Such a connotation renders the term "truth" useless for a reasonable and consistent debate about the structure, properties, and goals of science. The text below and the discussion of progress in Sect. 5.3 are based on an equivalence of truth and reliable knowledge (information, data, experimental facts).

The examples of problems discussed by philosophers of science should now be confronted with two aspects of scientific practice. The first aspect concerns the lowest level of scientific activity, a behavior any individual learns in the earliest stage of life and a behavior that humankind has practiced since its emancipation from apes. This activity is the observation of natural phenomena, their classification, and their extrapolation into the future. Natural phenomena are classified even in the subconscious into perfectly regular, more or less regular, and irregular events. The discovery of perfectly regular phenomena stimulates, in turn, an extrapolation into the future, either to avoid risks or to achieve an advantage, such as more or better food. The success of the extrapolation to the future is a consequence of the fact that the results of laws of nature must be reproducible irrespective of location and time (as long as their frame of validity exists). For instance, when a baby learns to stand and to walk upright, it is permanently confronted with the reproducible influence of gravity everywhere on the surface of the earth and at any time.

The discovery of a law of nature and the extrapolation of its effects on the time scale may be called inductive inference, but it does not necessarily involve the formulation of a hypothesis (and even less of a theory).

When a baby learns to cope with the problems of gravity it automatically assumes that gravity will persist in the future, otherwise a cumbersome time-consuming learning process would not make sense. The baby certainly does not develop a hypothesis about gravity, law of the lever, and the kinetics of motion. The same line of argumentation holds for the exploration of laws of nature by early humanity, as illustrated in Sect. 3.1.

The ability to discover and utilize laws of nature without elaborating hypotheses and theories is not confined to the human brain, it can also be observed for animals, because it is a fundamental principle of life. For instance, the brown bears living along the coast of British Columbia and Alaska gather in fall every year along the rivers that empty into the Pacific Ocean. They wait for the salmon coming up the river, because they need this calorie-rich food to survive the winter. Over a period of many thousands of years, the bears have learned to profit from this predictable event, but they have certainly not developed a theory about the yearly migration of salmons. In summary, any definition of science focused on the elaboration of theories and their qualities, such as absolute truth or truthlikeliness, is too narrow, one-sided, and may be misleading.

The second argument against a theory-laden view of science and against a fundamental lack of truth in scientific results comes from standard chemical research. Since the beginning of modern organic chemistry in the early nineteenth century a fundamental goal of chemists, decisive for the progress of chemistry, was and still is the elucidation of the chemical structure of all pure compounds. This kind of analytical work was difficult in the nineteenth century, in as much as a precise and generally accepted formula language was lacking before 1874. Yet,

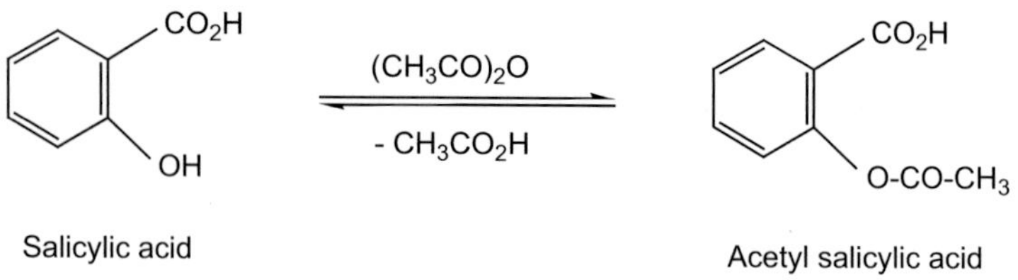

Fig. 2.1 Synthesis of aspirin from salicylic acid

today numerous analytical methods are available and allow rapid and unequivocal determination of chemical structures. Such analytical methods include elemental analyses (qualitative and quantitative determination of the elements that constitute a compound), UV-, IR- and Raman spectroscopy, several variants of nuclear magnetic resonance (NMR) spectroscopy, different mass spectrometry techniques (e. g., classical mass spectrometry, fast-atom-bombardment (FAB), and MALDI-TOF mass spectrometry), various types of chromatography, electron microscopy, and X-ray scattering. For the structural analysis of relatively simple molecules such as aspirin (Fig. 2.1) or procaine (see Fig. 7.1 in Sect. 7.3) the combination of four or five methods is usually sufficient. The structure of a compound is also confirmed by its synthesis from well-established precursor molecules. After confirmation by two or more different laboratories, the chemical structure of a molecule is a piece of reliable knowledge or absolute truth. The understanding of a synthetic procedure, such as the conversion of salicylic acid into aspirin (Fig. 2.1), is an additional piece of truth.

To illustrate the importance of this argument for the proper understanding of science, the following points need to be taken into account:

- The reliable knowledge of chemical structures is decisive for successful research in molecular biology, physiology, pharmacy, organic, inorganic and polymer chemistry, and in mineralogy.
- The rapid progress of human civilization over the past 150 years is based on the knowledge of chemical structures and chemical reactions (see Sect. 3.2). The production of local anesthetics (Sect. 7.3), medicaments, dyestuffs, and numerous polymers (see Sects. 9.4 and 9.5) serves as evidence.
- Pharmaceutical companies only produce and commercialize medicaments of well-known chemical structure (over the past 120 years nearly ten billion aspirin tablets were produced).
- Up to the year 2010 more than ten million chemicals of known structure were synthesized.

Of course, in several cases the initial structural analyses contained errors that were later revised. From this point of view chemical research is not different from research in other disciplines of science. Nonetheless, a body of far more than ten million "pieces of absolute truth" has been accumulated by chemists over the past 200 years. This is the second, the experimental, argument against the statements of Born, Heisenberg, and Laudan.

The third argument and the final answer to Laudan's statement is the fundamental law that all men must die. This is an absolute truth and its hypothetical falsification, an eternal life of skeptical philosophers or antiscientists, is a utopia.

2.6 Law of Nature Versus Model

The defamation of inductive inferences by Popper, Wittgenstein, and their followers; the negation of truth in science by Born, Laudan, and other skeptic philosophers; and the dramatic paradigm changes that took place in physics have stimulated some physicists to draw an extreme conclusion. The German physicist Thomas Millack declares in his 2009 book *Naturwissenschaft und Glaube im Gespräch* (Science and Religion in Dialogue) that he and other physicists prefer to use the term "model" to the term "law of nature." A model is an intellectual method, like hypothesis and theory, providing a preliminary explanation of complex natural phenomena that are influenced by numerous laws of nature and their interactions. Models need further modifications and improvements. Yet, what is the goal of these improvements, if not an as perfect as possible explanation of natural phenomena in terms of laws? Substitution of the term law of nature by model is a confusion of two dimensions because it puts an intellectual method on par with the properties of nature.

The substitution of law of nature by model is not only an issue for discussion among philosophers of science, it also has interesting and sometimes even amusing consequences for the conversation between scientists and among laics. Consider a child that is attempting to learn to stand upright and to walk. Does the child optimize its response to laws of nature, such as gravity or the law of the lever, or does it play with a model? When a motorcyclist has an accident because he ignored the fact that the stopping distance increases with the square of his speed, did he ignore a law of nature or a model? Finally it is instructive to look at the events following the birth of a human being. Immediately after the navel cord has been cut, a baby must start breathing, otherwise it will die within a shorter time than its birth required. Is this need to breathe a law of nature or a model?

This chapter should be closed with a modified aphorism of the German physicist Georg C. Lichtenberg (1742–1799), an aphorism he originally dedicated to chemists: "He who knows nothing but physics, does not even understand physics."

Bibliography

Achinstein P (1971) Laws and explanation. Clarendon, Oxford
Armstrong D (1983) What is a law of nature? Cambridge University Press, Cambridge
Asimov I (1984) Asimov's new guide to science. Basic Books, USA (and Penguin Books, 1987)
Beebee H (2002) The non-governing conception of laws of nature. Philos Phenomenol Res 62:571

Bird A (2005) The ultimate argument against Armstrong's contingent Necessitarian view of laws. Analysis 65:147
Carroll J (1994) Laws of nature. Cambridge University Press, Cambridge, NY
Dretske F (1977) Laws of nature. Philos Sci 44:248
Goodman N (1954) Fact, fiction, and forecast. The Athlone Press
Hacking I (1983) Representing and intervening. Cambridge University Press, London, NY
Harman G (1968) Knowledge, inference and explanation. Philos Quart 18:164
Hempel CG (1968) Maximal specificity and lawlikeliness in probabilistic explanations. Philos Sci 35:116
Feynman RP (1999) The pleasure of finding things out. Helix Books, Cambridge, MA
Kneale W (1950) Natural laws and contrary-to-fact conditionals. Analysis 10(6):121
Millack T (2009) Naturwissenschaft und Glaube im Gespräch. Oldenbourg, München
Mizrahi M (2013) What is scientific progress? Lessons from scientific practice. J Gen Philos Sci 44:375
Mumford S (2004) Laws in nature. Routledge, London
Popper K (1959) The logic of scientific discovery, 1st edn. Hutchinson, London (6th edn. 1972)
Schwartz N (1985) The concept of physical law. Cambridge University Press, Cambridge
Toulmin S (1961) Foresight and understanding. Hutchinson, London
Zagorin P (2009) Hobbes and the law of nature. Princeton University Press, Princeton, NJ

Chapter 3
Laws of Nature and Everyday Life

3.1 Laws of Nature and the History of Humankind

> If the human brain were so simple that we could understand it, we would be so simple that we couldn't.
>
> (Emerson M. Pugh)

Humans can approach nature in three different ways. The first way is to approach nature emotionally, admiring its beauty, for example, a wonderful sunset, or cursing nature, for example, because a catastrophe killed friends or relatives. The second approach is that of a businessman who looks for profit, for instance by mining or cutting trees to produce paper. The third approach is that of a scientist who tries to understand the origin, periodicity, and consequences of natural phenomena. Whenever humans try to understand natural phenomena, they are confronted with laws of nature, because laws are the language that nature uses for communication with the human brain (see Sect. 2.2).

The communication of an individual person with the laws of nature begins at the first second after birth, and the communication of humankind with the laws of nature began on the first day of emancipation from early apes. Hence, it is essential to have a correct understanding of the laws of nature (contrary to the opinion of many critics), that the laws of nature are not an invention of modern science. If the laws are properties of nature as defined in Sect. 2.2 and if nature uses laws for its communication with the human brain, humans must have been capable of identifying the laws of nature from the first day of their emancipation from apes. According to recent results in anthropology and paleontology, this emancipation began more than two million years ago, perhaps even more than three million years ago. Hence, perception and utilization of the laws of nature mainly occurred in the subconscious and certainly did not involve formulation of chemical or mathematical equations. In other words, perception of the laws of nature was intuitive and qualitative rather than quantitative. Nonetheless, it was decisive for the survival and expansion of humankind. Evolution of humans away from apes may be understood

H.R. Kricheldorf, *Getting It Right in Science and Medicine*,
DOI 10.1007/978-3-319-30388-8_3

as the increasing capacity of the human brain to identify and interpret the laws of nature and to use this knowledge as a tool or weapon in the struggle for life. The following text illustrates and supports these statements with concrete examples.

A law that humans (and animals) intuitively notice every minute, every hour, and every day is the permanent presence of gravity. Beginning at the earliest stage of the childhood, humans had—and still have—to learn how the confrontation with gravity can be optimized for any kind of bodily motion. More conscious actions were to throw stones at enemies and dangerous animals or to avoid falling from the top of a steep rock. Giving objects a firm stand on a table or on a shelf to prevent them from falling is an extension of this behavior.

People respond to gravity in a subconscious manner every morning when standing up after waking, and this subconscious response finally ends at night when falling asleep. In contrast, flying an airplane as a pilot is a far more sophisticated and conscious interaction with gravity. The subconscious interaction with laws of nature, whenever a frequent response is necessary, is a strategy favored by evolution to keep the consciousness free for new observations and special actions.

Humans were not only aware of the direct influence of gravity, as in the case of falling objects, they also noticed indirect effects. An important example is the observation that certain objects and materials float on water whereas others sink. Archimedes of Syracuse (287–212 B.C.) was apparently the first to discuss the origin of this phenomenon. He found that objects having a lower specific weight (including voids or the empty space inside a boat) than that of water float, whereas objects having a higher specific weight, such as stones, sink. The intuitive understanding of this law of nature prompted people long before Archimedes to construct boats that made fishing more efficient. The construction of boats and ships also enabled people to discover other countries across the sea, to establish trade connections across the sea, and to conduct wars across the sea.

Another law that is an indirect consequence of gravity is the rising of hot air, which has a lower density than the surrounding colder air. As soon as humankind was capable of making fire it observed this phenomenon and utilized it for roasting meat or vegetables above the fire and not in the lateral infrared radiation. Although cooking above an open fire is vanishing in western civilization and in most Asian civilizations, almost everybody knows the properties of a burning candle. Furthermore, the French brothers Joseph-M. and Jacques-E. Mongolfier invented flying with hot air balloons in 1783, an amusement that attracts an increasing number of participants, even in the twenty-first century.

Another combination of phenomena that humankind became acquainted with at its very beginning is the daily sunrise and the four seasons and their consequences for the temperature of the surrounding air. An intuitive understanding that these phenomena are a result of gravity and other laws of mechanics was certainly not possible, but early humans learned to profit from the reproducibility and predictability of these events, for example, for hunting and for optimization of their agricultural activities. The 24-h rhythm of the human metabolism is one more example of how evolution relies on the laws of nature and how humans subconsciously obey the laws of nature (see Sect. 7.3).

One more mechanical law of nature that humankind has intuitively noticed since its very beginning is the law of the lever. Application of this law for more efficient forward motion of animals was an evolutionary advantage as much as 550–600 million years ago. Primitive arthropods, such as the trilobites, were the first animals that used limbs (i.e., legs) for forward motion on a solid surface and for swimming (the term arthropod is derived from the Greek *arthros* meaning joint and *podos* meaning foot). The motion of any limb, including wings and flaps, is based on the law of the lever. The distance between the joint and the connection of muscle and bone is the short lever. Hence, humans used the law of the lever unconsciously whenever moving a limb, but they also soon learned a conscious, although intuitive, application when they began to use sticks as a weapon. Provided that the diameter and density of sticks are identical, the distance between the shoulder and center of gravity increases with the length of the stick. Therefore, the handling of a longer stick requires a greater force and, thus, stronger muscles. In the megalithic period of human civilization, men certainly used long wooden beams as nonsymmetrical levers to move heavy stones. Another early and intuitive application of the lever law concerns the lifting of stones with hand and arms. Lifting a relatively heavy stone requires shortening of the distance between the hand and body, because the maximum force that the shoulder muscles can exert is limited.

The first formulation of this law is usually attributed to Archimedes of Syracuse: "Magnitudes (M) are in equilibrium at distances reciprocally proportional to their distance (D) ($M_1D_1 = M_2D_2$)." Archimedes is reported to have illustrated this law by saying: "Give me a lever long enough and a fulcrum to place it and I will move the Earth." However, a similar formulation was published before Archimedes's birth in a work entitled *Mechanica* written by followers of Aristotle.

With the beginning of the bronze age, the progress of the human civilization gathered speed and numerous tools, instruments, and machines were invented, the function of which involves the law of the lever . Among the oldest tools of this kind, already used by the ancient Egyptians around 4000 years ago, are pincers and scissors. Further important applications in the antique world were catapults and lever-presses for the production of grape juice. In our modern world, bolt cutters, cranes, and recent models of water taps are applications of rigid levers. Pulley blocks, electric chain hoists, and the belts of cowboy saddles are applications of flexible levers.

The law of the lever is also applicable to rotation and twisting motions. Classical water taps and steering wheels of cars and ships are typical examples. A larger diameter of the steering wheel reduces the force needed for steering. A frequent application is also power transmission via cock-wheels in gear units or in the wheels and pinions of clocks and watches. They enable transformation of slow motion with high power into rapid motion with low power and vice versa.

Another aspect of linear motion is the impact of a moving object or body on a solid obstacle. From throwing stones or by falling from a galloping horse people learned thousands of years ago that the intensity of impact increases with speed. In the past, it was certainly difficult to learn intuitively that the kinetic energy increases with the square of the speed, whereas today every candidate for a driving license has to learn the consequences of this new insight.

One more mechanical law that was intuitively understood by people of the stone age was the relation between pressure (P) and area (A) on which a force (F) was applied: $P = F/A$. By sharpening the tip of an arrow or the edge of a stone ax, the area hit by the weapon was reduced and the hunters or warriors expected that their weapon would more easily penetrated the skin of a bear or the skull of an enemy.

An important experience of early humans was the perception of different temperatures. A brutal lesson man had (and still has) to learn from nature is that too high or too low a temperature can have lethal consequences. In connection with higher temperatures, humans had the opportunity to learn that in addition to light a second kind of radiation must exist, a radiation transporting heat, today called infrared radiation. Rock faces heated by the sun and radiating heat after sunset or the stones of a fireplace after extinction of the fire conveyed this experience.

Today people learn from handling radios, smart phones, or from physics lessons in high school that three types of radio waves exist: short, middle, and long waves that can propagate over long distances and can penetrate the walls of various buildings. After World War II an increasing number of people came into contact with microwaves and radar waves with positive and negative consequences. Heating food with microwaves is usually a positive experience, whereas traffic speed control using radar may be a negative experience.

Another kind of physical phenomena that early humans came across is electricity. From the lightning of a thunderstorm men learned intuitively that electricity can be a powerful phenomenon, sometimes even with lethal consequences. The impressive combination of lightning and thunder prompted all early civilizations to attribute a divine origin to these phenomena. As late as in the second half of the nineteenth century, a rational explanation on the basis of laws of nature was elaborated. The Europeans of the antique world also became acquainted with another kind of electric phenomena, namely with electrostatic charge. At least 5000 years ago, perhaps even earlier, people living along the coast of the Baltic Sea began to collect amber. Its high value, comparable to that of gold, allowed merchants to profit from trade over long distances, because amber was easy to transport. Therefore, small pieces of amber were found in the graves of Mycena and in the graves of Egyptian pharaohs. Amber has the conspicuous property of rapidly developing a high electrostatic charge when rubbed on wool or other materials. Electrostatic amber attracts or repels other objects, depending on their electric charge. Because of the electric properties of amber and because the Greek name for amber is *ελεκτρου* (elektron), this word became the origin of the modern terms electron and electricity.

Early humans were not only confronted with physical phenomena and the physical laws of nature, but also with biological and chemical phenomena. The three most fundamental biological laws are the inevitability of death, the necessity of breathing, and the necessity of eating and drinking. A particularly interesting aspect of human evolution is the human attitude towards its nourishment. Almost all apes are perfect vegetarians. Hunting and eating of other animals has been observed, but is a rare exception. The emancipation from apes was accompanied and perhaps even stimulated by the change from vegetarian to protein-rich food.

Humankind became aware that fish and meat are more efficient kinds of food and even accepted the risk of hunting dangerous animals such as bears or buffalos. Perhaps, men understood intuitively that meat is the same kind of material as the muscles in their own body and that the skin of animals is almost identical to their own skin. However, until the twentieth century men did not know the exact law behind their nutrition customs. This law says that the proteins making up muscles, skin, nails (claws), and tendons consist of 20 different α-amino acids, 12 of which can be synthesized by the human metabolism from various substrates. However, eight α-amino acids, the so-called essential amino acids, cannot be synthesized and need to be obtained from the proteins of plants or animals. When the supply of essential amino acids is insufficient, the body begins to consume its reserves, the muscles. The result can be seen in the horrible pictures of those who survived the Nazi concentration camps or the Russian concentration camps in Siberia.

As a result of protein-rich food, which is also calorie-rich food, humans did not need to spent all their time on retrieving food. Eventually, professions such as artists, philosophers, and politicians appeared on the scene that did not make a significant contribution to the survival of humankind.

Perception of all the natural laws mentioned above was based on causal interpretation of observed phenomena. However, in the second half of the nineteenth century, scientists had to learn that certain natural phenomena require statistical consideration and calculations. The first scientific problem that provoked a statistical approach was explanation of the gas laws on the basis of molecular motions. It was found that the motions of an individual molecule such as O_2 (oxygen), observed for a short time, do not allow prediction of the properties of 10^{22} O_2 molecules (approx. 1 L) observed for a longer time and vice versa. The same is true for molecules in a liquid. The upward, downward, or sideways motions of an individual H_2O molecule does not allow any prediction of flow rate or flow direction of a larger quantity, say 1 L of water (approx. 3×10^{25} H_2O molecules). The radioactive decay of 10^{23} uranium-235 atoms can be reliably described and predicted by a so-called first-order law, which says that the moiety of all atoms have decayed within the half-life time. When the half-life time is over, the remaining atoms again require a half-life time for the decay of 50 %. However, it is impossible to predict the moment of decay of an individual atom. Not all scientists were happy when the statistical view of natural phenomena appeared on the scene. The Nobel Prize laureate Sir Ernest Rutherford (see Sect. 10.3) is reported to have advised his students with the words: "If your experiment needs statistics you ought to have done a better experiment."

Interestingly, even humans of the stone age were capable of observing statistical phenomena and of intuitively evaluating their observations. This capability was a direct consequence (or perhaps a pre-requirement) of their change from strict vegetarians to carnivores. Men became hunters and most animals that were suitable as prey lived in groups or herds. Consider a herd of grazing antelopes that advances with a speed of 500 m/h. Not all members advance in exactly the same direction at any time and with identical speed. Older members may be slower and lag behind. Young animals may romp about, may move slower or faster than the herd, may move sideways or backwards and may even leave the herd for a short time. Those animals that leave the herd are the preferred prey of hunters.

These examples are, of course, not an exhaustive enumeration of all the natural laws that human kind has perceived during its long history. However, the examples presented above demonstrate that the interaction of men with the laws of nature occurred in three stages:

- Both the human body and brain obey the laws of nature and the vast majority of these actions are unconscious, such as the control of body temperature by the brainstem or the 24-h biorhythm.
- The second stage is conscious but intuitive perception and identification of the laws of nature, followed by inductive inferences, bare of hypotheses and theories, that have the purpose of supporting the survival of individuals or groups of humans.
- The third stage is a conscious and intentional search for the laws of nature, involving quantification of observations/measurements and condensation into chemical or mathematical formulas whenever possible. This stage is characteristic of modern science. Inductive inferences may be followed by hypotheses or theories and by technical inventions that fuel the progress of human civilization.

3.2 Laws of Nature and the Progress of Civilization

For almost two million years human civilization was based on only five materials: stone, wood, vegetable fiber, and animal bone and skin. Progress could only be achieved by improving the skills of carving, cutting, and fixing those materials and it was extremely slow. A new dimension, the biggest step forward in the history of human civilization, was opened when humans learned to produce new materials by means of chemical technology. Usually, the progress of civilization is described in terms of new tools, new instruments, and new machines. This restricted view ignores the fact that almost any kind of technical progress is based on new materials. To illustrate this point: even the invention of simple tools such as pincers or scissors was not possible in the stone age. Therefore, the following text emphasizes the role of new materials and the underlying invention of chemical technologies.

The fist chemical technology that allowed the production of new useful materials was "invented" about 30,000 years ago, namely the production of pottery by baking objects hand-formed from wet clay. The oldest items of pottery discovered so far were found in the Czech Republic. They were dated to an age of between 25,000 and 30,000 years. The oldest ceramic vessels found in China (Jiangxi) date back to 20,000 B.C. These dates clearly demonstrate that the production of pottery preceded agriculture and was not a consequence of agriculture, as is often assumed. Heating clay to temperatures above 900 °C is not a simple drying process; it involves various chemical reactions that may be summarized as rearrangements of silicon–oxygen and aluminum–oxygen bonds and it includes condensation of Si–OH groups with formation of new Si–O–Si bonds. The consequence is a transformation of soluble quasi band-like or sheet-like silicates into a three-dimensional network that is rigid and insoluble in water, even upon heating.

Pottery may be classified in several categories, which may partly be overlapping:

- Earthenware, stoneware, or porcelain
- Products used for dishes, decoration, or the construction of buildings
- Porous and nonporous products
- White/colorless pottery or colored pottery
- Pottery free of glaze or coated with glaze

Earthenware is the result of relatively low firing temperatures, typically in the range of 900–1000 °C. These low temperatures allow the production of more or less porous products, such as flower pots, normal bricks, and tiles. The penetration of water can be prevented by a waterproof glaze, a technique used for Dutch tiles, Majolica, and Faience. The firing temperature typically used for stoneware is in the range of 1100–1300 °C whereby nonporous, dense products such as clinkers (hard bricks), drain-pipes, and terracotta are obtained. Porcelain requires firing temperatures in the range of 1250–1450 °C so that a dense, colorless, and translucent pottery is obtained. Porcelain was first produced in China from an iron-free white kaolin, also called China clay.

The yellowish, red, and brown colors characteristic of the vast majority of pottery and ceramics result from the presence of fully oxidized iron ions (Fe_2O_3, when pure). A black color is based on partially reduced iron oxide (Fe_3O_4 when pure) and is obtained when the atmosphere around the heated pottery has a reducing character (i.e., deficient in oxygen). The chemical laws responsible for oxidation and reduction processes were elucidated by the French chemist Antoine Lavoisier (see Sect. 9.2). However, numerous early civilizations, notably the ancient Greeks, intuitively understood the relation between the color of their pottery and the reducing character of the atmosphere surrounding the pottery during the firing process. Therefore, they were able to produce (and to reproduce) pottery with red and black decorations without needing to use any glazes. Over the past 30,000 years probably around 50 billion people have lived on earth, which means that certainly more than 100 billion pieces of pottery and ceramic have been produced, a figure that illustrates the reliable reproducibility of the chemical reactions involved the production of pottery. In other words, over a period of 30,000 years potters and their customers have trusted in the reliability of these chemical laws of nature.

The second chemical technology, which was invented in the late Paleolithic, was probably of even greater importance: production of the first metal, copper. The first production of copper is documented in Anatolia around 7000 B.C. Knowledge of this technology migrated slowly from east to west. It was only around 5000 B.C. on the Balkan and around 4000 B.C. in valleys of the northern Alps that production of copper was introduced. Hence, for countries of the near orient and southern Europe a copper age preceded the bronze age.

The people of the Paleolithic noticed that the properties of copper, notably its hardness, varied with the origin of its ore (copper sulfide). Small amounts of arsenic, antimony, lead, or zinc present in the copper ore in the form of sulfides were responsible for the variation in properties. However, the early metallurgists were, of course, not able to discover this relation. Pure copper is a ductile metal allowing easy shaping by mechanical methods such as hammering, but it is not hard enough to be used for tools or weapons. Unfortunately, it is still not known how the early metallurgists discovered that mixing copper with a smaller amount of tin considerably enhances the hardness and enables systematic variation in hardness and other properties as a function of the tin content in the alloy.

The discovery and development of bronze was a complex process for three reasons. First, tin ores are relatively rare and were difficult to discover, because almost all tin minerals are colorless. Second, tin ores are usually not found in the immediate proximity of copper ores, so it was necessary to establish an international trade network to bring the tin and copper ores (or both pure metals) together. Third, the only tin mineral used in the Neolithic and in the antique world was the oxide cassiterite (SnO_2) and its transformation requires reaction with charcoal or another organic reducing agent ($SnO_2 + C \rightarrow Sn + CO_2$). This reaction involves the opposite concept to the process of copper production from copper sulfides. The latter requires heating with oxygen (air) to achieve oxidation of the sulfur and liberation of the copper metal ($CuS + O_2 \rightarrow Cu + SO_2$). Even without any understanding of the chemical background of these reactions, humankind learned to make use of them and to combine two quite different metals to obtain a useful alloy. Both chemical reactions are laws of nature that have been reproduced several million times by humanity over the past 5000 years.

The production of bronze alloys was first developed in Anatolia or in valleys of the Caucasus around 3500 B.C. and knowledge of this procedure spread in all direction. On the Balkan and in Egypt the oldest bronze objects date back to 3200–3100 B.C. and in Mesopotamia or China to 3000–2900 B.C. However, it took more than another 1000 years until this knowledge had completed its way through Europe and arrived in Scandinavia. The availability of bronze initiated numerous groundbreaking technical inventions. For example, tools such as pincers or scissors and simple but extremely useful items, such as nails, became available. Probably the most influential invention was the sword, which took its place as the most efficient weapon for more than 4000 years. It was still used by the French and German cavalry in the war of 1870/1871 (sabers made from steel at that time). After 1000 B.C. bronze was gradually substituted by the harder and cheaper iron/steel, but this substitution was never complete. In the twenty-first century bells are still cast from bronze and not from iron. One could even say that the copper age has never ended, because copper wires are widely used for their extraordinarily high electric conductivity to make electricity available all over the world and 24 h per day.

The invention of technology allowing the production of iron was made around 1400 B.C. in locations along the southern coast of the Black Sea. Depending on whether the iron ore used for the production was a sulfide (e.g., FeS_2) or an oxide (e.g., Fe_2O_3), either an oxidation process analogous to that of copper sulfide or a reduction process analogous to that used for tin oxide had to be applied. The main problem was the relatively high melting temperature (>1500 °C) of iron. It was difficult to achieve this temperature with the rather primitive furnaces of that time. Permanent improvement of the furnaces made melting and casting of iron possible in the fifteenth century, but the casting of cannon balls and cannon barrels together with the production of small firearms revolutionized warfare. This development had tremendous consequences for world history.

It took three more centuries before large production of steel (a modification of iron containing a little carbon) became feasible. Yet, the availability of large amounts of steel after 1800 entailed a technical revolution in the nineteenth century. Invention of the steam engine, the construction of steamships, and the development of railroad networks in Europe and in North America are symbols of this revolution.

Meanwhile, development of another metal began: aluminum. Small amounts (<1 g) of aluminum were prepared for the first time in 1827 by the German chemist Friedrich Wöhler (see Sect. 9.3). The aluminum–oxygen bond is particularly strong and, therefore, all natural minerals of aluminum contain Al–O bonds. Furthermore, reduction of aluminum oxides with the only widely available and inexpensive reducing agent, carbon, is impossible. Hence, it was the availability of cheap electricity at the beginning of the twentieth century that made technical production of aluminum possible. The cleavage of aluminum oxide by electrons ($2Al_2O_3 \rightarrow 2Al + 3O_2$) is a law of nature that enabled the reproducible production of aluminum over many centuries in numerous countries all over the world. The availability of aluminum stimulated, in turn, numerous inventions and new technical developments. To name only one example: Because of its useful mechanical properties in combination with its low specific weight (one third that of copper) aluminum is an ideal material for the construction of airplanes.

A fourth element, a so-called metalloid, that has had a conspicuous career in human civilization only for the past 100 years is silicon. This element was first prepared in small quantities in 1824 by the Swedish chemist Jöns J. Berzelius (1779–1848). After oxygen, silicon is the second most abundant element on (and in) the earth's crust. Its technical production is usually based on the reduction of silicon dioxide (sand, quarz) with carbon. Silicon can be obtained in the form of a gray amorphous powder or in the form of shiny pure crystals. Silicon is a semiconductor. Its moderate conductivity increases with temperature or upon irradiation with sunlight. Furthermore, the conductivity can be enhanced and systematically varied when extremely pure silicon is doped with small amounts of another element containing one electron more or less than silicon in its outer orbitals (for orbitals see Sects. 10.3 and 10.4). Doped semiconductors, the foundation of microelectronics, became the basis for the invention and development of numerous new instruments, electric devices, and gadgets, for example, allowing enhancement and variation of electronic signals. These devices are built into every single computer and into all mobile phones and smart phones, a true technical revolution.

Whereas the production of silicon from sand or silicates is based on chemical laws, the purification of crystalline silicon is based on a physical law of paramount importance. This law, at first glance very simple, states that impurities lead to a reduction in the melting temperature of a uniform compound or element. The melting temperature can, therefore, serve as a sensitive criterion for the purity of an element or compound, and the comparison of melting temperatures before and after a purification step informs the chemist about its success. Before World War II, analytical instruments and methods enabling scientists to check the purity of elements and chemicals did not exist. Determination of the melting temperature was the only universally applicable method and, hence, this method and its underlying law were decisive for the progress of any kind of chemistry and material science at that time.

The past 150 years may be understood as a period in which organic chemistry, including biochemistry, pharmacy, and polymer chemistry (see Sects. 9.4 and 9.5) have played a decisive role in the progress of human civilization. As already mentioned, more than ten million chemical compounds have been synthesized and their structures unambiguously determined. Many thousands of these play a role in everyday life, in medicine, and in almost every kind of scientific research. Even theoretical physicists would have difficulties in working under the conditions of the stone age. It is, of course, neither possible nor reasonable to try to itemize in this chapter all chemicals and materials used today, but one aspect of extraordinary importance should be highlighted here, namely the number of people living on earth.

Over a period of two million years until the Middle Ages the worldwide population stagnated at a level below 500 million. Until the middle of the nineteenth century, this number increased to about one billion. The main reasons responsible for this stagnation were infectious diseases such as plague, smallpox, and cholera; endemic diseases such as malaria, tuberculosis, and diphtheria; and last but not least, infected wounds. Another factor that contributed to the stagnation was the unreliable and insufficient supply of food. After 1850, an explosion-like increase in population occurred up to the present number of more than seven billion people. Where did this unexpected, unpredictable, and undesired (at least from the viewpoint of animals and environment) increase come from? Four factors, which were all somehow related to the progress of chemistry, played and continue to play important roles:

1. Around 1860 Louis Pasteur definitely proved that spontaneous generation of living organisms from dead matter does not exist (see Sect. 8.1). Pasteur's work stimulated a rapid expansion of the concept of hygiene. Physicians, such as I. Semmelweis and J. Lister, introduced antiseptics in hospitals (see Sect. 7.2) and the application of hygiene standards in everyday life. This achievement considerably reduced the outbreak of infections such as cholera and dysentery, and many other epidemics. All the chemicals needed to maintain a high level of hygiene (e.g., soaps, surfactants, and insecticides) are produced by the chemical industry on the basis of laws of nature.
2. Biologist and chemists discovered or invented numerous antibiotics. Large-scale production of these antibiotics has contributed to saving the lives of many millions of patients suffering from infectious diseases and still continues to do so.

3. Chemists invented and developed synthetic procedures for the production of millions of tons of fertilizers and agrochemicals. In the first half of the nineteenth century millions of Europeans were starving and many thousands died of hunger. Today, more than 80 % of the European population owe their survival to the food supply based on fertilizers and agrochemicals.
4. The availability of electricity prompted the construction of refrigerators and deep freezers that play an important role in broader and easier distribution and longer storage of all kinds of food. However, most people, including academics, who profit from this progress are oblivious of the fact that the application of electricity and construction of refrigerators or freezers requires materials produced by the chemical industry and that their function involves several physical laws.

In an article entitled "Why We Need to Understand Science" the astronomer Carl Sagan (1990) wrote: "We live in a society exquisitely dependent on science and technology, in which hardly anyone knows anything about science and technology."

In summary, any kind of philosophy criticizing or denying the reliability of inductive inferences or the reliability of laws of nature is in contradiction to 30,000 years of progress of human civilization based on chemical technology and an understanding of physical laws. Furthermore, it ignores the fact that this progress has enabled the growth in population and existence of billions of people, including skeptical historians, philosophers, and all kinds of antiscientists.

3.3 Laws of Nature and the Future

The question of if and to what extent scientific knowledge, notably the laws of nature, allows a forward look was discussed even at the early stages of modern science. The Scottish philosopher David Hume (1711–1776) held that the utility and benefit of science consists in teaching humanity how to control and regulate future natural events (e.g., potential catastrophes) by knowledge of their causes. In his *Treatise* he also said (Bk.1, Pt. III, Sec. II) that the only relation that enables us to infer from observed matters of fact to unobserved matters of fact is the relation of cause and effect. A similar stance was taken by the Austrian physicist Ernst Mach (1832–1916) and by the German Heinrich Hertz (1857–1894). They believed that it is the purpose of physics to "foresee future experiences." After World War I the so-called Vienna Circle organized and headed by the physicist and philosopher A. M. Schlick (1882–1936) advocated a similarly positivist but less extreme view. In contrast, the emigree Austrian philosophers Ludwig Wittgenstein (1889–1951, see below), K. Popper, (see Sect. 2.3), and P. Feyerabend (see Sect. 4.3) advocated a skeptical view on the predictive power of science.

The question of the extent to which the reproducibility of natural phenomena or experiments allows prediction of events, at least for the near future, was not and is not an object of permanent debate among scientists and philosophers of science; it was and is of permanent importance for the survival of humanity and for the development of its civilization, as discussed in the preceding sections. It is a triviality that all scholars and laics who reject science as a desirable, important, or useful human enterprise have a negative attitude towards predictions based on scientific results. More interesting and worth discussion is the criticism of the aforementioned skeptic philosophers. Wittgenstein's critique, like Popper's, starts from formal logic. In his work *Tractatus logico-philosophicus*, published in 1922, Wittgenstein wrote that "There is no compulsion making one thing happen, because another happened" and he presented a concrete example: "It is a hypothesis that the sun will rise tomorrow, and this means we don't know that it will rise." This latter statement combines two fallacies. The first is typical of many philosophers of science (and also shared by Popper), namely the incorrect understanding of the term "hypothesis," as already discussed in Sect. 2.4. All hypotheses result from inductive inferences, but not all inductive inferences are hypotheses. According to the original meaning of this Greek term, all hypotheses are, in principle, falsifiable but as demonstrated in Sect. 2.3 not all inductive inferences are falsifiable. All humans are mortal is an inductive inference but it is neither a falsifiable statement nor a hypothesis.

The second fallacy is application of formal logic to scientific research. In Sect. 2.3 it has already been demonstrated that Popper's defamation of inductive inferences based on formal logic is not justified. Wittgenstein's example provokes the following objection. Sunrise and sunset have been known to humanity since its emancipation from apes, and current astronomical and physical knowledge convincingly explains that the regular rotation of the earth around its axis dates back to the time when the earth–moon system was born about four billion years ago. Sunrise and sunset are based on laws of nature (e.g., on gravity) and it is inherent in the definition of natural laws that their effects are reproducible irrespective of time and place (within the limits of their validity). Reproducibility irrespective of time means, in turn, that the effects of natural laws are reliably predictable as long as the experimental (natural) conditions do not change. Therefore, it is reliably predictable that the sun will rise tomorrow, provided that the solar system maintains its current structure. The life cycle of the sun and similar stars is known. The sun will maintain its current structure and activity for another four billion years and during this period of time sunrise and sunset will be observable on earth, even though the 24-h rhythm will gradually change.

In summary, the positivist position holds that reproducible experiments allow reliable inductive inferences and the resulting universal statements represent laws of nature. These laws allow, in turn, reliable predictions for the future within the limits of their validity. The author shares this positivist or, better, empiricist view, which is not identical with an absolutely reductionist stance (see Sect. 4.4). This positivist position stands on two legs. The first leg is 50 years of experimental research (including temporary cooperation with surgeons), which allowed the author to observe numerous times the effect of the self-healing mechanism in his own working field (see also Sects. 9.4 and 9.5). The second leg is an "institution" that is bare of religious, ideological, or emotional prejudices and bare of linguistic ambiguities. This "institution" is the evolution of living organisms. Evolution could and can only proceed on the basis of trustworthy laws of nature. All living organisms are confident that the natural laws that have influenced their life yesterday and influence their life today will also exist tomorrow with the same reliability and efficacy as before. When the next day is over, the organism has learned that the expectation and confidence it had today was justified, and this process continues day by day. In other words, all living organisms, including skeptical philosophers, are inductionists in everyday life, irrespective of whether they are conscious of their behavior or not. This is not a new insight of the author. This stance was first formulated by the British philosopher John S. Mill in 1843 (see Sect. 4.2), and in the twentieth century numerous evolutionists and philosophers of biology have supported this view with results of their evolutionary research.

In this connection, it may be useful to have a short and very simplified look at the influence of natural laws on the ontogenesis of a human. The life of an individual begins with the fecundation of an egg and the formation of a complete genome by combination of the male and female chromosomes. This step is counted here as the first law. The fecundated egg is then fixed in the mucous membrane of the uterus, which controls the nutrition of the growing fetus (second law). Then follows a division of the fecundated egg cell into two daughter cells (third law) that undergo an analogous division, so that four cells having identical genomes are formed (fourth law). After birth, the separation of mother and baby requires cutting of the navel cord (fifth law). Immediately after cutting the navel cord, the baby must begin to breathe (sixth law). Although the fetus was kept sterile inside the body of his mother, it becomes "contaminated" during the birth with bacteria, notably those of the family *Lactobacillus*, which the baby needs for the digestion of milk (seventh law). Later, the child needs teeth for gnawing and digestion of solid food (eighth law). Furthermore, any motion of a limb involves the law of the lever (see Sect. 3.1).

When all the details of metabolism are taken into account, it seems that thousands of laws control the growth and survival of the child. All the laws of metabolism are likewise operating in the body and brain of an adult individual, but humans are usually unconscious of the numerous natural laws that are involved in all functions of their body. For example, the brainstem regulates the temperature of the human body, the concentration of glucose in the blood, and keeps the pH of blood and cerebrospinal fluid within the limits of 7.35–7.45, because otherwise the nerves of the brain would be damaged. These physiological activities happen in the brain and body of all humans, including antiscientists and skeptic philosophers. Therefore, all laics and scholars who take an antiscience stance and all those who criticize inductive inferences and the laws of nature owe their existence to the wisdom of evolution which ignores their criticism or skepticism and considers formal logic as irrelevant for nature and its laws.

The final comment on Wittgenstein's statement about sunrise concerns the fact that most animals and all humans, including Wittgenstein himself, have a metabolism that obeys a 24-h biorhythm. If this biorhythm is ignored for a long time, the organism becomes sick and finally dies. This is one more example of the fact that living organisms trust in the reproducibility of natural phenomena in the future, provided that these phenomena obey the laws of nature, and the daily sunrise is no exception. The British poet Lord Byron (1788–1824) is known for the insight: "The best prophet of the future is the past." A complementary approach was formulated by the American scientist Alan Kay: "The best way to predict the future is to invent it."

Furthermore, critical statements by certain physicists about the predictability of scientific measurements and results need discussion. The German physicist and Nobel Prize laureate Werner Heisenberg (1901–1976) is widely known for his "uncertainty principle." This principle states that it is impossible to determine all the properties of elementary particles (e.g., impulse, spin, location) simultaneously. Moreover, any experiment designed to determine the properties of elementary particles modifies their properties at the moment of measurement. In other words, the properties of a single particle are not all predictable for a certain place at a certain time. In his own words: "The more precise the measurement of position, the more imprecise the measurement of momentum and vice versa"; "Every experiment destroys some of the knowledge of the system which was obtained by previous experiments"; and (as cited in Sect. 2.4) "Therefore, it will never be possible by pure reason to arrive at some absolute truth." The problem with Heisenberg's and Born's statements is that they presented their experience with elementary particles as if it was valid for any kind of natural phenomenon and experiment. In this way, Born, Heisenberg, and their followers have contributed to a negative image of natural sciences, as if any kind of scientific result is uncertain and unreliable. Today, it is a triviality for scientists that the laws of nature typical for the macroscopic world (the world of human dimensions) are not applicable to elementary particles, but it also true that the properties and laws typical for elementary articles may not be extrapolated to the macroscopic world. Three examples should illustrate this statement.

In contrast to an electron, a moving car allows simultaneous characterization of several properties, such as color, length, form, speed, and location, by combination of three techniques: photography, radar, and GPS measurements. Far more important than this example is the "behavior" of large quantities of elementary particles, such as 10^{23} electrons. The metallic element sodium contains one reactive electron in its 3s orbital (in the ground state).

An amount of 3.7 g of sodium contains 10^{23} atoms and, thus, 10^{23} 3s electrons. When heated in close contact with the gaseous element chlorine, the 3s electron jumps from sodium to chlorine and the net result of this (nearly) quantitative reaction is the formation of sodium chloride, commonly known as table salt. This experiment is reproducible and predictable. All the thousands and millions of chemical reactions known so far, have in common that the electrons in the outer orbitals of atoms and molecules react in an uniform manner with electrons of other atoms or molecules. A striking example is the worldwide production of more than one billion aspirin tablets over a period of 120 years.

A much larger number of reproducible chemical reactions are performed by nature itself via photosynthesis. Photosynthesis is the synthesis of glucose from carbon dioxide and water by means of activated electrons. The green chlorophyll molecules absorb photons from sunlight and use their energy to activate electrons. Hence, synthesis of a glucose molecule involves the cooperation of two sorts of elementary particles, photons and electrons. Because all green organisms, cyanobacteria, grasses, and the green leaves or needles of bushes and trees perform trillions of photosynthetic reactions per year, and because photosynthesis was "invented" by evolution about 3 billion years ago, the number of photosyntheses that have occurred on earth is not seriously computable. The glucose and its reaction products, starch and cellulose, serve as nutrients for almost all species from microbes to humans. Furthermore, it is essential for plants as a building block for polysaccharides. In other words, over the past three billion years evolution has relied on the reproducibility and predictability of all reaction steps making up successful photosynthesis.

Finally, the discrepancy between the mental world of antiscientific scholars, skeptic physicists, or philosophers, on the one hand, and their daily life, on the other, needs a short discussion. Coming back to Heisenberg, one of his first duties after World War II was to develop a concept for the construction of the first two nuclear power plants in the Federal Republic of Germany. It is known that Heisenberg was a religious man and not a cynic. If so, how could he dare to develop and recommend the construction of nuclear power plants if he was not convinced that the atoms and elementary particles react in a calculable and predictable way. Worldwide more than 400 nuclear power plants exist today and their explosion would extinguish humankind. Obviously, all the physicists who contributed to the construction of these power plants are also convinced that their calculations are reliable and that the reactions of uranium and plutonium atoms are predictable.

Considering everyday life, it is worth mentioning that probably all antiscientific scholars, skeptic scientists, and philosophers like to live in houses. They work in libraries or in institutes and they deliver lectures in lecture halls. Obviously, they all trust in the stability of these buildings that were constructed by means of static considerations and calculations. Regardless of the extent to which an architect has used mathematical formulas, all static calculations are based on the laws of nature, such as the law of the lever, the relation between force (weight) and pressure, or Newton's law of gravity. Newton's law is only a limiting case of Einstein's law of gravity, but Einstein's theory of general relativity is not needed for the construction of buildings. Some of the most complex buildings ever erected, the Gothic cathedrals, were constructed without knowledge of Newton's law of gravity and without support from computers. Physicists plan and equip huge research centers, such as the CERN in Switzerland, expecting that they will be stable and useful at least for the next 100 years. Furthermore, physicist and philosophers plan conferences for the coming two or three years and expect that the hotels and lecture halls they will use are safely constructed and will be available at that time.

Another, even more important example, concerns the daily diet. The biographies of antiscientists and skeptic philosophers suggest that they eat normal food, just like other people and positivist scientists. Yet, why is this so if the laws of nature either do not exist or are unreliable and their consequences unpredictable?

The vast majority of people eat food based on starch, such as rice, wheat, rye, corn, and potatoes. Starch is a biopolymer built up of glucose units, and the human digestion degrades the starch back to glucose (which is finally oxidized by oxygen and yields almost all the chemical and thermal energy needed by the human body). Cellulose, the main component of grasses, bushes, and trees, is also built up of glucose but the chemical bond connecting the glucose units is different from that in starch. Why do antiscientists and skeptical philosophers not eat cellulose, which is inexpensive and available everywhere in the form of cotton or paper? The answer is simple. Although their frontal cortex does not know much about biopolymers, both their subconscious and their body know that only starch can be digested. It is a law of nature that the human body produces enzymes enabling the digestion of starch, and a complementary law that it cannot produce enzymes for the digestion of cellulose. Both laws are examples of laws of nature that do not need a mathematical formula and that are non-falsifiable truths because they are constituents of the biological and chemical definition of *Homo sapiens*.

More generally speaking, antiscientists and skeptical philosophers, like almost all people eat normal food because their body and subconscious are convinced that the biochemical and physiological laws responsible for digestion and metabolism of food will do their job in the future as they did in the past. Therefore, antiscientists and skeptical philosophers who criticize the existence of laws of nature or their reliability (and predictability of consequences), but obey numerous biological, chemical, and physical laws in everyday life, demonstrate a kind of intellectual schizophrenia. This phenomenon has not been identified and classified by physicians or psychologists yet and the author proposes the abbreviations AS for "antiscientist schizophrenia" or SPS for "skeptic philosopher syndrome."

3.4 Rules, Laws, and Their Importance

In the *Encyclopedia Americana* (6th edition) the keyword "law of nature" does not exist and under the keyword "law in science" the following description may be found: "A scientific law is a general statement that purports to describe some general fact or regularity of the universe. For example, Newton's law of gravitation. ... Any such regularity may be termed a law of nature."

The philosopher John W. Carroll wrote in the *Stanford Encyclopedia of Philosophy* (2010): "Two separate (but related) questions have received much recent attention in the philosophical literature surrounding laws. Neither has to do with what it is to be a law. ... First, does any science try to discover exceptionless regularities in its attempts to discover laws? Second, even if one science—fundamental physics—does, do others?

The German physicist T. Millack asks in his book *Naturwissenschaft und Glaube im Gespräch* (Science and Religion in Dialogue): "Should any minor chemical rule be called law of nature?"

Schrödinger said in his frequently published and cited lecture (see Sect. 2.2) that "Laws of nature are obviously nothing more than a sufficiently confirmed regularity of phenomena."

These and similar statements raise two questions that deserve discussion. First, are the meanings of rules and laws almost identical, and, if not, what is meant by regularity? The above statements reflect the traditional prejudice of physicists and physics-infected philosophers that that laws of nature are a property of physics, whereas biological, chemical, and geological laws are a kind of rule (see Sect. 2.2). Second, is it feasible to establish a classification of laws of nature according to their importance for a scientist or even better for the entire human race?

In contrast to definitions of scientific laws or laws of nature, a definition of scientific rules is difficult to find in encyclopedias. However, the common understanding of rules as the principles underlying regular phenomena or observations includes temporary or local irregularities and includes the experience that exceptions confirm the rule. In contrast, laws of nature have limits with regard to temperature, time, and location (e.g., most laws detectable on earth are not effective in black holes), but within those limits laws do not have exceptions. Consider the existence of gravity on earth over the past four billion years and suppose that gravity was absent at a tiny spot for only a few months. The atmosphere would vanish within a few hours and the resulting vacuum would attract the neighboring atmosphere so that eventually the atmosphere of the entire earth would disappear within a few weeks. All plants and animals would explode and eventually all water would vanish. In other words, if there had been even the faintest temporary and local exception to the law of gravity it would have been detrimental for all life on earth. Therefore, scientific rules and laws should not be confused. Terms such as regular and regularity do not allow a clear distinction between rules and laws unless they are more precisely specified. For example, terms, such as "strictly regular" or "regularity without exceptions" indicate that a law is meant.

An interesting example of the confusion between rule and law concerns the terms "law of the lever" and "lever rule." Apart from the different meaning of both terms, the term lever rule is incorrect. The lever rule is a method for determining the percentage weight of a liquid or solid phase in an equilibrium of a given binary system and temperature. In an biphasic alloy of the elements A and B, the weight percentage of the α-phase can be calculated according to the lever rule via the formula: $X_\alpha = (c - b)/(a - b)$, where a is the percentage of b in the α-phase, b its percentage in the β-phase, and c the percentage of b in the entire alloy. This formula emphasizes that the lever rule is in fact a law of nature.

Millack's question and Kuhn's definition of paradigm changes (see Chap. 5) draw attention to if and how the importance of laws of nature and scientific knowledge in different disciplines of science can be determined and compared. Within a discipline, comparisons can be relatively easy. For example, most, if not all, chemists agree that the knowledge required for synthesis of aspirin is more important than the knowledge needed for synthesis of the dyestuff Bismarck Brown. Yet, even between closely neighboring disciplines, such as organic and polymer chemistry, comparisons resulting in clear decisions may be difficult. For example, which is more important, invention and development of styrene-containing polymers or the synthesis of aspirin? Styrene is a component of all tires and of the widely used engineering plastic ABS. Pure polystyrene is used as solid foam (Styropor), as engineering plastic, and for food packaging. In most applications it is difficult to find a suitable substitute for styrene, whereas aspirin can be replaced by paracetamol, metamizole (novaminsulfon, dipyrone), or ibuprofen.

As demonstrated below, a reasonable comparison and classification across the imaginary borderline between different branches of science is impossible. As outlined in Sect. 2.1, physics has and will continue to provide important contributions to a better understanding of the universe and, thus, to the world view of humankind. For the past 150 years, chemistry has not made such a contribution, but chemistry is responsible for most of the improvements in daily life and for decisive contributions to the progress of medicine. Approximately 95 % of all medicaments are produced by the pharmaceutical industry and all instruments and equipment in hospitals are based on materials produced by chemical companies (including producers of metals).

According to the historian and theoretician T. Kuhn (see Chap. 5), knowledge acquired in chemistry and in physics is incommensurable. Nonetheless, it is informative to study the following four fictitious examples:

First example: A group of physicist is asked what knowledge they consider to be more important for them, Max Planck's discovery and calculation of the Planck constant or the knowledge required for the synthesis of Bismarck Brown. It is no risk to predict that all physicists will vote in favor of Planck's calculations.

Second example: Various scientists with severe toothaches are asked what they think is more important for humanity, the knowledge required for the synthesis of local anesthetics (with the consequence of painless dental treatment) or more knowledge about quarks. In this case it is unlikely that all scientists will vote for the same alternative and the result is unpredictable.

Third example: A group of people with cancer and awaiting a surgery are asked what is more important for humanity, the invention and production of narcotics or Heisenberg's uncertainty principle. It can be assumed that most of these people will vote for narcotics.

Fourth example: The entire population of a European country is carefully instructed that only 5–10 % of the population would survive if the chemical industry did not know how to synthesize fertilizers and agrochemicals, such as herbicides, pesticides, and insecticides. The question asked is: What is more important for the country, the production of fertilizers and agrochemicals or knowledge of Einstein's theory of special relativity? Now, it is foreseeable that most, if not all, people prefer to survive without knowing anything about Einstein's special relativity. In the modified words of the German playwright and theater director Berthold Brecht: “First the food and afterwards theoretical physics.”

These examples allow at least the inference that it is impossible to establish a list of increasing (or decreasing) importance of scientific knowledge that is acceptable for all people.

Bibliography

Millack T (2009) Naturwissenschaft und Glaube im Gespräch. Oldenbourg, München

Sagon C (1990) Why we need to understand Science. Skept Inq 1990 14(3)

Wittgenstein L (1922) Tractatus logico-philosophicus. Translated by C. K. Ogden from the German original: Logisch-philosophische Abhandlung (1921) Annalen der Naturphilosophie 1921:14

Chapter 4
Antiscience and Antireductionism

4.1 Antiscience

The first principle is that you must not fool yourself, and you are the easiest person to fool.
(Richard Feynman)

Antiscience is a mental position that criticizes science in general and certain scientific methods in particular. This chapter complements the discussions of critical and skeptical theories about science presented in Sects. 2.3, 2.4, 2.5, 3.3, 4.1 and 4.2. Antiscientific feelings, thoughts, and theories have many sources, including religions, ideologies, and political positions and economic, sociological, or psychological scenarios. Of course, this chapter cannot provide a full account of all the numerous aspects and authors of antiscientific literature. Therefore, the following text is focused on antiscientific critiques published by scientists, historians, and philosophers of science. The history of antiscience can be traced back to Thomas Hobbes (1588–1679) and his *Six Lessons to the Professors of Mathematics* and to Jean-Jacques Rousseau (1712–1778) and his essay "Discourse on the Arts and Sciences." Hobbes underlined the nonrational in human behavior and was skeptical about an objective description and explanation of natural phenomena. He argued: "We can only know the causes of what we make. So geometry is demonstrable because the lines and figures which we reason are drawn and described by ourselves. ... But we can only speculate about the natural world, because we know not the construction, but seek it from the effects." Rousseau criticized that progression of science caused a corruption of virtues and morality. He warned against the dangers involved in political commitment to scientific research and he was concerned about the ways in which a happy future of humanity might be secured. The authors and statements discussed in the following text were selected with the purpose of illustrating the diversity of antiscientific positions and arguments.

Under the ironic title *Science: The Glorious Entertainment*, the French-born American historian Jacques Barzun (1907–2012) described, discussed, and criticized the influence of modern science on the present culture and the difficulties

H.R. Kricheldorf, *Getting It Right in Science and Medicine*,
DOI 10.1007/978-3-319-30388-8_4

most people have with a proper understanding of science. The most serious critique is presented on p. 110: "I called it [science] earlier man's stupendous and unexpected achievement, yet, if one accepts the conclusion of the most advanced physicist-philosophers, science has no stability and yields no truth and deals with anything but the concrete. Besides which, it now feeds and inspires to exacerbate our desires and torment body and mind. What can justify the word 'achievement' for an undertaking so perverse and prolonged." This superficial and crazy statement requires at least three answers. First, if the anonymous physicist-philosophers are Popper, Wittgenstein, Feyerabend, Laudan, and Kuhn, their misleading or wrong statements are revised in Sects. 2.4, 2.5, 3.3, Chap. 5, and below. Second, that science has no utility and deals with nothing concrete is in contradiction to 30,000 years of civilization (see Sect. 3.2) and in contradiction to the immense progress of medicine (see Sect. 5.3 and Chaps. 6 and 7). Third, what would an experienced surgeon, who has performed more than 5000 successful operations based on numerous laws of nature and numerous technical inventions, say after reading Barzun's statement? Most likely he would recommend sending Barzun to a mental institution.

Typical of Barzun's superficial understanding of science is its definition as a body of rules, while the term "law of nature" is missing. His superficial view of science is also evident from his comment on the relation between science and technology (p. 3): "It is not true that technology is the offspring of science." On p. 19, he states that "science was not initiator it was the beneficiary of mechanical inventions" and on p. 20: "The steam engine, the spindle frame, and power loom, the locomotive, the cotton gin, the metal industries, the camera and plate, anesthesia, the telegraph and telephone, the photograph and the electric light, these and other familiar wonders were the handiwork of men whose grasp of science was slight or at least empirical." Part of these statements is wrong (see Part II) and they ignore what should be a triviality for any scientist, namely that the relation between scientific discoveries and technical inventions is not a one-way street. Scientific discoveries stimulate technical inventions, and new instruments and machines support rapid progress in fundamental research. Wilhelm Röntgen's discovery of X-rays and their properties entailed the construction of effective and useful X-ray devices, which enabled, in turn, further progress in chemistry, physics, and medicine. The elucidation of crystal structures and the detection of bone fractures or tumors serve as examples.

Barzun's strange view on science culminates in two statements (p. 110): "The activities of science are properly a game, because in it convention and chance divide the interest" and "Science is a play, because of the very meaning of the inquiry: nor this nor that urgent result, but laissez faire, laisser jouer." On p. 3, he writes: "Yet science is an all pervasive energy, for it is not once a mode of thought, a source of strong emotions, and a faith as fantastic as any in history." All these statements demonstrate the inconsistency in the thoughts of Barzun and leave the reader with the choice of considering science as a stupendous achievement, a body of rules, a game without utility, an unstable enterprise bare of any truth, a pervasive energy, or a fanatical faith. What a fantastic bouquet of attributes!

The American psychologist Robert A. Wilson (1932–2007) was a successful writer, poet, novelist, essayist, and futurist and recognized as an Episkopos and saint of discordianism. A major goal of all his activities was “to try to get people into a state of generalized agnosticism, not agnosticism about God alone but agnosticism about everything.” In his book *The New Inquisition* Wilson advises the reader to be skeptical about any perception because no perception and, thus, no inference or theory based on perception is certain.

In his introduction (p. i) Wilson explains the title of his book: “By the New Inquisition I mean to designate certain habits of repression and intimidation that are becoming increasingly commonplace in the scientific community today.” No explanation follows this obscure statement and the author must confess that he does not understand what kind of repression and intimidation in the scientific community is meant. However, in several paragraphs of his book Wilson mentions that his work is also meant as an attack on fundamentalist materialism. Apparently, he considered this ideology as typical of the mentality of scientists and repression and intimidation as typical of fundamentalist materialism. In this connection, he declares on p. i: “I am criticizing what I call Fundamentalist Materialism—a term I coined over ten years ago, and have used in many articles and a few books—I am opposing the Fundamentalism not the Materialism. (This point will be clarified as we proceed).”

In Chap. 1, Wilson explains his fundamental critique of any kind of perception (observation, experiment) on the basis of the eye–brain system. He argues that the physical part of the vision (see Sect. 10.5) is interpreted by the brain and it is this interpretation that reaches the consciousness and not the physical signal itself. In his own words (p. 4): “Aristotle without knowing the modern laws of optics, understood this general principle well enough to point out once that ‘I see’ is an incorrect expression and really should be ‘I have seen.’ There is always time, however small, between the impact of a signal on our eye and the perception and or image in our brains. In that interval the brain imposes form, meaning, color, and a great deal else. What is true for the eye is true for the ear and other senses.”

On the one hand, this comment is correct, but it is a triviality for most scientists and even for laics with interest in science and medicine. On the other hand, this argumentation demonstrates that Wilson is not familiar with the purpose of the brain activities. If all chemical and physical signals that reach the body and brain immediately reached the consciousness, the huge amount of information would flood the brain within a few minutes. Because of the limited capacity of animal and human brains, it is the primary job of the brain to suppress signals that are not relevant for survival. The selection of signals requires, of course, an interpretation and evaluation process. Errors may occur and may be revised. However, the sheer existence of several billions of species and many trillions of individuals having a central nerve system proves that this evolutionary strategy was and is successful and enables even the existence of skeptical philosophers.

What is amazing and incorrect is Wilson’s ensuing conclusion: “On the face of it, once this has been pointed out, there seems no escape from an at least partial agnosticism, i.e., from recognition that all ideas are somewhat conjectural and inferential. Aristotle escaped from that conclusion, and until recently most philosophers and scientists have escaped it, by asserting or assuming or hoping that a method exists whereby the uncertainty of perception can be transcended and we can arrive at certitude about general principles" (universal statements in the language of K. Popper, see Sect. 2.4).

Wilson supports his conclusion by describing examples of optical illusion or visual hallucination. Furthermore, he illustrated the fallibility of perception with the "warm water experiment" (p. 5): "The Greeks as we say or the ancient Greeks ... were well aware of the fallibility of perception and an illustration well known in Athens in the Golden Ages went like this: take three bowls of water. Make one of them quite hot, one of medium temperature and the third quite cold. Put your right hand in the hot bowl for a while and your left hand in the cold bowl. Then put both hands in the medium bowl. The same water feels cold to your right hand and hot to your left hand."

Because scientists have been aware of the problem of unreliable perceptions for more than 2000 years, they have developed four methods to avoid this problem. The first method consists of the invention of instruments that allow precise measurements that can be reproduced by other scientists. Temperature measurements are performed with thermometers or electronic devices using thermoelements. The second strategy consists of the repetition of the observations/experiments by the same scientist who made the first observation/experiment to check the reproducibility in the hands (and in the brain) of the same scientist. The third method begins with publication of the observation/experiments, so that the international community of scientists has access to the new results. Other scientists then check the reproducibility at another time and at another location. With a positive outcome of all three methods, new experimental facts are established. However, this process does not immediately yield reliable results when new theories are under investigation. The history of science demonstrates that even a large number of experts can adhere to the same fallacy. However, the fourth method, the self-healing mechanism of scientific research described at the end of Chap. 1 and in all examples of Part II, can overcome this problem. It is characteristic of the texts of Wilson and many other philosophers that the role of reproducibility is never discussed.

Finally, Wilson cites the numerous paradigm changes that have happened in the past as evidence for the failure of scientists to elaborate reliable theories. For example, on p. 8, he states the following:

> There remains, of course, Scientific Method (SM), the alleged source of the certitude of those I call the New Idolators. SM is a mixture of SD (sense data, usually aided by instruments to refine the senses) with the old Greek PR (pure reason). Unfortunately, while SM is powerfully effective, and seems to most of us the best method yet devised by mankind, it is made up by two elements which we have already seen to be fallible—SD (sense data) and SPR (pure reason) can both deceive us. Again: two fallibilities do not add up to one infallibility. Scientific generalizations which have lasted for a long time have high probability, perhaps the highest probability of any generalizations, but it is only Idolatry which claims none of them will ever again have to be revised or rejected. Too many have been revised or rejected in this century alone.

Wilson like all philosophers who use this argument ignores the fact that science possesses an innate self-healing mechanism. As already mentioned at the end of

Chap. 1, any step into a new working field is grounded on knowledge, methods, materials, and instruments invented and developed in the past. Thereby, all previously elaborated knowledge is permanently reviewed and scrutinized. This is exactly why numerous examples of paradigm changes are described in Part II of this book. These examples demonstrate that the revisions of these mistakes, errors, and fallacies were not made by philosophers, writers, or journalists, but by the scientists themselves.

A critique of science guided by a quite different world view was published by the American ecologist and Marxist (his own self-characterization) Richard Levins in two essays entitled "Whose Scientific Method? Scientific Methods for a Complex World: New Solutions" and "Ten Propositions on Science and Antiscience." Three of the ten propositions listed in the second article deserve to be mentioned and discussed here.

Proposition (2): "To call something scientific does not mean that it is true. … Therefore, we have to consider the notion of the 'half-life' of a theory as a regular descriptor of the scientific process and even be able to ask (but not necessarily answer), under what circumstances might the second law of thermodynamics be overthrown?" At the end of proposition (5) Levins adds: "While all theories are eventually wrong, some are not even temporarily right." This view suffers from the fact that Levins, like most philosophers and historians of science, exclusively consider science as a theory-generating enterprise. He has ignored the millions of reliable, reproducible experimental facts, ignored the huge number of non-falsifiable inductive inferences (see Sects. 2.4, 3.2, and 3.3), and ignored the fact that statements such as "an absolute truth can never be achieved" negate themselves.

Proposition (3): "Science has a dual nature. On the one hand, it really does enlighten us about the interactions with the rest of the world, producing understanding and guiding our actions. … On the other hand, as a product of human activity, science reflects the conditions of its production and the viewpoints of its producers and or owners." It is trivial that theoretical knowledge, experimental methods, and instruments used at a certain time are characteristic of that time, because they are the fruit of the preceding research. Yet, it is a strange conclusion that the natural sciences are formed and owned by an individual society. For example, at the time of Aristotle no Greek imperium existed and at least 99 % of all Greeks (who were mainly merchants, seamen, fishermen, or craftsmen) were not interested in his insights and theories. However, all scientists and philosophers of the Roman Empire knew of them and after the appearance of Mohamed, Aristotle's work was picked up by Arabian scientists and disseminated in every part of the Islamic world. Furthermore, which society owned the work of Copernicus, Kepler, and Galilei (see Sect. 10.1) and which society owned (and owns) polymer science (see Sects. 9.4 and 9.5)?

Proposition (6): "Modern European/North American science is a product of the capitalist revolution. It shares with modern capitalism the liberal progressivist ideology that informs its practice and that it helped to mold. Like bourgeois liberalism in general it is both liberated and dehumanized." This statement is too short-sighted and narrow-minded to merit a detailed response. Yet, this author is tempted to ask two questions: When R. Levins has severe toothache and visits a dentist, does he reject a pain-killing injection because local anesthetics are a bourgeois invention (see Sect. 7.3)? When R. Levins needs an operation, does he reject narcosis, because narcosis (and narcotics) are a traditional, conservative, and dehumanized technique?

Furthermore, the interesting and informative book *Science and Antiscience* by Gerald Holden (1993) should be mentioned. It reports various pro-science and antiscientific positions and theories published in the past. Of particular interest is here the Chap. 5 entitled "The Controversy over the End of Science." This chapter begins with the following text:

> Even while science has been asserting ever greater success in its aim of encompassing the understanding of all natural phenomena, antithetical forces have been gathering outside the laboratory in what amounts to an effort to delegitimate science as we know it. At different times in modern history this challenge to the role of science in culture has assumed different forms; but its roots are ancient and robust. ... This chapter focuses on the confrontation between the two main, thematically opposed positions: one claims that the sciences are by their nature subject to eventual decay; the other argues that the sciences are designed to merge eventually into one coherent body of understanding of all phenomena.
>
> To most scientist today, the first of these choices seems too unreasonable to be taken seriously, they are unlikely to pay any attention to currently fashionable writings claiming that science, traditionally the source of new insights, of material progress, and of intellectual emancipation, may now be coming to a close—to its end—not merely to a recognition of the limits on the power of science, limits of which the scientists themselves on the whole are quite aware. To the historian of science, a debate on the possible decay and death of science is neither a contradiction nor a novelty. The idea has been proposed many times in the past. To give one example, towards the end of the nineteenth century, a number of new problems could not be solved by the then-current mechanistically based physics. In disappointment, the European scientist Emil Du Bois-Reymond wrote that science had at least come up against unbreakable barriers of understanding, beyond which we shall always remain ignorant.

In this connection Holden presented two footnotes:

- The Nobel Conference XXV, held in October 1889, contained the following agenda-setting paragraphs in its letter of invitation to the participants of the conference: "As we study our world today, there is an uneasy feeling that we have come to the end of science, that science, as a unified, universal, objective endeavor, is over. Even the consensus that science is a recently formed alliance, a consensus that has led to the grand methodologies of science, is in fragments."
- In the same spirit, a conference was held in December 1991 at the Massachusetts Institute of Technology under the title "Progress: An Idea and Belief in Crisis"; the letter of invitation remarked that "the idea of progress is predicated on the belief in reason and material advancement. The value and validity of both of these have now been seriously called into question. It is this situation that has produced a crisis in belief."

What is remarkable in this review are the footnotes that indicate that the discussion about the end of science was particularly vivid towards the end of the nineteenth century, fueled by the physicists (see Sect. 2.2). However, at the same time chemistry was flourishing. Both fundamental research and the transformation of chemical inventions into technical production were booming and this boom is still continuing at the end of 2015 when the manuscript of this book was completed. During the last three decades of the nineteenth century, the Periodic Table was elaborated, the chemical formula language was completed, syntheses of important dyestuffs such as indigo (still used today for blue jeans) were discovered, the first useful local anesthetic was synthesized (see Sect. 7.3), the synthesis and technical production of aspirin were developed, and several other useful medicaments were invented. The situation of physics and the paradigm changes induced by Planck and Einstein did not cause any changes in the course of chemical research. From the viewpoint of a chemist, any discussion on the end of science looks strange, not to say ridiculous.

This historic scenario illustrates what has already been mentioned before, namely that all the scientists, philosophers and historians whose knowledge of science is limited to physics and its history have a high tendency to develop wrong or misleading theories on the nature of science.

The Austrian philosopher Paul Feyerabend (1924–1994) was disciple of Popper in 1952, but he developed a harsh critique of Popper's rationalism and falsificationalism (see Sect. 2.4). He was proud to be nicknamed an "epistemological anarchist." Feyerabend rejected any prescriptive methodology and argued that the progress of science is based on making use of any and all available methods supporting new theories. He also rejected any reliance on scientific method. Together with the Hungarian philosopher Imre Lakatos (1922–1974) he postulated that the demarcation problem of distinguishing science from pseudoscience on objective grounds is impossible and, thus, deadly for the understanding of science as an enterprise following fixed universal rules. His book *Against Method* is the most widely known work of a skeptic philosopher (at least five editions). Characteristic of his work is the absence of any definition of science, but from the text it may be inferred that his view of science includes historical research and sociology in addition to the natural sciences. Furthermore, terms such as the law of nature or reproducibility of experiments are not discussed (the same is true for all authors mentioned in this chapter).

A positive aspect of this book (the fourth edition is cited here) is the so-called analytical index, which lists the many arguments that head and summarize the 15 chapters of the first edition (20 chapters including new appendices in the fourth edition). The two most important statements in the introduction are "Science is an essentially anarchic enterprise" and "Theoretical anarchism is more humanitarian and more likely to encourage progress than its law-and-order alternatives."

Feyerabend did not provide any justification for these statements. The first statement suggests that he extrapolated the anarchy in his brain to the brains of all scientists, a severe deficit of self-critique. Obviously, Feyerabend had never heard or read anything about reproducibility of observations and experiments, the basis of any inductive inference and any law of nature. When science is anarchic, medicine is anarchic too, because about 90 % of its diagnostic and therapeutic methods are based on scientific research (see Chap. 6). Feyerabend was severely wounded in World War II and healed by physicians trained in traditional medicine. Why did he trust in traditional western medicine when methodical research and laws of nature do not play any role or are even disadvantageous?

Concerning the second of his statements, the first question to ask is: What is meant by "humanitarian"? If happiness and euphoric feelings are meant, his statement is absurd because no sane scientist has claimed that science has the purpose of producing happiness. If, however, improvement in health and reduction in pain and mortality are meant, science in combination with medicine has been far more successful than any other human enterprise.

The main arguments of the first and second chapters of *Against Method* expand on the statements given above:

Chapter 1: "This is shown both by an examination of historical episodes and by an abstract analysis of the relation between abstract idea and action. The only principle that does not inhibit progress is: anything goes."

Chapter 2: "For example, we may use hypotheses that contradict well-confirmed theories and/or well-established experimental results. We may advance science by proceeding counterinductively."

The second argument raises the question: What is meant by "counterinductively"? Feyerabend offers the following explanation: "Examining the principles in concrete detail means tracing the consequences of counterrules which oppose the familiar rules of the scientific enterprise. To see how this works, let us consider the rule that it is experience or the facts or experimental results which measure the success of our theories, that agreement between a theory and the data favors the theory, while disagreement endangers it and, perhaps, even forces us to eliminate it. ... It is the essence of empiricism. ... The counterrule corresponding to it advises us to introduce and elaborate hypotheses which are inconsistent with well-established theories and/or well-established facts. It advises us to proceed counterinductivity."

In science, established facts mean reproducible results free of experimental error and it is the fundamental law of all natural sciences that correct and reproducible experiments never lie. What may be wrong is their interpretation. An example from the history of chemistry is given in Sect. 9.2. The alchemist G. Stahl performed experiments (combustion of metals) that let him develop the "phlogiston theory." These experiments were perhaps reproduced by Stahl himself in his laboratory, but Antoine Lavoisier was not able to reproduce them and found different results. In contrast, other scientists were able to reproduce Lavoisier's experiments and, therefore, Lavoisier's "oxidation theory" survived and not the phlogiston theory.

An elaboration of new hypotheses may proceed on two tracks. First, they can be developed without considering established facts. Such a mental process is outside science; it is a kind of dreaming or reeling around in a pool of anarchic or chaotic thoughts. Second, new hypotheses are developed on the basis of established facts with the purpose of challenging older hypotheses. This process is a normal scientific procedure and Feyerabend does not propose anything new. In other words, counterinductive thinking either means developing alternatives grounded on fact, which means the prefix "counter-" is wrong, or, counterinductive thinking means developing ideas in a way contrary to inductive inferences based on facts. This second interpretation means intellectual anarchism, which has nothing to do with science. In either case, Feyerabend's anarchic philosophy does not make any useful contribution to better science or a better understanding of science, but it was and is certainly welcome by those who are not able or willing to learn what science means.

In 2013, the British biologist Rupert Sheldrake published a book entitled *The Science Delusion* in which he describes his position (p. 6): "In this book I agree that science is being held back by century-old assumptions that have hardened into dogmas. The science would be better off without them: more interesting and more fun." The irony and inconsistency behind this declaration is the fact that Sheldrake himself proposed a dogma, the existence of morphic fields, in a previous book entitled *The Presence of the Past.*

On pages 7 and 8 of *The Science Delusion* Sheldrake formulated the basis of his critique:

> Here are the ten core beliefs that most scientists take for granted:
>
> 1. Everything is essentially mechanistic.
> 2. All matter is unconscious.
> 3. The total amount of matter and energy is always the same.
> 4. The laws of nature are fixed. They are the same today as they were at the beginning and they will stay the same forever.
> 5. Nature is purposeless and evolution has no goal or direction.
> 6. All biological inheritance is material.
> 7. Minds are inside heads and are nothing but activities of brains.
> 8. Memories are stored as material traces in brain.
> 9. Unexplained phenomena like telepathy are illusory.
> 10. Mechanistic medicine is the only kind that really works.
>
> Together, these beliefs make up the philosophy or ideology of materialism.

Sheldrake devotes a chapter to each of the ten doctrines. A critical reviewing of all chapters is, of course, outside the scope of this book and, therefore, the following comments are focused on points (1) and (4) (an answer on point (10) has already been given in Chap. 1).

Point (1) represents a striking example of Sheldrake's strategy of using a few negative examples for a generalization that defames all scientists and any kind of science. Sheldrake's reproach that all scientists are materialists who try to explain all aspects of nature in terms of mechanics may be acceptable as a response to the physicists of the nineteenth century (see Sect. 2.2) or to Dawkins theory of the egoistic gene (see Sect. 8.4). However, Dawkins and those physicists are not representative of the millions of scientists who have worked in the past or are still at work today. For, instance, the author himself has demonstrated in Sects. 2.2and 8.4 that he does not agree with them.

For scientists, the discussion of point (4) is perhaps the most interesting aspect of Sheldrake's critique. In contrast to most philosophers, he has at least understood that laws of nature play a decisive role for a proper understanding of science but he ignores the consequences for evolution, human civilization, and medicine. The following quotation illustrates Sheldrake's view (p. 84): "Most scientists take it for granted that the laws of nature are fixed. They have always been the same as they are today and will be forever. Obviously this is a theoretical assumption not an empirical observation. On the basis of 300 years of earth-bound research, how can we be sure that these laws were always the same and always will be everywhere?"

First, Sheldrake ignores any kind of space research. The minerals found on the Moon and Mars obey the same laws of chemistry and physics known from experiments on earth. He also ignores the numerous chemical analyses of meteorites. Furthermore, he ignores the examination of the universe via spectroscopic methods ranging from radio waves to γ-radiation and he ignores the fact that the observation of distant galaxies is a look into the history of the universe (see Sect. 10.2). Second, although the term "modern science" is limited to the past 500 years, the discovery and utilization of laws of nature is as old as humanity itself (see Sect. 3.1). Third, Sheldrake ignores the results of paleontology and, thus, the course of the evolution over the past three billion years.

Sheldrake continues on p. 84: "As soon as we begin to question them, eternal laws become problematic for two main reasons. The very idea of a law is anthropocentric. Only humans have laws." This is one out of numerous incorrect statements. As outlined in Sect. 3.1, humanity has explored the laws of nature from the very beginning, and, therefore, at the same time or even before it ordered its social life by human laws. Furthermore, the hierarchy in a clan of *Pithecanthropus erectus* was not the consequence of a consciously elaborated set of laws, it was the sheer consequence of a law of nature. The evolution of multicellular organisms, like that of a social group, is grounded on three principles: specialization of cells (individuals), cooperation of specialized cells, and hierarchical order of all activities.

On p. 85 Sheldrake asks the following: "If everything else evolves, why don't the laws of nature evolve with nature?" This statement discloses a total misunderstanding of the laws of nature. Constancy is inherent in the definition of laws of nature because laws of nature are responsible for the reproducibility of natural phenomena and experiments irrespective of time and location. He continues on the same page: "There is no reason to assume that the laws that govern molecules, plants, and brain were all present at the moment of the Big Bang, long before any of these systems existed." Obviously Sheldrake has never learned that all laws of nature are limited to a certain frame of validity (as outlined in Sect. 2.2) and may be undiscoverable outside this frame. A more detailed answer to Sheldrake's argument is given in Sect. 2.3.

Sheldrake continues on p. 85: "I suggest an alternative to eternal laws: evolving habits." This is a nice idea for spiritualists but far from reality. This suggestion means that the inevitability of death is nothing more than a bad habit. Unfortunately, Sheldrake does not offer advice how to get rid of this bad habit. However, the necessity of dying has at least two satisfactory aspects: because it applies to everybody, it will prevent "megacriminals" (such as Hitler, Stalin, Napoleon, or Genghis Khan) from continuing with evil deeds for all eternity, and it will also prevent certain stubborn antiscientists from continuing to publish nonsense for all eternity.

4.2 Reductionism, Antireductionism, and the Origin of Life

Reductionism is an intellectual and philosophical position that interprets a complex system as the sum of its parts or, in other words, an account of a system can be reduced to accounts of individual components. According to the reductionist concept, which plays a decisive role in scientific research and for the structure of scientific theories, a full understanding of all parts of a complex system allows the prediction of all properties and functions of the complex system under investigation. Reductionism does not preclude the existence of emergent phenomena, but it does include the ability to understand these phenomena thoroughly in terms of the constituents and processes from which they are composed. Therefore, the reductionist concept is very different from the usual understanding of the term "emergence," which intends to indicate that a complex phenomenon is more than the sum of the components and processes from which it emerged.

The opposite position is called "holism," a term derived from the Greek word *όλοσ* (holos) meaning entire or complete. Like almost all fundamental ideas and concepts, the origin of reductionist and holistic views on theories or on living and non-living objects can be traced back to the ancient Greeks. Demokritos of Abdera (460–370) may be seen as the first prominent advocate of the reductionist concept, because he explained all phenomena (irrespective of whether dead matter or living organisms) by the combination (or aggregation) of atoms that differed in form, size, and number. Aristotle (384–322) reports that he believed "the whole being more than the sum of its constituents."

Most philosophers differentiate between three types of reductionism: ontological, methodical, and theoretical reductionism.

- Ontological reductionism is the assumption that reality consists of a minimum number of entities or materials. This position is usually metaphysical and frequently the basis of monism, which claims that all events, objects, and properties may be reduced to one primary medium or substance.
- Methodological reductionism is the strategy to reduce explanations to the smallest possible entity. In an essay entitled "Is the Universe a Computer?" the American physicist Steven Weinberg (born 1933, Nobel Prize 1979) wrote the following: "but in seeking the laws of nature it is the essence of the art of science to avoid complexity."
- Theoretical reductionism is the concept that one general theory absorbs a more special theory. The Slovakia-born American philosopher Ernst Nagel (1901–1985) presented in his book *The Structure of Science* the following definition: "In a reductionist explanation, a theory or situation, TH, is shown

to be deductively derivable from a more basic theory with a smaller ontology, TB, in a manner analogous to deriving a theorem from a set of axioms in a logical proof." For example, the theory that the formation of table salt (sodium chloride) by reaction of the metal sodium with chlorine proceeds by electron transfer from sodium to chlorine is part of the broader theory that all metal halide salts may be formed by electron migration from metal to halogen. This broader theory is, in turn, embedded in the universal concept of oxidation–reduction reactions (see Sect. 9.2).

Reductionist thinking is the fundamental strategy in most branches and disciplines of the natural sciences, in as much as it is directly correlated with causal thinking. The American philosopher Thomas Nagel (born 1937) remarked that "Reductionism has been a creative driving force in the history of modern science." However, reductionist thinking is not limited to science and also exists in other field such as philosophy, linguistics, sociology, mathematics, and even in theology in connection with the discussion of "free will." In all fields, reductionist thinking has attracted much critique and in the fields of ecology and sociology the American ecologist Fritjof Capra may be mentioned as a representative of a holistic approach. In philosophy, the terms reductionism and reductionist have recently acquired a pejorative or defaming connotation. The Korea-born American philosopher Jaegwon Kim wrote in a 1998 essay: "Reduction and reductionism have become common epithets thrown at one's critical targets to tarnish them with intellectual naiveté and backwardness." The Canada-born American philosopher Patricia Churchland (born 1943) wrote: "Reductionism has come in some quarters to be used as a general term of insult and abuse. . . . Sometimes it is used as a synonym for behaviorism ... or as a symbol for such diverse sins as materialism, bourgeois capitalism, experimentalism, vivisectionist, communism, militarism, sociobiology, and atheism."

Well-balanced views of both positions were contributed by the American philosopher Richard H. Jones in his book *Reductionism: Analyses and the Fullness of Life* and by Todd Jones in his article "Reductionism and Antireductionism: Rights and Wrongs." Todd Jones outlines his goal as follows: "In this article, I want to try to clarify what our attitudes towards reductionism should be. I will make three points. First, contrary to standard rhetoric in the humanities, reductionism is a very good epistemic strategy. Second, contrary to the standard arguments in some areas of philosophy, reduction to physics is always possible for all causal properties. Third, there are, nevertheless, reasons why we want science to discover properties and explanations other than reductive physical ones."

An antireductionist position rejecting any kind of reductionist thinking is identical to total rejection of natural sciences and does not deserve further discussion at this point. Almost all antireductionist scientists, sociologists, and ecologists agree that reductionist thinking is useful for a proper understanding of natural phenomena involving non-living matter. Their critique is directed towards reductionist thinking as the only intellectual strategy in science and also encompassing all aspects of life. They wish, as suggested by Todd Jones, to complement reductionist analyses with a holistic approach whenever the origin of life, the function of single cells, the behavior of individual animals or humans, and the behavior of organs and groups of individuals are concerned. All aspects of this broad and intensive discussion cannot, of course, be discussed in this section and, therefore, the following text focuses on two particularly interesting questions:

– Is it possible to explain the origin of life exclusively on the basis of a reductionist theory and to what extent is this theory supported by experimental facts?
– Is it possible to explain evolution via natural selection by a reductionist theory?

The properties of organic molecules can be traced back to the Big Bang (see Sect. 10.2) in a reductionist scheme as follows: In the first moment after the Big Bang, sheer energy at an incalculable high temperature was generated. This energy obeyed laws of nature, which enabled the formation of elementary particles such as electrons, protons, and neutrons after a few seconds of cooling. The nucleons were, in turn, capable of forming atomic nuclei.

These atomic nuclei automatically created orbitals in the surrounding space (or relativistic ether, see Sect. 10.4). With decreasing temperature, these orbitals were filled with electrons step by step according to the Pauli principle. At temperatures below 3000 °C more and more atoms became capable of forming salts via ionic bonds. Below 1000 °C, simple organic molecules, such as hydrogen cyanide or acetylene, were formed via covalent bonds. Below 500 °C and notably below 300 °C complex organic molecules can form and below 200 °C the building blocks of biopolymers, such as α-amino acids, saccharides, and nucleobases can exist (all the steps following the formation of electrons and protons required millions and billions of years). Each stage may be understood as a necessary consequence of the properties of the preceding stage. Yet, it should be mentioned, with the example of the British chemist and writer Eric Scerri, that not all chemists and philosophers accept an entirely reductionist explanation of all aspects of chemistry via quantum mechanics.

At this point the question arises whether the atoms and small organic molecules are endowed with the "talent" to produce those biopolymers required for the formation of a living cell. Charles Darwin was the first to formulate a pertinent speculation: "It is often said that all the conditions for the first production of a living organism are now present, which could ever have been present. But if (and oh! what a big if!) we could conceive in some little warm pond, with all sorts of ammonia, phosphoric salts, light, heat, electricity, etc. present, that a protein compound was chemically formed ready to undergo still more complex changes, at the present day such matter would be instantly devoured or absorbed, which would not have been the case before living creatures were formed." About 50 years later the Russian chemist A.I. Oparin and the American chemist J.B.S. Haldane independently published hypotheses about the origin of life as a consequence of molecular evolution under prebiotic conditions. They assumed a warm or hot "prebiotic soup" as the arena of molecular evolution and birthplace of biopolymers.

After World War II numerous research groups began to study those hypotheses in various directions. The following discussion concentrates on proteins, because proteins due to their sheer mass and their function as enzymes are the most important group of biopolymers in all living organisms. These proteins are built up from 20 different α-amino acids. The first big success was reported by the American chemist Stanley L. Miller in 1953. He found that exposure of a hypothetical prebiotic atmosphere to electric discharges yielded numerous organic compounds, including four protein amino acids (prAAs), a result for which he was awarded the Nobel Prize. In later experiments, conducted with variations of the gas mixture, 11 prAAs were detected, less than the hoped for 20 prAAs. Other research groups performed similar experiments using different sources of energy, such as UV light or radiation of radioactive elements. However, to the best knowledge of the author, no model experiment has been reported to yield all 20 prAAs. Even worse was the finding of the Miller group that the prAAs were accompanied by more than 15 non-protein amino acids and imino acids. These amino and imino acids are not only obsolete they are in fact deadly for any evolution of proteins. Their reactivity is identical with or similar to that of the prAAs and, hence, any polymerization process will incorporate these non-protein amino acids in the growing peptide chains. Any explanation of why, when, where, and how these detrimental amino acids disappeared is so far lacking. The third negative result obtained by the Miller group was a large fraction of reactive byproducts, such as amines, alcohols, carboxylic acids, and hydroxyl acids, which can stop or prevent any polymerization process. Again, there is no explanation of why the byproducts did not hinder the evolution of biopolymers. The fourth negative result was the absence of oligopeptides (the simple precursors of proteins) in the reaction products of almost all research groups studying the prebiotic synthesis of prAAs.

In addition to model experiments, there is another and even more important source of information on prebiotic chemistry, namely meteorites carrying organic materials. These so-called carbonaceous chondrites only make up a small percentage of all meteorites. The most intensively studied representative of this group is the Murchison meteorite, which fell on 28 September 1969 near Murchison, Victoria, in Australia. The total mass of all fragments collected immediately after the fall amounted to 100 kg. The analyses reported by various research groups revealed a pattern largely resembling the results of the model experiments. Only around 50 % of the 20 prAAs were detected. No peptides were discovered, but more than 70 non-protein AAs were found, far more than in Miller's experiments. Furthermore, Philippe Schmitt-Köpplin et al., authors of a careful analysis reported: "Here we demonstrate that a nontargeted ultrahigh-resolution molecular analysis of the solvent-accessible organic fraction of Murchison extracted under mild conditions allows one to extend its indigenous chemical diversity to tens of thousands of different molecular compositions and likely millions of diverse structures. This molecular complexity, which provides hints on heteroatoms chronological assembly, suggests that the extraterrestrial chemodiversity is high compared to terrestrial relevant biological- and biochemical-driven chemical space."

These results indicate:

- The prAAs and saccharides needed for molecular evolution of biopolymers represent only a minute fraction among thousands and possibly millions of useless or detrimental chemicals.
- All extraterrestrial chemistry is based on random reactions of radicals, exactly contrary to a targeted or at least predominant synthesis of molecules relevant for the molecular evolution of biopolymers and living organisms.

Furthermore, it should be mentioned that model reactions designed to prepare polypeptides or proteins under allegedly prebiotic conditions were all conducted with pure prAAs and in the absence of any chemicals that might interfere with the polymerization under investigation. Such experiments are useless as evidence for a molecular evolution of proteins. Another unsolved problem is the role of chirality. All results taken together are far from supporting a reductionist explanation of the origin of biopolymers. They even point in the opposite direction. The research activities of the next 100 years may be more favorable for the reductionist hypothesis, but it will be difficult to disprove that the random character of extraterrestrial chemistry is highly unfavorable for molecular evolution everywhere in space.

The hypothesis of spontaneous formation of living cells from dead matter raises a second question: Provided that all the biopolymers and small molecules (e.g., vitamins) needed for the metabolism and self-reproducibility of a living cell were present in a "warm pond" (Darwin's terminology) somewhere on the early earth, was their simultaneous presence sufficient for self-organization of a living cell? Since World War II numerous studies have been devoted to the self-organization and self-assembly of various types of molecules. At this point it is worth mentioning that both terms are not identical, although many authors do not care about the difference. Specific folding of a protein chain with the consequence of a special biological function is a kind of self-organization but not an assembly, which requires two or more protein chains. Crystallization of organic substances may be understood as the simplest kind of self-assembly, although with little or no relevance for the origin of life. An intensively studied example of self-assembly with high relevance for the hypothesis of spontaneous generation is the formation of a double helix from two complementary strands of poly(deoxyribonucleic acid), DNA. Another group of intensively studied self-assembly phenomena is the formation of micelles and vesicles from soap molecules and similar surfactants (detergents).

All these self-organization processes are thermodynamically controlled, which means that the driving force is a gain of free energy and that the organized state is energetically more stable than the preceding random conformation or distribution of molecules. However, living organisms also involve a second group of self-organized systems that are formed under kinetic control, meaning that they are the result of the fastest process. Examples of such systems are biosynthesis, virus assembly, formation of beehive and anthill or swarm intelligence, which are formed under enzymatic, genomic, and/or evolutionary control. All these studies have considerably improved knowledge of the structural order, metabolism, and self-reproducibility of living cells. However, any experiment demonstrating that the combination of all these ingredients needed for the construction and function of a living cell will automatically generate a living cell is missing so far. Therefore, the antireductionist view that the principle of life is an "added value" that is not a direct consequence of the sum of the self-organized components of the cell has not yet been refuted.

Nature itself presents an interesting insight into this problem with the existence of viruses. A virus may be understood as an assembly of all the DNA molecules needed as information for the structure of its "descendants." Yet, the virus alone is nothing but dead matter, which requires the protein-based machinery of a living cell for its reproduction. The viruses demonstrate, on the one hand, that a highly sophisticated self-organized assembly of structural information can exist apart from the metabolism and reproduction machinery of a cell. Yet, on the other hand, they demonstrate that without information for the construction and function of such a machinery the structural information is just dead matter. In other words, the existence of viruses proves rather than disproves that an added value is necessary for the generation of a living organism.

At this point another interesting question comes into sight. Assuming that an added value—resembling the *vis vitalis* of the vitalists (see Sect. 9.3)—is absolutely necessary for the principle of life, is this added value a law of nature inherent in the universe since the Big Bang or is it the consequence of a separate creational act (the Big Bang of life) that occurred on earth about four billion years ago? This question could be answered by the detection of living organisms on another planet, regardless of how primitive they are. However, the discovery of extraterrestrial life does not necessarily answer the primary question of whether a reductionist approach alone can explain the origin of life.

In this context, Darwin's principle of evolution via natural selection is intensively discussed. Darwin himself considered it an indispensable property of evolution and life: "This presentation of favorable individuals, differences, and variations and the destruction of those which are injurious I have called Natural Selection or the Survival of the Fittest." Now three questions deserve an answer:

First, is evolution without natural selection possible?
Second, if natural selection is an absolutely necessary property of evolution, is it a law of nature?
Third, if evolution grounded on natural selection is a law of nature, can it be explained by reductionist theory, that is, by chemical and physical laws alone?

The vast majority of scientists agree that evolution without natural selection is unproven and not imaginable. A detailed discussion of the second and third questions was published in 2005 by the American philosophers Alexander Rosenberg and D. M. Kaplan in an article entitled "How to Reconcile Physicalism and Antireductionism about Biology?" They state on p. 49:

> In considering the relation between the PNS [principle of natural selection] and physics, three alternatives suggest themselves:

(a) The PNS is an underived law about biological systems, and is emergent from purely physical processes. This alternative would vindicate the autonomy of all biology, following Dobzhansky's (1973) dictum, but leave biological phenomena physically unexplained and/or emergent.
(b) The PNS is derived law; it is drivable from some law of physics and/or chemistry. This alternative would vindicate the reductionist's vision of a hierarchy of scientific disciplines and theories, with physics at the foundations.
(c) The PNS is an underived law about physical systems (including non-biological ones), and from it the evolution of biological systems can be derived so that the principle we recognize operating at the biological level is also an underived basic law of physical science. This is an alternative no one has canvassed, and one which we shall defend.

To end this section, the author concludes that a final decision on whether the origin of life can be explained by a reductionist theory alone, or whether combination with a holistic concept is the better alternative, cannot be made at the current state of knowledge. As long as quantum mechanics cannot explain the origin of life, the evolution of living organisms and its later results, such as Beethoven's symphonies, Picasso's paintings, and antiscientist philosophies, it is justified to assume that life involves principles and laws that cannot be reduced to quantum mechanics.

Bibliography

Barzun J (1964) Science the glorious entertainment. Harper & Row, New York
Capra F (1982) The turning point. Simon & Schuster, New York
Churchland P (1986) Neurophilosophy. MIT Press, Cambridge, MA
Diethelm P, McKee M (2009) Denialism. What is it and how should scientists respond? Eur J Public Health 19(1):2
Feyerabend P (2010) Against method, 4th edn. Verso, London, New York (first published by New Left Books, 1975)
Holden G (1993) Science and antiscience. Harvard University Press, Cambridge, MA
Jones RH (2000) Reductionism: analyses and the fullness of life. Bucknell University Press, Lewisburg, PA
Jones T (2004) Reductionism and antireductionism: rights and wrongs. Metaphilosophy 35(5):614
Kim J (1998) Mind in a physical world. MIT Press, Cambridge, MA
Klir GJ (1991) Facets of system science. Springer, Heidelberg
Levins R (1996) Ten propositions on science and antiscience. Social Text 14(46/47):101–111
Nagel E (1961) The structure of science. Harcourt, Brace & World, New York, Burlingame, CA
Pizzarello S (2006) The chemistry of life's origin: a carbonaceous meteorite perspective. Account Chem Res 39(4):231–237
Rosenberg A, Kaplan DM (2005) How to reconcile physicalism and antireductionism about biology? Philos Sci 72:43
Scerry E (2008) Collected papers on philosophy of chemistry. Imperial College Press, London
Schmitt-Köpplin P et al (2009) Proc Natl Acad Sci 107(7):2763–2768
Sheldrake R (2013) The science delusion. Hodder & Stoughton, London
Wilson RA (1999) The new inquisition. New Falcon Publications, Scottsdale, AZ

Chapter 5
Paradigm Change and Progress

5.1 What Is a Paradigm?

> I think mistakes are the essence of science and law. It's impossible to conceive of either scientific progress or legal progress without understanding the important role of being wrong and of mistakes.
>
> (Alan Dershowitz)

In 1962, the American historian and philosopher of science Thomas R. Kuhn (1922–1996) published a book entitled *The Structure of Scientific Revolutions*. This book attracted a great deal of attention among historians, philosophers, and theoreticians of science and among sociologists, but it found relatively little attention among scientists. In their appreciative or critical response to Kuhn's work most historians and theoreticians concentrated on the question of the extent to which theories or historic stages of scientific progress are comparable or incomparable, commensurable or incommensurable. Little attention was paid to the more important question of the extent to which Kuhn's view of science agrees with reality. Kuhn's book is here discussed in some detail for three reasons. First, it is probably the most famous twentieth century book on the structure and history of science. In the words of the reviewer Nicholas Wade in the journal *Science*: "A landmark in intellectual history which attracts attention far beyond its own immediate field." Second, its discussion complements Chaps. 2 and 4. Third, Kuhn's most famous invention, the term "paradigm change" or "paradigm shift" is now widely accepted by scientists, although its definition is not quite clear, as demonstrated below.

Prior to a review of Kuhn's book and a description of his background, it should be mentioned that the term "paradigm," which plays a fundamental role in Kuhn's book, is derived from the Greek word *paradigma* meaning example, model, or pattern. For correct interpretation of Kuhn's inferences and theories, it is necessary to have some information about his mental background. Kuhn studied theoretical physics and never performed experimental research. During his Ph.D. work he became interested in the history of science and finally devoted his professional life

H.R. Kricheldorf, *Getting It Right in Science and Medicine*,
DOI 10.1007/978-3-319-30388-8_5

to this topic. In contrast to most other philosophers of science, he cared about the meaning of the term "science" and defined it as a human enterprise that, in contrast to politics, arts, and the humanities, is capable of producing measurable progress (see Sect. 5.3). The concrete examples mentioned in his book concern the natural sciences, above all physics and its history.

However, Kuhn never used the term "law of nature," but in several chapters he spoke of Kepler's laws or Newton's laws, as if those laws were the properties of those scientists. This terminology is a consequence of Kuhn's view of science as a theory (or paradigm)-producing enterprise, which in the optimum case approaches truth about the properties of nature. Kuhn explains his understanding of science and the structure of scientific revolutions by means of six terms, which are correlated in the following way: the period of "normal science" is characterized by "ordinary research" under the dictate of a "paradigm." A scientific revolution occurs during the (short) period of "anomalous science" based on "extraordinary research" and ending with a "paradigm change." About the long periods of normal science, Kuhn says (p. 10): "In this essay normal science means research formally based upon one or more past of scientific achievements, achievements that some particular scientific communities acknowledge for a time supplying the foundation for its further practice."

At the time when Kuhn studied physics, textbooks were up to date and contained all important information concerning a certain discipline. This point needs to be taken into account for a proper understanding of Kuhn's definition of paradigm (p.10), as follows:

> Today such achievements are recounted, though seldom in their original form, by science textbooks, elementary and advanced. These textbooks expound the body of acceptable theory, illustrate many or all of its successful applications, and compare these applications with exemplary observations and experiments. Before such books become popular early in the nineteenth century (and until even more recently in the newly matured science), many of the famous classics of science fulfilled a similar function. ... These and many other works served for a time implicitly to define the legitimate problems and methods of a research field for succeeding generations of practitioners. They were able to do so because they shared two essential characteristics. Their achievements were sufficiently unprecedented to attract an enduring group of adherents away from competing modes of scientific activity.
>
> Simultaneously, it was sufficiently open-ended to leave all sorts of problems for the redefined group of practitioners to resolve. Achievements that share these two characteristics I shall henceforth refer to as paradigms, a term that relates closely to normal science.

This somewhat tortuous definition is supplemented and specified later on p. 12: "To be accepted as a paradigm, a theory must be better than its competitors, but it need not and in fact never does, explain all the facts with which it can be confronted." On pp. 23 and 24 Kuhn states the following:

> The success of a paradigm ... is at the start largely a promise of success discoverable in selected still incomplete examples. Normal science consists in the actualization of the promise, an actualization achieved by extending the knowledge of those facts that the paradigm displays as particularly revealing, by increasing the extent of the match between those facts and the paradigm's prediction and by further articulation of the paradigm itself.

Kuhn describes three classes of experimental research that are characteristic of normal science dominated by a paradigm (pp. 25–27): "First is that class of facts that the paradigm has shown to be particularly revealing of the nature of things. By employing them in solving problems, the paradigm has made them worth determining, both with more precision and a larger variety of situations. ... A second usual but smaller factual determination is directed to those facts that, though often without much intrinsic interest, can be compared directly with prediction from the paradigm theory. ... A third class of experiments and observation exhausts, I think, the fact-generating activities of normal science. It consists of empirical work undertaken to articulate the paradigm theory, resolving some of its residual ambiguities and permitting the solution of problems to which it had previously only drawn attention." Finally, he concludes (p. 34): "These three classes of problems—determination of significant fact, matching of fact with theory, and articulation of theory—exhaust, I think, the literature of normal science, both empirical and theoretical."

It is characteristic of Kuhn's concept of normal, paradigm-dominated science that in this phase of scientific activities important discoveries and theoretical novelties do not occur. In his own words (p. 35): "Perhaps the most striking feature of the normal research problems we have just encountered, is how little they aim to produce major novelties." On p. 52 (chapter VI), he states: "Normal science, the puzzling-solving activity we have just encountered, is a highly cumulative enterprise, eminently successful in its aim, the steadily extension of the scope and precision of scientific knowledge. In all these respects it fits with great precision the most usual image of scientific work. Yet, one standard product of the scientific enterprise is missing. Normal science does not aim at novelties of fact or theory and, when successful, finds none."

The following examples demonstrate that Kuhn's fundamental characterization of what he calls "normal science" is far from reality. First, all physicists who lived in the second half of the nineteenth century or later agree that James Clerk Maxwell's concept of electromagnetic phenomena is one of the greatest steps forward in the history of physics. In the words of the Nobel Prize laureate Feynman: "From a long view of the history of mankind—seen from, say, 10,000 years from now—there can be little doubt that the most significant event of the nineteenth century will be judged as Maxwell's discovery of the laws of electrodynamics." However, Maxwell's theory was not a revolutionary paradigm change, it was immense progress born out of Kuhn's normal science.

The second example is the invention of laser light (light amplification by stimulated emission of radiation) and its application in fundamental research, medicine, and material science. The theoretical foundation of this working field was laid down by Einstein in his relativity theories, but Einstein did not initiate any research on laser or maser radiation and never published any contribution to this field. The development of this research area 30–50 years later was not an intellectual revolution but entailed enormous progress in the understanding of light and other electromagnetic radiation. It also entailed numerous practical applications, part of which may be understood as revolutions in material science and medicine.

Such applications include laser pointers, optical disk drives, barcode scanners, measurements of range and speed of vehicles and airplanes, cutting or welding of metals, bloodless surgery with laser scalpels, lithotripsy of kidney stones, tissue ablation, and cosmetic skin treatment. A particularly conspicuous example of a medical application is fixation of a loose retina across the intact eye.

The third case is invention and development of nuclear magnetic resonance (NMR) spectroscopy after World War II. Many nuclei, such as the proton (^{1}H) or the carbon isotope ^{13}C possess a magnetic spin $I = ½$. In a strong magnetic field these nuclei preferentially adopt positions parallel or antiparallel to the field The energy difference, which can be measured with high precision, depends on the neighborhood of these nuclei, that is, it depends on the entire structure of a molecule and its mobility. After the invention of cryomagnets around 1960, NMR spectroscopy became the most versatile and powerful analytical tool in chemistry and had a major impact on the progress of biochemistry, organic and inorganic chemistry, pharmacy, and polymer science.

The same basic phenomenon, the orientation of protons in a strong magnetic field, can be "translated" into pictures that display the different mobilities of protons in the tissues of living organisms. This magnetic resonance tomography (MRT) has become the most powerful diagnostic method in medicine, particularly well-suited for analysis of soft tissue. Only the high cost prevents broad application in every medical office (i.e., in the year 2015).

The fourth example concerns the evolution of genetics (see Sect. 8.4). Its origin, the work of the Augustinian monk Gregor Mendel, included a partial paradigm change of the established theory of inheritance, but for nearly 40 years almost nobody noticed Mendel's studies. After 1900 the evolution of genetics proceeded rapidly but in a stepwise manner through the contributions of numerous scientists and physicians and culminated in complete analysis of the human genome. Within the field of genetics an immensely important paradigm change occurred after 1950, which entailed a paradigm change in the Darwinian theory of evolution (see Sect. 8.4). Yet, the emergence of genetics as a whole cannot be considered as a revolution in biology; it represents the rise of a new discipline from the "normal science" of biology.

Chapter IX of Kuhn's book discusses the nature and necessity of scientific revolutions. Kuhn himself concedes (p. 95): "In principle a new phenomenon might emerge without reflecting destructively upon any part of past scientific practice ... a new theory does not have to conflict with any of its predecessors. It might exclusively deal with phenomena not previously known, as quantum theory deals with subatomic phenomena unknown before the twentieth century." This is not the only point where Kuhn contradicts his own concept. In summary, Kuhn's view of normal science does not agree with the reality of scientific research and the following chapters of this book (5.2 and 5.3) demonstrate that other aspects of Kuhn's model of science are also inconsistent or disagree with the history of science.

5.2 Paradigm Change

According to the title of his book, Kuhn interpreted a paradigm change as a crisis in the course of normal science and as a revolutionary process. He wrote several pages devoted to a comparison of revolutions in science and in politics. His own words give again the best insight into his concept (p. 52):

> New and unsuspected phenomena are, however, repeatedly observed by scientific research, and radical new theories have again and again been invented by scientists. History even suggests that that the scientific enterprise has developed a uniquely powerful technique for producing surprises of this kind. If this characteristic of science is to be reconciled with what has already been said, then research under a paradigm must be a particularly effective way of inducing paradigm change. That is what fundamental novelties of fact and theory do. Produced inadvertently by a game played under one set of rules, their assimilation requires the elaboration of another set. After they have become parts of science, the enterprise, at least of those specialists in whose particular field the novelties lie, is never quite the same again.

On p. 62 he says about discoveries and paradigm change: "Those characteristics [of discoveries] include: the previous awareness of anomaly, the gradual and simultaneous emergence of both observational and conceptual recognition, and the consequent change of paradigm categories and procedures often accompanied by resistance" and continues on p. 65: "Anomaly only appears against the background provided by the paradigm. The more precise and far-reaching the paradigm is, the more sensitive an indicator it provides of anomaly and, hence, of an occasion for paradigm change." On pp. 66–67, Kuhn writes: "Discoveries are not, however, the only source of these destructive-constructive paradigm changes. In this section [chapter VII] we shall begin to consider the similar, but usually far larger, shifts that result from the invention of new theories. ... In taking up the emergence of new theories we shall inevitably extend our understanding of discovery as well. ... If awareness of anomaly plays a role in the emergence of new sorts of phenomena, it should surprise no one that a similar but more profound awareness is prerequisite to all acceptable changes of theory."

In chapter XI, Kuhn explains the title and design of his book by comparison of political and scientific revolutions (pp. 92–93): "Why should a change of paradigm be called a revolution? In the face of the vast and essential differences between political and scientific development what parallelism can justify the metaphor that finds revolution in both. ... Scientific revolution, as we noted at the end of Section V, need seem revolutionary only to those whose paradigms are affected by them. To outsiders they may, like the Balkan revolutions of the early twentieth century, seem normal parts of the developmental process. ... The genetic aspect of the parallel between political and scientific development should no longer be open to doubt. The parallel has, however, a second and more profound aspect upon which the significance of the first depends. Political revolutions aim to change political institutions in ways that those institutions themselves prohibit. Their success, therefore, necessitates the partial relinquishment of one set of institutions in favor of another, and in the interim, society is not fully governed by institutions at all." He concludes on p. 94: "The remainder of this essay aims to demonstrate that the historical study of paradigm change reveals very similar characteristics in the evolution of science. Like the choice between competing political institutions, that between competing paradigms proves to be a choice between incompatible modes of community life."

Kuhn's comparison of paradigm changes with social or political revolutions is an informative and fascinating approach to a deeper understanding of the structure and history of science, but it is only half truth. Political/social revolutions are quasi-explosive, short-term events with a typical duration of 1–5 years. Such duration is an extremely short period of time compared with the written history of humanity. Such short-term paradigm changes also exist in science, and a comparison with political events and the term "revolution" may indeed be justified. Examples are Lavoisier's introduction of the oxidation theory (Sect. 9.2), Max Planck's quantum theory of energy, and Einstein's publications on special relativity (1905) and general relativity (1917). However, other paradigm changes required a much longer time. Copernicus's work *De revolutionibus*, published in 1543 (see Sect. 10.1), was ignored or rejected by almost all experts of that time for nearly 50 years and only Kepler's calculation of planetary motions and Galilei's new astronomical observations promoted the change from the geocentric to the heliocentric view of the universe in the course of the seventeenth century.

The origin of the wave theory of light can be traced back to the work of the Dutch physicist Christian Huygens (1629–1695), but his theory was for a long time in conflict with Newton's corpuscular theory, published in 1704. Completion of wave theory required almost 250 years, with Einstein's publication on general relativity in 1917 as the final contribution. The paradigm change of male homosexuality from sinful and criminal behavior to a nervous and psychical disorder required several decades and the second paradigm change from a disease to a natural property of many (if not all) animal species (including humans), is still going on and will presumably take more than a century (see Sect. 7.7). These slow paradigm changes can certainly not be classified as revolutions in the sense described by Kuhn. They have instead an evolutionary character and, thus, it makes sense to distinguish between evolutionary and revolutionary paradigm changes, although a sharp borderline between these categories does not exist.

Another important characteristic Kuhn attributes to revolutionary paradigm changes is a change in world view. In his own words (chapter X, p. 111): "Examining the record of the past research from the vantage of contemporary historiography, the historian of science may be tempted to exclaim that when paradigms change, the world itself changes with them. . . . Nonetheless, paradigm changes do cause scientists to see the world of their research engagement differently. In so far as their only recourse to their world is through what they see and do, we may want to say that, after a revolution, scientists are responding to a different world." On p. 128, Kuhn says: "By the same token, the Copernicians who denied the title 'planet' to the sun were not only learning what 'planet' meant or what the sun was. Instead they were changing the meaning of 'planet' so that it could continue to make useful distinctions in a world where all celestial bodies, not just the sun, were seen differently from the way they had been seen before." The final conclusion of chapter IX is summarized in the following statement (p. 135): "The data themselves had changed. That is the last of the senses in which we may want to say that after a revolution scientists work in a different world." The latter statement is misleading, not to say wrong. In the case of the phlogiston theory (see Sect. 9.2) frequently cited by Kuhn, Lavoisier demonstrated that G. Stahl's experiments were incorrect and, thus, their interpretation too. Lavoisier developed his new theory on the basis of new correct experiments. The new paradigm followed new experimental data and not vice versa. It is the normal course of normal science that improved experimental methods and better instruments yield new experimental data that either confirm the existing paradigm or stimulate the formulation of a new paradigm. The vast majority of paradigm changes follow this pattern, as can be learned from all the paradigm changes in medicine reported in Sects. 7.1–7.7, from biology in Sects. 8.1–8.4, from chemistry in Sects. 9.3 and 9.5, and from astronomy in Sect. 10.2. Kuhn's view that revolutionary paradigm changes change the world view in general or at least that of an entire branch of science is also a half-truth. Kuhn's interpretation may be acceptable for paradigms such as those induced by Copernicus, Newton, Einstein, and perhaps Lavoisier (i.e., for those examples frequently cited by Kuhn). Yet, his view does not fit in with the paradigm changes caused by Hermann Staudinger and Wallace Carothers (see Sects. 9.4 and 9.5). Their work neither changed the world view of humankind nor that of most scientists and even the vast majority of chemical laws and theories were not affected. However, Staudinger's and Carother's work had and still has an immense influence on the everyday life of men and on the progress of medicine. Their work and Einstein's relativity theories represent two extremes in the spectrum of paradigm changes: a change in world view with little influence on everyday life and vice versa. This aspect was ignored by Kuhn and it is ignored by most physics-infected philosophers of science.

In chapters IX and XII, Kuhn presents another conclusion that is not acceptable because it deviates from real science (p. 94): "To discover why the issue of paradigm change can never be unequivocally settled by logic and experiment alone, we must shortly examine the nature of the differences that separate the proponents of a traditional paradigm from their revolutionary success." On p. 148, he states that "The competition between paradigms is not a sort of battle that can be resolved by proofs" and on p. 122 that "Paradigms are not corrigible by normal science at all."

Kuhn derived these strange conclusions from the historic (and psychological) fact that scientists adhering to a seemingly established paradigm are quite often reluctant to accept a new theory within a short time. He presented several examples, such as the fierce debate between the French chemists Joseph L. Proust (1754–1826) and Claude L. Berthollet (1748–1822). Proust argued that chemical compounds consist of two or more atoms, but was not able to convince Berthollet. Kuhn also mentions that Priestley never accepted Lavoisier's oxidation theory, although he was the first to prepare oxygen (see Sect. 9.2). Furthermore, Kuhn quoted Charles Darwin, who wrote at the end of his work *On the Origin of Species*:

"Although I'm convinced of the truth of the views given in this volume . . . I by no means expect to convince experienced naturalists whose minds are stocked with a multitude of facts all viewed during a long course of years, from a point of view directly opposite to mine. . . . But I look with confidence to the future—to young and rising naturalists who will be able to view both sides of the question with impartiality." Finally, Kuhn cites Max Planck who sadly remarked in his scientific autobiography: "A new scientific truth does not triumph by convincing its opponents and making them see the light but rather because its opponents eventually die, and a new generation grows up that is familiar with it."

What Kuhn and the cited scientists could not know are the results retrieved over the past 50 years by neurologists and physiologists about evolution and the structure of the human brain. It is a reasonable and successful strategy of evolution that the majority of animals and men of a given population are reluctant to leave the home and environment that have proven satisfactory for years and decades. An expedition into a new area is understood as a risk for the survival of individuals and species (extraordinary individuals are an exception from this rule). What is true for the body is also true for the mental situation. Religious people who are satisfied with their religion, politicians who are satisfied with their ideology, and scientists who are satisfied with a paradigm are unwilling to leave their mental home and risk conflicts with unpredictable consequences. The decision-making institution in the human brain is an area below the *gyrus cinguli*, which applies emotional standards. This middle part of the brain represents an early stage of evolution and dominates over the relatively new frontal cortex, which provides logical arguments. Hence, whenever a logical argument comes into conflict with strong emotions it will lose the battle. It is not the quality of an argument but the strength of the emotions that is decisive. Scientists may be more prone to accept logical arguments than the average person, but evolution and the structure of their brain is the same. Hence, the reluctance of many (but not all!) scientists to rapidly accept a new theory or paradigm change cannot serve as argument against the quality and logic of new experiments and their correct interpretation.

Kuhn's statement that the competition between paradigms cannot be solved by logic and experimental facts alone is in contradiction to the historical examples given in Part II of this book and it is in contradiction to other parts of Kuhn's own work:

– Kuhn attributes science to be the only human enterprise for which progress may be objectively defined (see Chap. 5.3 of this book). How will Kuhn define progress if it is fundamentally impossible to decide whether a new paradigm is more correct and reliable than the previous one?
– If experiments and their logical interpretation never allow for a final decision on which inductive inference and theory is right or wrong, then science as it is understood by scientists cannot exist at all and Kuhn's entire work is obsolete.

In summary, Kuhn's work is a useful study containing numerous original insights and informative comparisons. However, his "biphasic model," in which normal science is from time to time interrupted by evolutionary paradigm changes, is an over-simplification of real science. As already mentioned at the end of Chap. 1, science produces a permanent flow of new facts, hypotheses, and theories including mistakes, errors, and fallacies. These "flaws" range from simple experimental errors to wrong interpretations of the universe or of the evolution of life. Kuhn's focus on a few mega-paradigm changes, such as those of Copernicus, Newton, and Einstein, means that he only sees the outermost tip of an iceberg of errors and fallacies. Because he has never performed any experimental research, he has never seen the size of the iceberg and has overinterpreted and misinterpreted the few historic examples he frequently cites.

Finally, it should be mentioned that Kuhn's biphasic concept of science is not highly original. A similar scheme was designed by the British philosopher John S. Mill (see Sect. 4.2) about 100 years before Kuhn's book was published. The following quotation originates from an article the Canadian philosopher Fred Wilson (University of Toronto) contributed in 2007 to the *Stanford Encyclopedia of Philosophy* (p.11): "Mill's picture of the inductive method of inquiry and the research that it guides is remarkably close to T. Kuhn's picture of 'normal science.' What Mill calls a law about laws, Kuhn calls a 'paradigm,' but that is a terminological difference. For both are theories that guide research. . . . Mill also allows for something like what Kuhn calls 'revolutionary science,' inquiry undertaken when the paradigm or background theory no longer leads to the discovery of specific laws. . . . When a theory is not falsified but fails to be a successful guide in research, scientists begin to search for a new theory. But this research is not guided by a theory: the research guiding theory is no longer available. . . . When a new theory is located by this research and it replaces the old theory then what Kuhn calls a scientific revolution occurs: the practice of normal science is restored guided by a new theory."

5.3 What Is Scientific Progress?

The term "scientific progress" (abbreviated here to sc. pr.) does not exist in standard encyclopedias such as *Encyclopedia Britannica* or *Encyclopedia Americana*, which focus their comments on social progress. The *Oxford Bibliographies* give a short outline of the history of sc. pr., but without exact description or definition of the term. However *Wikipedia* offers an interesting definition: "Scientific progress is the idea that science increases its problem-solving ability through application of scientific methods."

A comprehensive article with a detailed discussion of sc. pr., including 106 references, is offered in the *Stanford Encyclopedia of Philosophy*. The author of that article, Ilkka Niiniluoto, avoids a short definition and introduces the topic with the following words: "Science is often distinguished from other domains of human culture by its progressive nature: in contrast to art, religions, philosophy, morality, and politics there exist clear standards or normative criteria for identifying improvements or advances in science. For example, the historian of science, George

Sarton, argued (in 1936) that the acquisition and systematization of positive knowledge are the only human activities which are truly cumulative and progressive."

The scope of the discussion of sc. pr. by almost all theoreticians may be described by the following categories:

Methodical progress: invention of new methods of research, refinement of scientific instruments
Cognitive progress: increase or advancement of scientific knowledge
Technical progress: increased effectiveness of tools and techniques
Economical progress: increased funding of scientific research
Educational progress: increased skill and expertise of the scientists
Professional progress: increasing status of scientist and scientific institutions

It is conspicuous that in the entire debate on the meaning and definition of sc. pr., one category, perhaps the most important, is missing, namely progress in medicine.

As explained in Chap. 6, medicine on the whole, may not be classified as a natural science, but more than 95 % of its diagnostic and therapeutic activities are based on scientific discoveries and inventions and, therefore, progress in medicine is directly related to sc. pr.

In the past, numerous historians and philosophers of science have formulated fundamental criticism of science and its capability to retrieve absolute truth or reliable knowledge. This criticism includes, of course, the question of whether science may be progressive. The most prominent representatives of these skeptics and their central statements are mentioned and discussed in Sects. 2.3, 2.5, 3.3, 4.2, 4.3, and 4.4 and repetition of this discussion is not needed here. The main direction of the ongoing debate is illustrated by the following questions (*Stanford Encyclopedia of Philosophy*):

- What is meant by progress in science?
- To what extent and in which respect is science progressive?
- How can we recognize progressive developments in science?

and by the titles of pertinent books and articles:

"Scientific Progress and Peircean Utopian Realism," Almeder R (1983) Erkenntnisse 20:253
Realism Rescued: How scientific progress is possible, Aronson JL et al. (1994) Duckworth, London
"What is Scientific Progress?" Bird A (2007) Nous 41:92
"Scientific Progress as Accumulation of Knowledge: A Reply to Rowbottom," Bird A (2008) Studies in History and Philosophy of Science 39:279
"Progress as Approximation to the Truth: A Defense of the Verisimilitudinarian Approach," Cevolani G, Tambolo L (2013) Erkenntnis 78:921
Scientific Progress: A Study Concerning the Nature of the Relation Between Successive Scientific Theories, Dilworth C (1981), Reidel Dordrecht
"Pure Science and the Problem of Progress," Douglas H (2014) Studies in History and Philosophy of Science (Part A) 46:55

"On Relative Progress in Science," Jonkisz A in: *On Comparing and Evaluating Scientific Theories*, Jonkiz and Koj, eds. (2000), Rodopi, Amsterdam
Progress and its Problems: Towards a Theory of Scientific Growth, Laudan L (1977), Routledge and Kegan Paul, London
"Is There Genuinely Scientific Progress? Moulines CU in: *On Comparing and Evaluating Scientific Theories*, Jonkisz and Koi eds. (2000), Rodopi, Amsterdam
"Scientific Progress," Niinoluoto I (1980), Synthese 45:427
Is Science Progressive? Niiniluoto I (1984), Reidel, Dordrecht
History of the Idea of Progress, Nisbet R (1980), Heinemann, London
Change and Progress in Modern Science, Pitt JC, ed. (1985), Reidel, Dordrecht
Progress and Rationality in Science, Radnitzky G., Andersen G. eds. (1978), Reidel, Dordrecht-Boston
Realism and the Progress of Science, Smith P (1981), Cambridge University Press
"Progress Metaphysical and Otherwise," Wachbroit R (1986), Philosophy of Science 53:354

Almost all philosophers, with the exception of a few biologists who define sc. pr. in the light of Darwinism (see below), discuss sc. pr. in relation to the quality of theories and in relation to the question of if and to what extent scientific theories can approach absolute truth. For example, the English philosopher Alexander Bird defined and compared in his 2008 essay an epistemic (E) and a semantic (S) version of sc. pr. as follows:

(E): An episode constitutes sc. pr. precisely when it shows the accumulation of scientific knowledge
(S): An episode constitutes sc. pr. precisely when it either shows accumulation of true scientific belief, or shows increasing approximation to true scientific belief

Concrete examples are rare in philosophic discourses and almost exclusively concern the history of astronomy and physics, from Ptolemy to Einstein, but ignore all other branches of science. As already commented in Chaps. 1 and 2, this one-sided understanding of science runs a high risk of misinterpretations if definitions and conclusions are extended to all natural sciences. A concrete example should substantiate this critique. It has been (and still is) an important goal of chemical research since the beginning of modern organic chemistry 200 years ago to elucidate the chemical structure of compounds produced by plants or animals. Originally the main driving force was curiosity, but meanwhile potential applications of natural products as aromas, food additives, or medicaments has greatly stimulated these research activities. Today, numerous analytical methods (see Sect. 2.5) enable rapid and unambiguous elucidation of the chemical structure of not only simple molecules such as aspirin, but also of complex molecules such as hemoglobin. Traditionally, synthesis of the compound under investigation from simple, well-known, precursor molecules completes the study. At this point, the chemist can say that the goal of the project has been reached and the absolute truth about the structure of the molecule under investigation is known; this knowledge represents significant progress.

The historically important example of penicillin allows the author to illustrate the meaning of absolute truth and progress for another typical aspect of chemical research, namely optimization of a synthetic procedure. Identification of the structure of penicillin was followed by synthetic variation of the structure with the goal of finding variants with different or more intensive antibiotic effects. Prior to technical production of penicillin and its useful variants their syntheses need optimization. The yield of the product depends on several parameters, such as temperature, time, and pressure. Traditionally, the optimization is performed in such a way that all but one parameter are fixed and only one parameter (e.g., the reaction time) is systematically varied. Figure 5.1 is an illustration of the variation in yield with reaction time for a fictitious synthesis. Figure 5.2 illustrates the temperature dependence of the yield with fixed reaction time. This synthesis is reproducible everywhere in the world and at any time, as expected from a law of nature, and thus the measured course of this synthesis represents again an absolute truth for chemists. The optimized synthetic procedure serves, in turn, as basis for the technical production of penicillin or its variants. The progress resulting from this kind of research is evident from the survival of millions of patients who

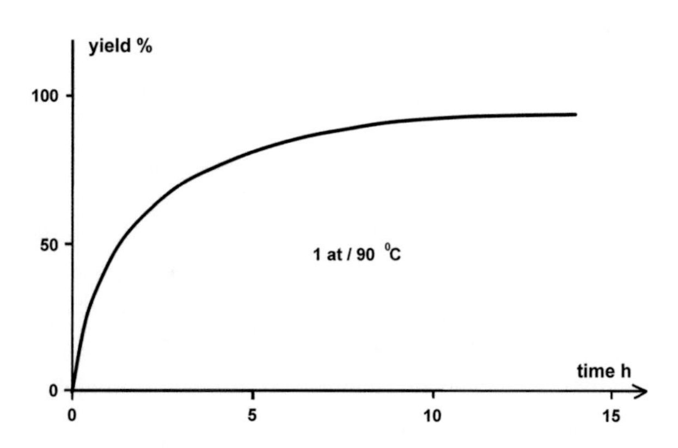

Fig. 5.1 Dependence of the yield of a fictitious antibiotic on the reaction time

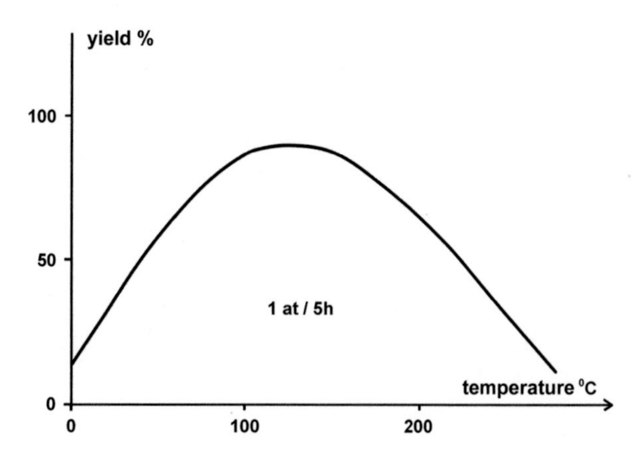

Fig. 5.2 Dependence of the yield of a fictitious antibiotic on the temperature

suffered from potentially deadly infections. The author is himself a "piece of evidence," because he contracted diphtheria at the age of 6 and had a narrow escape from death with the help of penicillin. These examples and the arguments presented in Sect. 2.3 demonstrate that terms such as "truthlikeliness" and "verisimilitude" are obsolete or even misleading when applied to all kinds of scientific research.

The classical definition of sc. pr. as accumulation of knowledge can be traced back to Francis Bacon (1561–1626) and George Sarton (1884–1956). It is, in principle, best suited to cover all branches of science and any kind of research. However, this definition was not only attacked by skeptic philosophers (see Sect. 2.3) but also by philosophers who applied a narrow and one-sided interpretation of sc. pr. as simple compilation of data. Yet, this problem can be eliminated by a slight modification: "sc. pr. is accumulation of knowledge in a quantitative and in a qualitative sense." The term qualitative sense may include:

- Better understanding of consequences and interactions of laws of nature (see Chaps. 2 and 3)
- New insights resulting from paradigm shifts
- Increased problem-solving capacity of theories

In other words, this modified definition includes all semi-definitions having in common that reliable knowledge (absolute truth) may, in principle, be achieved.

Two treatises on sc. pr. are commented on here in more detail, because one of them relates to Kuhn's work and because both are related to Darwin's theory of evolution. Kuhn devoted the last chapter of his book to his understanding of sc. pr. On page 166 he writes: "In its normal state then a scientific community is an immensely efficient instrument for solving the problems and puzzles that its paradigm defines. Furthermore, the result of solving these problems must inevitably be progress. ... Once again, there is much to be learned by asking what also the result of a revolution may be. Revolutions close with the victory of one of the opposing camps. Will the group ever say the result of the victory has been something less than progress?" However, on p. 170 he denies progress towards a final truth: "In the sciences there need not be progress of another sort. We may, to be more precise, have to relinquish the notion, explicitly or implicitly, that changes of paradigms carry scientists and those who learn from them, closer and closer to the truth." This statement is in contradiction to his aforementioned conclusions and in contradiction to chapters VII–XII, where he argues that in science the big steps forward are the consequence of revolutionary paradigm shifts.

Finally, Kuhn compares sc. pr. with Darwin's evolutionary theory, embarking on the absence of any final goal (pp. 171–173): "We are all deeply accustomed to seeing science as the one enterprise that draws constantly nearer and nearer to some goal set by nature in advance. ... and the entire process [of research] may have occurred, as we now suppose, biological evolution did, without benefit of a goal, a permanent fixed scientific truth, of which each stage in the development of scientific knowledge is a better exemplar." This view contradicts the definition of science and various arguments presented in Chaps. 2 and 3 and suggests that science is nothing more than an intellectual game inside the community of scientists.

Another extreme interpretation of sc. pr. based on modern Darwinism (see Sects. 8.3 and 8.4) was published by E. Volland, professor of philosophy of biology. The title of his German article "Fortschrittsillusion" (Illusion of Progress) indicates the basic concept: progress is nothing but a fiction of the human brain. In his own words (translated by the author): "Brains resemble dogmatic egocentrics which have difficulties to learn to change the perspective and which don't understand why they should do it with respect to themselves. The idea of progress is also one of these useful constructions of the brain. … Finally, progress is not a feature of evolution even when in sloppy speech sometimes terms such as 'Höherentwicklung' (evolution towards a higher cognitive capacity) or progressive evolution or similar suggestive terms are used. … Perhaps evolution means increasing complexity, but higher complexity is not identical with progress and progress is not a biological category."

Possibly, many scientists and interested laics can agree with this line of argumentation, but the following statement deserves critique: "Why this at all? Why has the brain constructed the concept of progress and cares for it until the end of life (although with varying intensity depending on age)? Well, it is established that natural selection works with the evaluation of differences, and from this simple matter of facts it may be concluded that Darwin's survival of the fittest automatically and necessarily favors evolutionary competition.

The advantage of one individual (or species) is quite often a disadvantage for another one and therefore men live in comparatives. Stagnation entails elimination from the evolutionary game, and therefore, the Darwinian properties 'higher,' 'larger,' and 'faster' are necessarily inherent in all organisms. … Hence, the idea of progress belongs to the class of constructions that were mentioned before. Our brain creates an idea which does not have an objective equivalent outside the consciousness. The standards we use for measuring progress result from our individual goals and desires in the actual consciousness of a fully developed, strategically egocentric brain. It is obviously home-made and remains fixed in the world of subjectivity."

Even if it is correct that the course of evolution may not be interpreted in terms of progress, it cannot be inferred that any idea of progress is nothing but a subjective feeling and objective measurements are never possible. As outlined below, categories of progress exist outside evolutionary theory and these categories may include a high degree of intersubjectivity. Volland correctly answered his critics that a determination of progress requires standards and methods that allow comparison of two different states. However, Volland's theory ignores the fact that in everyday life comparisons with intersubjective relevance are feasible, even when progress cannot be measured in exact numbers. The following examples illustrate this point.

At least since humankind settled down and began to develop agriculture, most, if not all, men ascribe progress to any kind of invention, such as new technologies, new tools, new machines, or new materials, that alleviate or reduce strenuous bodily work. In the twenty-first century a complementary definition can be added: progress is any invention or development of hard- and software that make data transfer, data storage, and evaluation of data more efficient.

The second and more important example, which is relevant for all humans including (skeptic) philosophers, concerns progress in medicine. It can be safely assumed that all sane and psychically intact people agree that any discovery or invention that contributes to safer diagnosis, shorter and more efficient therapy, less dangerous surgery, reduction in pain, and longer life may be called progress. A concrete example is given by toothaches. It would be difficult to find any adult who has never suffered from toothache and, therefore, comparisons concerning toothaches represent a high level of intersubjectivity. Anybody who suffers severe toothache will visit a dentist, given the opportunity and money. The German poet and artist Wilhelm Busch (1832–1908) illustrated what a visit to the dentist (the fictitious Dr. Schmurzel) looked like in around 1870 (Fig. 5.3). For comparison, Fig. 5.4 displays the office of the real Dr. N. Dudeck in 2015, in which the author has enjoyed several almost pain-free dental treatments under the influence of local anesthetics (see Sect. 7.3). Discovery and technical syntheses of local anesthetics (and narcotics) represent a special and extraordinarily important contribution to the

Fig. 5.3 Visit to a dentist in around 1870 (Wilhelm Busch in "Balduin Bählamm")

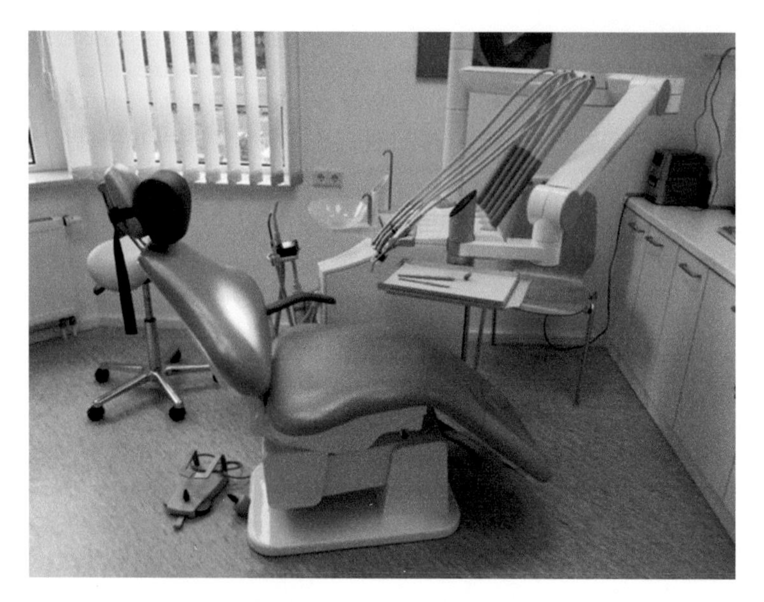

Fig. 5.4 Office of Dr. N. Dudek, 2015

progress in medicine. The chemical structure and synthetic procedure of local anesthetics are absolute truths, and progress can reliably be determined for all individuals by comparison of painful and pain-free states.

The difference between the mental world of philosophers and the reality of scientific progress may be illustrated and summarized by the following insights:

The French philosopher Gaston Bachelard (1884–1962) wrote that "The characteristic of scientific progress is our knowing that we did not know."
The patient of Dr. Schmurzel (Fig. 5.3) certainly thought that "Progress is when future dental treatment is less painful."

Bibliography

Harré R (ed) (1975) Problems of scientific revolution: progress and obstacles to progress in science. Oxford University Press, Oxford
Hempel CG (1965) Aspects of scientific explanation. The Free Press, New York
Laudan L (1984) Science and values: the aims of science and their role in scientific debate. University of California Press, Berkeley
Laudan L (1987) Progress or rationality? The prospect of normative naturalism. Am Philos Quart 24:19
Mizraki M (2013) What is scientific progress? Lessons from scientific practice. J Gen Philos Sci 44:375
Niiniluoto I (2011) Revising beliefs towards the truth. Erkenntnis 75:165

Niiniluoto I (2014) Scientific progress as increasing verisimilitude. Studies in history and philosophy of science. Science (A) 75:7377
Rantala V (2002) Explanatory translation: beyond the Khunian model of conceptual change. Kluwer, Dordrecht
Stegmüller W (1976) The structure and dynamics of theories. Springer, Berlin, Heidelberg

Chapter 6
How Much Science Is There in Medicine?

6.1 Introductory Remarks

Whoever can heal is right

The term "medicine" is derived from the Latin term *medicus* meaning physician. Both *Encyclopedia Americana* and *Encyclopedia Britannica* do not offer a definition. *The text of Wikipedia* begins with the words: "This article is about the science and art of healing." It continues with the definition: "Medicine is the science and practice of the diagnosis, treatment, and prevention of disease." This is an interesting description because no other human activity is called science, practice and art. The author has asked several physicians and surgeons, if they consider medicine to be a kind of natural science and the uniform answer is NO! Because this answer is the rule and not a law, exceptions may exist.

This chapter explores whether this answer is justified and to what extent medicine is founded on scientific methods, scientific discoveries, and related technical inventions.

6.2 Medical Diagnosis

When a person visits a doctor because of pain or illness, the doctor's first activity is diagnosis. Medical diagnosis is the process of determining which wound or disease is responsible for the pain and symptoms of the patient. The first step of the diagnosis is usually an interview, which has the purpose of retrieving information about the following points:

- Chief complaints, that is, the reasons for the current visit
- History of the current symptoms and illness
- Previous medication and past medical history

H.R. Kricheldorf, *Getting It Right in Science and Medicine*,
DOI 10.1007/978-3-319-30388-8_6

- Family history
- Social background and history

This interview is frequently followed by a physical examination of the patient using instruments, the production and application of which involve laws of nature. Consider first a dentist: He uses a dental probe made of a hard elastic metal, which enables him to detect caries. The caries-infected material in a cavity is softer than the intact dentin or cement of a tooth. This simple diagnostic method involves two physical laws: the law of the lever and the proportionality between force and pressure. If, as is usual, a dental drill is used to remove the caries, mechanical laws are again involved and the electricity needed for the drilling machine obeys the laws of electricity. Regardless of whether a dental probe or a drill is used, both instruments have been applied many billions of times over the past 100 years, which illustrates the reproducibility of these diagnostic methods.

Another quite simple instrument used by physicians is a tongue depressor, the application of which involves the same physical laws mentioned above. The most widely used diagnostic instrument is, however, the stethoscope, which has become the symbol of doctors and medicine. It consists of two rubber tubes that connect a small disc-shaped resonator to the ears of the doctor. It transports sounds from the interior of the body, protected against noise from outside. It is an acoustic medical device that allows the physician to listen to the heart beat and to the sound of the lungs, intestines, veins, and arteries.

A more sophisticated method designed to study the function of the heart is electrocardiography (ECG or EKG). Electrodes are placed on the skin of the breast and connected to a receiving device that projects curves onto a screen, indicating voltage changes. This method allows physicians to record the change in voltage resulting from contraction and depolarization of the heart muscles over time. The regularity, shape, and amplitude of the recorded voltage curves carry a large amount of information about the correct or incorrect function of ventricles and heart muscles. It is trivial to say that this method, like all the physical methods described below, relies on the laws of electricity. Hospitals and the offices of most surgeons and orthopedists are equipped with ultrasonic devices. These instruments use ultrasonic waves to distinguish between organs or tissues of different density. A probe head is placed on the skin above the organ under investigation and ultrasonic waves of a frequency between 1 and 40 MHz are emitted with an intensity of approximately 100 mW/cm^2. Bones, tissues rich in fat or rich in water scatter these waves in different ways and the receiver "translates" this pattern of scattered waves into a black, gray, and white picture.

Probably the most widely applied diagnostic method relying on physical laws is based on X-ray scattering. X-rays are high-frequency electromagnetic waves (see Sect. 10.4), the diagnostic usefulness of which was discovered by the German physicist Wilhelm C. Röntgen in around 1895. The high frequency (n) of the X-ray radiation corresponds to a short wavelength (l) according to $c = nl$ (where c is the maximum speed of light), with the consequence that X-rays "see" the tiny atomic nuclei and not the electronic clouds around atoms and molecules as visible light does. The efficiency of X-ray scattering increases with the size (or weight) of the atomic nuclei. Therefore, soft tissue (i.e., proteins), which mainly consist of the light nuclei hydrogen, carbon, nitrogen, and oxygen, are barely visible, whereas bones and teeth, which are made up of the heavier atoms calcium and phosphorous (in the form of phosphate ions), scatter the X-rays quite well. Titanium or tantalum bars and screws, frequently used in surgery, are also easy to detect.

A more modern diagnostic method that has rapidly gained (and still gains) importance despite high cost is nuclear magnetic resonance tomography (MRT). The term tomography is derived from the Greek words *temnein* (cutting) and *graphein* (writing or drawing) and indicates that the MRT method scans cross sections of a limb or of the entire body. Magnetic resonance means that the magnetic properties of the hydrogen nucleus (^1H = proton) serves as a source of information. As mentioned in connection with NMR spectroscopy (see Sect. 5.1), the proton can occupy two different energy levels when exposed to a strong magnetic field and the exact energy level depends on the chemical neighborhood, notably on the mobility of the protons. The protons of water are highly mobile in contrast to protons attached to the protein or polysaccharide chain in solid tissue. The mobility of most protons in soft tissue varies between these extremes, and diseases such as infections can modify the temperature and water content of a tissue and, thus, the mobility of protons. The computer of the MRT system "translates" the information provided by the MRT machine into a black-gray-white picture and the physicians learn how the 50 shades of gray may be interpreted in terms of the healthy or sick states of a tissue. Because protons are rather rare in bones, the MRT method is particularly suitable for the diagnosis of soft tissue and, thus, it is complementary to X-ray computer tomography.

A very expensive method that is rarely applied is the positron emission tomography (PET). Positrons are the positively charged counterparts of electrons and have a short life time because of their rapid reaction with the electrons surrounding all atoms and molecules. The application of PET requires injection of positron-emitting radioactive chemicals with the purpose of elucidating anomalies in the metabolism of an organ, but they do not provide pictures of an organ.

All these diagnostic methods have in common that their function and application are grounded in the physical laws of nature. Therefore, they yield reproducible results irrespective of time and location, otherwise the numerous instruments and machines would not be applicable worldwide and around the clock. Furthermore, it is the predictability of their performance and usefulness that makes their production profitable for many companies.

Medical diagnostics profit not only from physical methods and physical laws but also—with steadily increasing success—from chemical methods. A standard procedure practiced for almost a century is the blood count. A complete blood count may reveal the presence or absence of nearly 30 different components, such as inorganic ions, glucose, fat, hormones, and proteins. Quite often, the blood count is focused on a few components, such as sodium (Na^+) and potassium (K^+) ions, which are important for the function of the heart and nerves, or calcium (Ca^{2+}) and phosphate (HPO_4^{2-}) ions that are needed for the growth of bones and teeth. Perhaps most frequently, the concentration of glucose is measured (diabetes test) and the concentration of cholesterols, which is related to the risk of high blood pressure and brain strokes. Special rapid chemical reactions yielding colors or precipitates underlie these tests. Specialized laboratories perform these tests and the modern analytical methods enable accurate determination of low concentrations of the blood components of interest. Further important tests based on chemical and biological laws have been designed to identify certain species or families of infectious bacteria.

A steadily increasing number of chemical tests, often called flash tests or rapid tests, can be conducted by the patients themselves. Pregnancy tests and glucose tests are the most widely applied tests of this kind. These flash tests usually work using paper strips or polymer films coated with materials that undergo a color reaction when the tested hormones or compounds are present in blood or other body fluid. Such tests are developed by pharmaceutical companies and the optimization of their reliability and applicability requires many years. In summary, more than 90 % of all diagnostic methods used by physicians in the twenty-first century are the result of scientific discoveries and technical inventions.

6.3 Therapy

Therapy is another a term derived from ancient Greek, *θεραπεια* means healing or medical treatment. A therapy is the usual and logic consequence of a diagnosis. When averaged over all disciplines of medicine, approximately 90 % of therapies involve medicaments and, thus, profit from biological and chemical discoveries and inventions. Surgery and orthopedics are those disciplines that most frequently apply "nonchemical" therapies. Such therapies encompass physiotherapy (exercises), massage, acupuncture, and local heat treatment (e.g., by means of infrared radiation or mud packs).

An operation is mainly skilled manual work involving numerous instruments, such as scalpels, scissors, pincers, clips, saws, and medical drills. However, all operations also need support from chemistry. The "chemicals" used in this connections encompass special soaps, disinfectants, antibiotics, pain-killing drugs, local anesthetics, and frequently narcotics. Furthermore, hemostatic agents (styptics) or hormones (e.g., adrenalin) may be used, notably in dentistry. Moreover, synthetic materials that play a mechanical role are involved, for example, elastic polymers used for flexible tubes and cables, or special rubbers for gloves that do not induce latex allergy.

The application of local anesthetics is widespread in dentistry, but it is also the method of choice for minor surgery near the surface of the body (see Sect. 7.3). Over the past 120 years, the effect and success of local anesthetics have been reproduced more than a billion times; they do not affect the brain and have the additional benefit of few, usually negligible, side reactions. In contrast, narcotics have a strong influence on the mental capacity of the patient and strongly interfere with important functions of the brain. The risk of side reactions that may even have lethal consequences is much higher and, therefore, operations conducted under the influence of narcotics require artificial respiration and control of heart rhythm and blood pressure.

All cancer therapies also involve physical and chemical methods. Irradiation of tumor cells with laser light or α-particles (see Sect. 10.3) and superheating of cancer tissue all belong to the physical part of therapy. Administration of cytostatic medicaments stops growth of tumor cells, either by affecting their metabolism and proliferation or by reducing the access of fresh blood. The treatment of all other major diseases, such as bacterial or viral infections, diabetes, or heart–circulation disorders, is almost exclusively based on medication or, in other words, on chemistry.

If the administration of a drug is considered as a chemical experiment and compared with an experiment in a chemical laboratory, the following similarities and differences are worth noting. Exact reproducibility of an experiment, such as the synthesis of aspirin (see Fig. 2.1), has been proven a billion times and is predictable if all experimental parameters, such as temperature, reaction time, concentration, pressure, and purity of the reactants, are reproduced and kept constant. When these parameters are varied, the results (e.g., the yield of aspirin) scatter around a mean value and the extent of scattering increases with more extensive variation of the experimental parameters.

However, when a drug is administered to a patient, it is not possible to reproduce and maintain the external parameters (temperature, time, etc.), because the "reaction vessel," the human body, varies over a wide range. If variables such as sex, age, size, weight, and content of fat and water were identical for all patients (a fiction), the effect of the medicament would show less scattering and the treatment with medicaments could approach the situation of a chemical experiment in a laboratory. Yet, considering the heterogeneity of patients, it is amazing that a physician is usually confronted with the good reproducibility of most therapies based on drugs and remedies. Reproducibility is particularly high for injections of local anesthetics, which have been reproduced in dentistry more than a billion times. The reason for the good reproducibility of therapies with chemicals is the fact that the metabolism of all humans obeys the same chemical laws. A relatively broad scattering of the therapeutic effect is observable and predictable when psychological disorders are involved and when emotions play a role in the intensity of the symptoms (e.g., headache).

Excluding surgery and dentistry, if therapies based on medicaments are considered, three categories may be defined:

- Homeopathy
- Treatment with (allegedly soft) natural medicine
- Treatment with (allegedly hard) synthetic medicaments

Homeopathy has frequently been described, debated, and criticized. Nonetheless, one more discussion is needed here for comparison with the other two categories. The Greek name for this medical strategy and philosophy indicates that medicaments and methods are used that intensify the symptoms of a disease or cause similar symptoms in healthy people (*similia similibus curentur*, which can be translated as "like cures like").

The modern interpretation of this concept, which was developed around 1796 by the German physician Samuel Hahnemann (1755–1843), assumes that the drugs and remedies provoking symptoms resembling those of the disease activate or intensify the immune response, thereby accelerating the healing process. Furthermore, it has been demonstrated in extensive studies that the placebo effect plays a significant role. Homeopathy has attracted numerous skeptical or critical comments, but it is unquestionable that for a limited number of diseases homeopathy has a healing success above the probability limit of 50 %. A typical case where homeopathy is usually more successful than traditional western medicine is neurodermatitis. The author has himself witnessed an impressive case in his family. A boy was born with neurodermatitis and in addition to this problem his tear ducts were not wide enough to allow rapid drainage of tears and eye fluid. The consequence was repeated eye infections. The first doctor who was consulted administered an antibiotic to heal the infection. The antibiotic did its job and after 3 weeks the infection had vanished, but 3 months later the infection came back. The second physician proposed another antibiotic and the same sequence of events occurred. The third physician followed the same strategy as his colleagues but recommend a third new antibiotic. None of those physicians cured the neurodermatitis. Now, the parents of this boy visited a homoeopathist who had a good reputation for healing neurodermatitis. Using the usual *globuli*, he was able to heal the neurodermatitis in the course of 6 months. Furthermore, the homoeopathist stopped the infection of both eyes. He was, of course, not able to widen the tear ducts but this process occurred automatically during the following 2 years, simply as consequence of the continuous growth of the boy. By the age of 4 years, the boy had got rid of all his former health problems.

The fact that at the beginning of the twenty-first century none of the natural sciences can satisfactorily explain how and why homoeopathy can successfully treat certain diseases does not justify criticism of all homoeopathy as nonsense or sheer fiction. Usually, the "placebo effect" and great empathy of the healer are considered to be responsible for the success of homoeopathy and this explanation may be partly true. Yet, it is unlikely that the placebo effect is responsible for the healing of children aged two or less. Furthermore, healing of animals can certainly not be explained by the placebo effect, but the author has himself witnessed healing of horses by homoeopathic methods. Nobody can predict what the scientists will know in 100 or 500 years from now and perhaps a scientific explanation of most aspects of homoeopathy will be available in the future.

Nonetheless, the dilution method, which is a characteristic part of the homoeopathy, deserves a critical comment. In the second half of his life, Hahnemann began to dilute his drugs and remedies to reduce the intensity of side reactions and to reduce the costs. Initially, he worked with dilution factors in the range of 10–100 but during the last years of his life he worked with dilution factors up to 10^4 (meaning a drug to water ratio of 1:10,000). Because he observed healing effects even at such low concentrations, he finally speculated that it was not the medicaments themselves but their "souls" that were responsible for the observed success. This hypothesis was related to the concept of vitalism, which was still *en vogue* at that time (see Sect. 9.3).

However, in the course of the past 150 years, people have flushed all kind of drugs and remedies into the rivers and finally into sea, mainly into the Mediterranean, the Baltic Sea, the North Sea, and the Atlantic. Although those medicaments have been degraded by hydrolysis and/or oxidation, their "souls" should have survived. Yet, it is not documented that repeated drinking of sea water is a satisfactory cure for most diseases. Furthermore, it has been reported several times during the past 10 years that most European lakes and rivers contain significant concentrations of many medicaments. In the case of diclofenac, the concentration is so high that the fertility of certain fish is affected. If, alternatively, we assume that it is not the souls of the drugs but the diluted drugs themselves that are responsible for their efficacy in homeopathy, repeated drinking of water from lakes and rivers should heal almost any kind of illness. But, once again, any evidence for this assumption is lacking. Ironically, it is the failure of the dilution method that guarantees that homoeopathists have enough patients.

In the case of natural medicine, the view that it is "soft" and preferable to the medicaments of traditional scientific medicine is an emotional rather than a rational, evidence-based view. To avoid misunderstanding and to shorten the discussion, the term natural medicine is used here as a generic term for application and administration of medicinal herbs, plant extracts, and animal products. Unfortunately, almost all laics in chemistry and pharmacy—including most academics—do not know that many allegedly "hard" synthetic medicaments of traditional western medicine are natural products or derived from them. Two widely known medicaments serve as examples, aspirin and penicillin. The chemical name for aspirin is acetyl salicylic acid (ASS) because it is prepared from salicylic acid by acetylation (see Sect. 2.5). Low concentrations of ASS can be found in several plants such as *Filipendula ulmaria*, the former name of which, *spirea*, was the origin of the trademark aspirin. The true active agent, salicylic acid, has been found in numerous plants, usually in higher concentrations than ASS, for example in the bark of willow trees. The pain-killing effect of willow bark extracts was known to the ancient Greeks as long as 2500 years ago. However, salicylic acid has a bitter taste, stimulates vomiting, and attacks the mucosa layer and inner walls of gullet and stomach. These disadvantages are considerably reduced when an acetic acid (acetyl) group is attached to salicylic acid. The human metabolism removes the acetyl group by hydrolysis and the liberated salicylic acid (now in extremely low concentration everywhere in the body) can do its job. No other drug has found such extensive application worldwide. Over the past 120 years, approximately 5–10 billion aspirin tablets have been produced. It is obvious that such a large quantity cannot be obtained by extraction from willow trees and that only technical production can satisfy the worldwide demand. The technical procedure was elaborated by the German chemist Felix Hoffmann, of Farbenfabriken F. Bayer, in 1897.

The second example is penicillin. The original compound (benzyl penicillin or penicillin-G) was discovered in 1928 by the Scottish biologist Alexander Fleming (1861–1955) as a secretion of the mold *Penicillium notatum*. Extensive application—including veterinarian medicine and animal breeding—has had the consequence that numerous species and strains of bacteria have become resistant to penicillin-G. In order to circumvent the resistance, and to broaden the spectrum of infections that can be treated, chemists have modified the structure of penicillin. These modifications and sufficient supply of the original penicillin-G required technical production.Hence, the sharp borderline between natural and synthetic medicaments only exists in those cases where the natural remedies are extracts of plants or animal products with a complex composition, whereas the medicaments of traditional scientific medicine are pure chemicals with established structure. The complex composition of natural remedies could be an advantage, because more than one component may have a healing effect. Furthermore, two or more components might undergo a synergistic interaction. Nonetheless, all components of natural medicines are chemicals that somehow interact with human metabolism via biochemical reaction pathways. The fact that the chemical structures of most components of natural medicines and their biochemical reaction pathways are unknown does not justify ascribing them with extraordinary efficacy or mystifying them. In other words, lack of knowledge is not an advantage per se.

In this context, it is worth mentioning that over the past two decades an increasing number of pharmaceutical companies have launched research programs designed to discover new pharmaceutically active substances in plants and (primitive) animals from all areas of the world. These new natural drugs are hoped to enable new therapies and, in the case of success and after elucidation of their chemical structure, technical production will follow.

What is meant by "soft" and "hard" medicine was and still is emotionally debated. However, the core issue of such a comparison may be discussed on the basis of scientific terms and standards. For this purpose, it makes sense to define the adjectives soft and hard in terms of the therapeutic index. In the past, this term was defined as the quotient LD_{50}/ED_{50}. The median lethal dose (LD_{50}) is the dose that entails the death of 50 % of tested laboratory animals (usually mice). The median effective dose (ED_{50}) is the dose at which 50 % of animals show a healing effect. For reasons of safety, today the ratio LD_{25}/ED_{75} is usually determined. Another important change concerns the testing method. Today, the toxicity of chemicals, in general, and that of drugs, in particular, is evaluated via special tissue tests and not using laboratory animals. It is particularly difficult to determine ED values for pain-killing medicaments and psychotropic drugs because the sex and the emotional situation of the patient play a role in the effect of these medicaments and animals or tissue tests cannot be used to evaluate the efficacy of psychotropic drugs.

The assumption that natural medicine is soft is partly based on the low concentrations used, which entails a low level of side reactions. But, the assumption that natural medicaments have few side reactions is not necessarily true. As already mentioned above, the natural drug salicylic acid is highly aggressive against the mucosa layers and inner walls of gullet and stomach. Furthermore, it suppresses blood clotting and, thus, favors long lasting bleeding in the case of injuries. Other examples are the digitalis glycosides, which were used for many decades to enhance the efficiency of the heart muscles. These drugs, originally extracted from foxglove (digitalis) are known for their particularly low therapeutic index. A minor overdose may have lethal consequences. In contrast, an absolutely artificial synthetic drug with an unusually high therapeutic index is bismuth subsalicylate (BSS). This drug is based on the metallic element bismuth and has been known and commercialized in the USA for almost 100 years (e.g., under the trade mark Pepto-Bismol). It is administered against a variety of stomach problems ranging from insufficient digestion to stomach ulcers. The recommended daily dose is 1.0–1.5 g (depending on the weight of the patient). At such a high dose most natural medicaments cause intensive side reactions. For all those reasons, digitalis glycosides have found a place in thrillers and police stories whereas the bismuth salts have not.

At this point, the question arises of how to define the term "poison." For most laics in chemistry and pharmacy, notably for members of green parties and journalist of newspapers, the world consists of two groups of chemicals and materials: poisons, on the one hand, and harmless products, on the other. It was recognized about 500 years ago by the German physician and naturalist Theophrastus B. von Hohenheim (known as Paracelsus, 1493–1541) that this simple view cannot be correct. His definition of poison is still valid in the twenty-first century. Using Latin, the language of philosophers and scientists until the eighteenth century, he coined the short definition: *dosis sola facit venenum* (the quantity alone makes the poison).

Two examples illustrate this definition. The D group of vitamins are needed in all periods of human life for growth, healing, and strengthening of bones. Vitamin D_3 and its precursor molecules are supplied by normal food in sufficient quantity, and irradiation of skin by sunlight transforms the precursor molecules into vitamin D_3. Yet, in a much higher dose vitamin D_3 is toxic and it is well documented that crews of fish trawlers have died after eating tuna liver, which may contain vitamin D_3 in a 1000–10,000 times higher concentration than normal food.

The second example concerns carbon dioxide (CO_2), a colorless and tasteless gas that is formed upon combustion of any kind of organic materials (e.g., fat, sugar, wood, and diesel).

Carbon dioxide is heavier than air and fills a room like water, beginning on the bottom and finally reaching the ceiling. In a room filled with carbon dioxide a human will die by asphyxiation and, thus, murders and suicides have been committed in a closed garage by means of the emissions of a running engine. Hence, newspapers and other public media usually call carbon dioxide a poison. However, the human body and brain produce carbon dioxide in every cell every minute, because this gas is the end product of the "combustion" of all food with inhaled oxygen. A small amount of it remains in the body and yields carbonic acid and hydrogen carbonate ions by reaction with water. The combination of both reaction products forms a buffer system that maintains a pH of around 7.4 in blood and brain. The excess of CO_2 is permanently removed by breathing to prevent acidification of the brain, because a lower pH affects the nerve membranes and can finally stop the brain from working. The intensity of breathing and, thus, the pH is controlled by the brainstem. In other words, the human body produces a chemical that at a low and constant concentration is necessary for survival, but which is lethal when present in excess. The entire system of CO_2 production, partial storage, and permanent elimination (by breathing) may be understood as a self-regulatory process obeying a combination of various laws of nature.

Finally, it may be said that today the capacity of the pharmaceutical industry and the knowledge of physicians provide good medicaments for most diseases in a dose that optimizes the healing effect and minimizes side reactions. Over the past 150 years, therapies based on natural and synthetic medicaments have made such tremendous progress that the following cynical comment of the French poet Moliere (1622–1673) is no longer valid: "Most humans are killed by their remedies and not by their diseases." However, the progress in medicine has not yet reached the stage described by the German writer Hans H. Kersten (1928–1986) with the words: "The progress in medicine is so immense that you cannot be sure anymore that you will die."

6.4 Psychosomatic Medicine

The prefix "psycho-" is derived from the Greek term *ψυχη* (psyche) that, according to standard dictionaries, may be translated by the term "soul" although it had several meanings for the ancient Greeks, such as breath, spirit, or soul. These different meanings are a first indication that soul is not necessarily an adequate translation of psyche. In a recent book entitled *Erasing Death*, the author, Sam Parnia, a physician and resuscitation expert, goes one step further and declares that the psyche, consciousness, and soul all mean the same. In his own words: "Most people are unaware that the terms soul and psyche actually correspond with the something that is really the 'self.'" For several reasons, the author of the present book does not share this view.

First, even the early Greek philosophers had different opinions about the fundamental properties of the psyche. Pythagoras and Plato believed that the psyche includes a transcendent component that survives after the death. For Demokritos and Aristotle, the psyche was a special kind of matter that vanished with the death of a human.

Second, the term soul has decidedly religious connotations related to the monotheistic religions Christianity, Islam, and Judaism. However, the most important roles of God/Allah are as the sole creator of the world, as moral instance and advisor, as judge, and in the case of Jesus Christ as redeemer. This understanding of God/Allah is totally different from any Greek religion and philosophy and, therefore, it is wrong or misleading to identify the psyche as soul. In all monotheistic religions, the soul somehow connects an individual with God/Allah. If this connection is understood as divine touch, as an emanation of God/Allah directed to an individual or as a spiritual "navel-cord," it is not relevant to this discussion.

Third, it is absolutely unlikely that a passionate communist and atheist would identify their "self" or "I-feeling" with a soul having the afore-mentioned religious connotation.

Fourth, it is incorrect or at least questionable to identify self and I-feeling with consciousness. The I-feeling certainly exist in dreams. In dreams feelings such as fear, frustration, or sympathy are often involved, but such emotions do not make any sense without an I-feeling. Emotions and I-feeling also exist during a coma. The author has had the occasion to make this observation several times when visiting a friend who was kept for 5 weeks in an artificial coma. But, neither the state of sleeping and dreaming nor coma are identical with full consciousness.

For the purpose of this chapter, the differentiation between psyche and soul is particularly important. Only a few people consider a human to be exclusively a chemical machine bare of any transcendent component. As an example, the ironic comment of the Danish writer Karen Blixen (1883–1962) can be cited (translated by the author): "What else is a man—if we consider it right—than a precisely working, admirable machine, capable of transforming a first class Bordeaux wine into urine?"

For about 2500 years in Europe, and even longer in Egypt, the vast majority of people had (and still have) a dualistic view of human beings, usually designated as the "body and soul" concept. Without differentiation between psyche and soul, emotions (notably positive ones) such as compassion, empathy, and love are usually considered to be expressions of the soul. This dualism is plagued by the problem that these properties (which for religious people are somehow connected with God/Allah) directly depend on the chemical and electrical activities of the human brain. Because of this dependence, it is possible to influence the emotions and quality of consciousness by means of chemicals such as alcohol, drugs, narcotics, or psychiatric medications.

A "trialistic" view based on body, psyche, and soul can offer an escape from this dilemma. The soul is here considered as an exclusively transcendent entity, certainly in contact with the psyche, but bare of any direct connection with chemical or electrical reactions in the human brain. All other properties of the soul are left to religions or ideologies for characterization and interpretation. With such a definition of soul, the psyche may be defined as the sum of all states of consciousness and emotions. Decisive for a proper understanding of the trialistic concept is the point that any change in emotions or thoughts is directly correlated with chemical and physical processes in the human brain. Descartes' famous statement "*Cogito ergo sum*" needs to be complemented by the statement "*Cogito ergo chimia est.*" This definition of psyche is now in perfect agreement with the standard terminology of medicine and related disciplines, which includes terms such as psychology, psychotherapy, psychosomatics, or psychiatric medication.

Psychosomatics is the tenet of the interaction between psyche and body (from the Greek *soma*). As correctly indicated by the term "interaction," the body has an influence on the psyche and the psyche has an influence on the body. For example, a serious injury, such as amputation of a limb, may cause depression. On the other hand, long lasting stress or a traumatic event, such as the loss of a beloved person, may trigger bodily pain or the outbreak of a disease, such as an infection, stomach ache, or heart rhythm disorder.

The interaction between body and psyche serves as the basis for numerous therapeutic concepts. For example, psychiatric counseling of patients suffering from chronic pain resulting from traumata or intensive stress can reduce the pain or even eliminate it. Such counseling and psychotherapy does not necessarily require a trained psychologist or psychiatrist. It has been well known for thousands of years that the empathy of a trustworthy doctor has a positive influence on the psyche of the patient and on the course of his healing process. Empathy, counseling, and help from family members and friends are other sources of nonprofessional psychotherapy. Over the past decades the "therapeutic touch," healing by laying-on of hands, has become widely accepted in the nursing profession, mainly in the USA. Because of the lack of efficient medicaments, the psychic aspect of therapies played a much greater role in previous centuries and for primitive races than in the western civilization of the twenty-first century. More than 50 % of the successes of medicine men and shamans may be attributed to their influence on the psyche of the patients. A transcendental aspect may be ascribed to this part of the therapy that goes beyond the scientific basis of the entire therapy.

The following diseases are the most frequent type of psychological illness that may also entail bodily symptoms and pain:

- Compulsory states of anxiety
- Compulsory motions, such Tourette syndrome or tics
- Depressions
- Burnout syndrome
- Dementia
- Psychosis, such as hysteria, schizophrenia, loss of self-critique or self-control, and loss of grasp of reality
- Drug addiction and manias

Depression has become the most frequent psychological illness (approximately 4.5 % of the population worldwide) and is therefore commented on here in more detail. Several types of depression are known. A temporary low mood due to sickness, loss of money, or traumatic event is called mood disorder. It is not identical to major depressive disorder (MDD), which encompasses the above-listed psychological diseases. Depending on the origin, three major groups of MDD may be classified:

- Endogenous depressions (most of which result from anomalies of the metabolism in the brain, based on genetic factors)
- Neurotic depressions (usually triggered by stress)
- Reactive depressions (i.e., long lasting or chronic reactions to an external traumatic event)

In addition to these "monopolar" depressions, so-called bipolar depressions are known that are characterized by a repeated change from a depressed mood to an enthusiastic mania.

All cases of MDD have in common that the efficacy of the hormones and neurotransmitters serotonin, dopamine, and noradrenaline in the human brain is somehow disordered (a neurotransmitter is a molecule that transports a nerve signal between two synapses and regulates it intensity). Depending on the type of depression, the disorder has different aspects. The concentration of the neurotransmitter may be too high or too low, or the receptor site that reacts with the neurotransmitter is not sensitive enough.

According to the different types of biochemical anomalies, various types of antidepressant medications have been developed to reduce at least the intensity of the symptoms. The art of the doctor consists of two steps. First the correct diagnosis must be found and, second, the optimum antidepressant drug must be selected. Quite analogous to the treatment of any other psychological disease, the best result is achieved when the chemistry-based therapy is complemented by psychiatric counseling and psychotherapeutic treatment. Successful treatment of most depressions occurs when the therapy is accompanied by intensive communication and discussion with the physician in charge. It is, of course, helpful if additional activities are involved, such as sport therapy, ergotherapy (occupational therapy), group therapy, and empathic care from the family. In summary, the success of psychosomatic medicine and psychiatric treatment depends roughly 50 % on the application of medicaments and 50 % on the psychological components of the entire therapy.

Finally, the two important questions that underlie this chapter should be answered: How much science is involved in medicine? Is medicine a branch of the natural sciences? An averaging over all disciplines of medicine (with exclusion of homoeopathy) allows the inference that more than 90 % of all diagnostic and therapeutic activities are grounded on scientific discoveries and technical inventions. Instruction and training of students of medicine in disciplines such as anatomy, pathology, physiology, and pharmacology is based on scientific insights and results.

In modern medical research, regardless of the discipline, about 98 % of all activities involve scientific methods, instruments, and machines. Examples are the development of new medicaments for all disciplines, the development of new implants, improved contact lenses, new applications of laser light, and the implantation of chips that can measure the pressure in joints or monitor the course of a healing process. Nonetheless, the question “Is medicine a science?” deserves a negative answer for two reasons. First, the psychic influence of the doctor and medical staff is a dimension outside the natural sciences. Even more important is the second aspect, namely the different goals of the natural sciences, on the one hand, and medicine, on the other. As defined in Sect. 2.1, science has the purpose and goal of learning more about nature and its phenomena and to explain all observations on the basis of laws of nature. However, the purpose and goal of medicine is to reduce pain, prevent diseases, restore health, and enhance the life span. Scientific results and inventions are a vehicle on the route to the final goal. In other words, the decisive difference between the natural sciences and medicine does not consist in different methods but in totally different goals.

However, because today about 98 % of all medical research involves scientific methods, it is justifiable to ask if and to what extent medical research produces mistakes and fallacies. How efficient is revision of mistakes and fallacies and how reliable are medical theories? From this point of view, medical research is not different from any other kind of scientific research.

Bibliography

Deurasch N, Abu Talib M (2005) Mental health in Islamic medical tradition. Int Med J 482:76
Levenson JL (2006) Essentials of psychosomatic medicine. American Psychiatric Press, Washington, DC
Melmed RN (2001) Mind, body, and medicine. Oxford University Press, Oxford
Parnia S (2013) Erasing death. Harper One (Harper Collins), New York

Part II
From Fallacies to Facts: Prominent Mistakes and How They Were Revised

Chapter 7
Medicine

7.1 The Circulation of Blood

> Courage to err is what makes a successful researcher
> (Charles Tschopp)

Both practical and theoretical knowledge in medicine were dominated in Europe for almost 2000 years by the writings of two extraordinary and famous physicians, Hippokrates and Galenos. Hippokrates was born around 460 B.C. on the island of Kos in the Aegean Sea as son of Heraclides, a physician, who was his first teacher. After the death of his parents, he went to Athens for further studies and died around 377 B.C. in Larissa, the capital of Thessaly. He was teaching at a time when all kinds of arts were flourishing in Greece and reached their climax. At this time he became the most famous physician of the ancient world. One reason for his fame and reputation in all areas of the Mediterranean world was the fact that he traveled a lot, teaching and practicing his "art" at various places in Greece and Macedonia. Unfortunately no text written by Hippokrates himself has survived, and all that is known about him and his work is scattered over more than 60 writings of various authors who lived between the third century B.C. and the first century A.D. and who did not know him personally.

Most of the texts of this so-called *Corpus Hippocraticum* are written in the Ionian dialect and aim to explain all medical theories and therapies on the basis of rational argument and systematic study of nature. Thus, Hippokrates made an important contribution to the freeing of medicine from superstition, speculative hypotheses, and religious influences. In several writings, the origin of diseases is attributed to an imbalance in bodily humors such as blood, phlegm, and black or yellow bile. The Hippocratic theory of the four bodily humors influenced the treatment of various diseases for more than 1000 years, in as much as the second star in the firmament of ancient medicine, Claudius Galenus, accepted, taught, and improved this theory.

H.R. Kricheldorf, *Getting It Right in Science and Medicine*,
DOI 10.1007/978-3-319-30388-8_7

Claudius Galenus (*Κλαυδιοσ Γαλενοσ*, Galen in English and German) was born in 129 or 131 A.D. in Pergamon (Bergama, Turkey), a city known for one of the best libraries of the ancient world. His father Aelius Nicon was a wealthy architect who paid for the best teachers as soon as he detected the talents and extraordinary mental capacity of his son. As was usual at that time and in previous centuries, Galen was mainly taught by philosophers and acquired a fundamental knowledge in Aristotelian and Epicurean philosophy. Therefore, his entire work reflects his strong interest in both medicine and philosophy. Pergamon belonged to the Roman empire at that time and, thus, it was quite normal that Galen traveled to various cities of the Roman empire. On one of his early travels he visited Alexandria and its famous library. In the year 162 he finally settled down in Rome, but soon had to leave it for several years, because he had come into conflict with traditional Roman physicians. However, when a plague broke out in 169 A.D., the emperor Marcus Aurelius ordered him back to Rome, and eventually Galen became court physician of four emperors: Marcus Aurelius (121–180), Commodus (161–192), Septimius Severus (145–211), and Caracalla (188–217). Galen died in Rome, probably in the year 216 A.D.

Galen had a great interest in human anatomy, but dissection of human cadavers was not permitted in the Roman Empire at that time. Therefore, he concentrated on the dissection of dead or living animals, mostly with focus on apes and pigs, assuming that the anatomy of animals closely mirrors that of humans. In this way he acquired numerous correct and useful insights, but he also made mistakes and adhered to fallacies that survived almost 1500 years until they were revised by Andreas Vesalius (1514–1564) and other physicians of the sixteenth century. Galen's writings comprise more than 400 treaties, and the surviving texts represent about one half of the surviving literature from ancient Greece. It is known that he employed up to 20 scribes to write down his thoughts. His main work, entitled *Methodi Medendi*, comprises 16 volumes. However, it is not clear to what extent the entire body of his work is based on Galen's original formulations. Other schools of medicine have left their footprints in texts written centuries after his death.

An important part of Galen's work is devoted to the circulatory system. He was in fact not the first to study and describe part of this system. The existence of cardiac valves was known from the *Corpus Hippocraticum*, but their function was not understood.

Herophilus of Chalcedon (335–280 B.C.) already recognized that both arteries and veins were somehow engaged in the transportation of blood, but he did not understand their function. The physician Erasistratus (304–250 B.C.) believed that blood flowing out of injured arteries was supplied by the veins. In other words, his concept of the direction of flow was exactly opposite to that of the real flow. Because after the death of a patient blood becomes concentrated mainly in the veins, the early physicians suspected that the rather empty arteries were also involved in the transportation of air.

Galen was the first to recognize that a living body contained two different kinds of blood, the relatively dark blood of the veins and the red blood of the arteries. This observation made him believe that two different one-way systems contribute to the overall blood transportation. He considered the arterial blood as originating from the heart, from where it was distributed to all organs of the body. In the liver, the arterial blood was somehow transformed into venous blood, which was, in turn, distributed to the organs where it was consumed. Neither Galen nor any other physician of the ancient world detected the pump function of the heart. Galen believed that the arteries themselves were responsible for the pulse beat and systolic blood pressure. It took 1000 years before the Arabian physician Ibn al-Nafis (1210/1213–1288) revised Galen's concept and described blood circulation through the lungs. However, his discovery (which exists in the form of detailed drawings) remained unknown to Christian Europe for a long time. In the year 1552, Michael Servatius (1511–1553) again described blood circulation through the lungs (perhaps knowing the work of Ibn al-Nafis), but his theory was not accepted by the physicians of his time, because it was in contradiction to the established theory of Galen. To understand the predominance of Galen's work and theories over a period of nearly 1500 years, two aspects must be kept in mind. First, his work was the most comprehensive and most detailed summary of all knowledge in the field of medicine published so far, and most conclusions were based on experimental results. Second, his entire concept of medicine and his theories of individual working fields were supported by philosophical arguments and considerations, an important argument at that time. Despite several mistakes and fallacies detected much later, his work appeared to be absolutely consistent in all directions and represented the prevailing paradigm for all areas of medicine.

As late as 1628, the first description of the complete circulatory system was published by the British physician William Harvey (1588–1657). After completion of his book *De Motu Cordis* he traveled to Frankfurt (am Main) where he attended the most important annual European book fair to publicize his book. Like Galen, Harvey had acquired most of his knowledge from the dissection of animals, but in contrast to Galen he also studied living animals such as snails, fish, and birds. He discerned that the heart had the function of a central pump for the entire circulation of the blood. This insight was stimulated by his former teacher Girolamo Fabrizio (1537–1619) who had taught him at the University of Padua. Fabricio was known for his book *De Venarum Ostiolis*, in which he described how the cardiac valves had a hydraulic function for the transportation of blood. On the basis of vivisections, Harvey observed that the left and right ventricles acted in a cooperative fashion, and that the systolic pulse beat of the arteries was a consequence of the contraction of the left ventricle. In his book Harvey finally formulated the correct circulation, beginning with the contraction of the right ventricle, which propels blood into the lungs for the exchange of carbon dioxide for oxygen. From there the blood returns to the left ventricle, which pumps the blood with enhanced pressure into the head and body, where it exchanges oxygen for carbon dioxide. Finally, it returns via veins to the right ventricle. However, Harvey, who only had lenses of poor quality and no microscope of high resolution, was not able to detect the capillaries and their function. This missing link in the proper understanding of the entire circulation was later discovered by Marcello Malpighi (1628–1694). Yet, because Harvey's hypotheses and conclusions were sufficiently supported by experimental facts and observations, his entire concept was gradually accepted by other physicians despite its contradiction of Galen's paradigm.

William Harvey

William Harvey was born in Folkestone (UK) on 1 April 1657 as one of five son of Thomas Harvey. His father was a jurat and served the office of Folkstone's mayor. Thomas Harvey's income allowed him to finance a good education for his sons. His sons, in turn, loved, revered, and trusted their father and made him the treasurer of their wealth when they had acquired great estates. William Harvey first attended the elementary school in Folkestone, where he began to learn Latin, a language that every educated man in Europe was at least able to read. From there he moved to Canterbury where he entered the King's School. After 5 years he joined Gonville and Caius College in Cambridge in the year 1593. He graduated from this college with a bachelor of arts degree in 1597. He then traveled to Padua, which was known as one of the oldest and most highly reputed universities in Europe. Because of his good knowledge of Latin he was able to understand and speak Italian within a few weeks, as the Italian of the sixteenth century resembled Latin much more than the Italian of the twenty-first century. Harvey began to study medicine in 1599, and on 25 April 1602, he graduated from the University of Padua as a doctor of medicine. Immediately afterwards, he returned to England, where he obtained a doctorate in medicine from the University of Cambridge.

Harvey established himself in London and joined the College of Physicians in October 1604. Membership of this highly reputed institution was an excellent basis for his future career. A few weeks later he married Elizabeth Browne, the daughter of the Dr. Physic Lancelot Browne. In 1607 he was elected a Fellow of the College of Physicians and accepted a position at St. Bartholomew's Hospital, which he occupied for the rest of his life. In October 1609 he succeeded Dr. Wilkinson in the position of the Physician in Charge at St. Bartholomew's Hospital. In this leading position he earned more than 30 pounds a year, an income that allowed him to live in a small house in Ludgate and to raise two children. His work included administrational work, and a simple but thorough weekly analysis of patients who came to the hospital and the writing of prescriptions.

A further important step in Harvey's career was an appointment as Lumleian lecturer in August 1614. The Lumleian lectureship was founded in 1582 by Lord Lumley and Dr. Richard Caldwell with the purpose of increasing the general knowledge of anatomy throughout England. Harvey began the lectures in April 1616 and continued for a period of 7 years,. His growing reputation reached a first climax when he was appointed Physician Extraordinary to King James I in February 1618, a position in which he also served other aristocratic members of the court. As distinguished member of the College of Physicians he was elected Censor of the College in 1613 and elected a second time in 1625. His book *De Motu Cordis* (published in 1628) drew much criticism from traditional physicians, who adhered to Galen's theories. This criticism affected his reputation, but he was able to continue his career, and in 1629 he was reelected Censor of the College and even elected Treasurer.

A new chapter in Harvey's life began with King Charles I and the Civil War. In 1629, the King commanded Harvey to accompany the Duke of Lennox during his trip to the continent. They traveled in France and Spain during the Mantuan War and outbreaks of plague and witnessed the human and economic catastrophe, as documented in a letter to Viscount Dorchester:

> I can complain that by the way we could scarce see a dog, crow, kite, raven or any other bird, or anything to anatomize, only some few miserable people, the relics of the war and the plague where famine had made anatomies before I came. It is scarce credible in so rich, populous and plentiful countries as these were that so much misery and desolation, poverty and famine should in a short time be, as we have seen. I interpret it well that it will be a great motive for all here to have and procure assurance of settled peace. It is time to leave fighting when there is nothing to eat, nothing to be keep, and nothing to be gotten.

After his return in 1632 Harvey had to accompany the King as Physician in Ordinary on all his trips. Particularly interesting to Harvey were the hunting expeditions, because he had access to the cadavers of deer and other animals, which enabled new anatomic studies and stimulated him to develop new hypotheses. During the Civil War, Harvey suffered various severe losses. Citizen-soldiers fighting the King entered Harvey's home, stole his goods, and destroyed or scattered papers that contained his records of the dissections of animals and of comparative anatomy. Furthermore, he lost three brothers and his wife at that time. During the war, he spent most of his time in Oxford, which was the main stronghold of Charles I. He cared for the King and his family and he continued his anatomic studies whenever possible. In recognition of his merits, he was made Doctor of Physic and later Warden of Merton College.

After the surrender of Oxford in 1645 Harvey returned to London and gradually retired from public life and from all his duties. He was childless and decided to live with his brothers Eliab and Daniel, separately and at different periods of time. He died in Eliab's house in Roehampton on 3 June 1657. Descriptions of his death suggest that he died of a cerebral hemorrhage. According to his last will, his wealth was distributed throughout his extended family and he left a considerable amount of money to the College of Physicians. Harvey was buried in Hampstead, Essex on 26 June 1657, in the Harvey Chapel built by his brother Eliab.

Bibliography

A-Dabbagh SA (1978) Ibn Al-Nafis and the pulmonary circulation. The Lancet 311:1148

Adams F (1886) The genuine works of Hippokrates. William Wood., NY

D'Arcy P (1897) William Harvey: masters of medicine. T. Fischer Unwin. ISBN 978-1-4179-6578-6

Furley D, Wilkie J (1984) Galen on respiration and the arteries. Princeton University Press
Gregory A (2001) Harvey's heart. The discovery of the blood circulation. Leon Books, Cambridge
Harris P (2007) William Harvey, Folkestone's most famous son. Lilburne Press, Folkestone
Jones WHS (1868) Hippokrates collected works. Cambridge Harvard University Press
Matter SP (1999) Physicians and the Roman imperial aristocracy. The patronage of therapeutics. Bull Hist Med 73(1):1–18
National Library of Medicine (2006) Images from the history of medicine. National Institute of Health
Nutton V (2004) Ancient medicine. Routledge, London
Pinault JP (1992) Hippocratic lives and legends. Brill Academic, London
Rapson H (1982) The circulation of the blood. Frederick Müller, London
Wright T (2012) Circulation. Chatto, London

Internet Links
Boylen M Galen, Internet encyclopedia of philosophy. http://www.iep.utm.edu/g/galen.htm

7.2 The Origin of Childbed (Puerperal) Fever

> A mistake uncovered late, may in science cost tax money, but in medicine it may cost many lives.
> (Hans R. Kricheldorf)

Childbed fever is a dangerous infection of the female reproductive organs that can follow childbirth. Characteristic of the outbreak of this disease is a high fever (>38 °C/100 F) 3–4 days after the birth. Responsible for the first stage of most cases of childbed fever is the bacterium *Streptococcus pyogenes,* which frequently lives in the throat or nasopharynx of otherwise healthy persons. However, in later stages a polymicrobial infection develops, which may include bacteria such as *Staphylococcus, Chlamydia, Lactobacillus*, and *Escherichia*. If not immediately treated with antibiotics this infection is often fatal. It is assumed that in the USA today between 1 and 8 % of all births are followed by childbed fever. In three out of 100,000 births the sepsis is deadly, with cesarean section being the most important single risk factor. For the UK a somewhat lower percentage (0.5–0.8/100,000 maternities) is reported.

These figures do not look alarming but in the nineteenth century, and before, childbed fever was for most women the main cause of premature death. It is shown for Europe that in the eighteenth and nineteenth century the level of maternal deaths was 1–10 % and reached values as high as 40 % for individual hospitals in the seventeenth and eighteenth century. The percentage of deadly childbed fever was significantly lower for births at home, but in the seventeenth century births in a hospital had become popular, mainly for women of lower social classes. For three reasons, physicians and professors who were in charge of obstetrics at municipal hospitals showed surprisingly little interest in this phenomenon. First, the income from any kind of medical service for poor women was low, and the obstetrical department was usually located in a less attractive wing of the hospital. Second, according to the paradigm of that time infectious germs (and other microorganisms) were a result of spontaneous generation and, thus, unavoidable. Louis Pasteur's discovery that spontaneous generation does not exist was published after 1860 (see Sect. 8.1). Third, the transfer of deadly germs via the hands of a doctor was in conflict with the image doctors had of themselves. A statement by Charles Meigs, obstetrician and teacher at a hospital in Philadelphia is representative of the mentality of that time: "Doctors are gentlemen and gentlemen have clean hands."

However, since the end of the eighteenth century, a small, but slowly increasing number of physicians began to suspect that childbed fever was a consequence of contagion and lack of hygiene. The Scottish naval surgeon and obstetrician Alexander Gordon published in 1795 a book entitled *Treatise on the Epidemic of Puerperal Fever* with the warning that this infection is transmitted from one case to another by midwives and doctors. He wrote: "It is a disagreeable declaration for me to mention that I myself was the means of carrying the infection to a great number of women." In 1842, Thomas Watson, professor of medicine at King's College Hospital (London) recommended handwashing with chlorine solutions and changes of clothes for doctors and staff of obstetrical departments "to prevent the practitioner from becoming a vehicle of contagion and death between one patient and another." Oliver Wendell Holmes published in 1843 an article entitled "The Contagiousness of Puerperal Fever," in which he draw the same conclusions as Watson a year earlier.

Gordon's book was ignored in the first half of the nineteenth century and the conclusions and recommendations of Watson and Holmes were ridiculed by most of their contemporaries. Some comments that appeared in articles or letters may illustrate the situation: "pus was as inseparable from surgery as blood"; "surgeons operated in blood-stiffened frock-coats—the stiffer the coat, the prouder the busy surgeon"; and "cleanliness was next to prudishness." A visitor to a hospital in Philadelphia wrote: "There was no object in being clean. … Indeed, cleanliness was out of place. It was considered to be finicking and affected. An executor might as well manicure his nails before chopping off a head."

It was Ignaz Semmelweis who achieved the paradigm shift in a life-long fight against the ignorance and arrogance of contemporary physicians and obstetricians. Semmelweis had finished his studies of medicine at the University of Vienna with a doctoral degree in 1844. After a first position at the K.u. K. Hospital he was appointed assistant lecturer at the Vienna General Hospital. He worked in the First Obstetric Division under the supervision of Professor Klein, who was responsible for the training of medical students. Semmelweis was apparently not informed about the publications of Gordon, Watson, or Holmes, but he observed that the mortality rate from childbed fever in the First Division was substantially higher

Table 7.1 Childbed fever mortality rates at the Vienna General Hospital

	First clinic			Second clinic		
Year	Births	Deaths	Rate (%)	Births	Deaths	Rate (%)
1841	3036	237	7.8	2442	86	3.5
1842	3287	518	15.8	2659	202	7.6
1843	3000	274	9.0	2739	164	6.0
1844	3157	260	8.2	2956	68	2.3
1845	3492	241	6.0	3241	66	2.0
1846	4010	459	11.4	3754	105	2.8

than in the Second Division (see Table 7.1) where midwifery students were trained. He also noticed that the male students and obstetricians in the First Division conducted autopsies almost every morning on women who had died the day before. In contrast, midwives were not allowed to perform or assist in autopsies.

At that time, hygiene instructions or directions did not exist in hospitals and the students and doctors changed from autopsies to the treatment of women giving birth without washing their hands or changing their clothes. Although the reasons were not understood, the different mortality rates in the First and Second Division were known to many pregnant women, and Semmelweis reported that women fell down on their knees begging to be admitted to the Second Clinic. Furthermore, he knew that births at home and so-called "street births" had a lower rate of maternal death than births in hospitals. He found his suspicion that students and doctors transferred infectious germs to birth-giving women confirmed when his friend and colleague Dr. Jakob Kolletschka (1803–1847) died of sepsis, an infection almost identical to childbed fever. A few days before his death, Kolletschka had injured one of his hands while performing an autopsy.

Immediately after this experience, Semmelweis began experiment with various cleansing agents and reached the conclusion that washing hands, gloves, and instruments in an aqueous solution of chlorinated lime might be a suitable means of stopping transfer of infectious germs. He ordered that from May onwards all physicians, students, and staff should wash their hands in this solution before working in the ward and before each vaginal examination. The mortality rate from childbed fever fell in his Division from around 18 % in May 1847 to less than 3 % in November. Despite this conspicuous success, he was exposed to attacks, skepticism, and ridicule from most colleagues. Therefore, he left Vienna and continued his work at the St. Rochus Hospital in Pest (Budapest in Hungary). He made the same observations and had the same success as before in Vienna and published his insights and results in 1860. Yet, his work was ignored by most colleagues despite the convincing evidence for the success of improved hygiene.

The breakthrough for the concept of systematic hygiene and aseptic work in hospitals came after Pasteur had published his findings that spontaneous generation does not exist. The two men who succeeded Semmelweis in advocating aseptic work in hospitals were the Italian surgeon Enrico Bottini and the British surgeon Joseph Lister (first Baron Lister). Bottini was born in the village Stradella near

Pavia in the year 1835. He was appointed professor of surgery at the University of Pavia in 1863 and served later as surgeon at the Ospedale Maggiore in Novara. In 1867, he returned to his position as professor at the University of Pavia where he was active as teacher and surgeon until his death in 1903. He published in 1866 a treatise recommending an aqueous solution of phenol for disinfection of surgical instruments and for washing hands and gloves. Phenol was identified as a component of coal tar by the German chemist Friedlieb F. Runge in 1834 and had become commercialized in a somewhat impure form under the name carbolic acid. It is possible, but unproven, that Bottini was informed about Semmelweis work, because large parts of north-east Italy belonged to the Habsburgian monarchy at that time, and newspapers written in German were certainly available in most cities of that area.

Only 1 year after Bottini's treatise had appeared, Joseph Lister published a series of six articles in *The Lancet* describing the successful suppression of infections by treatment of wounds with diluted aqueous solutions of carbolic acid. He was certainly not informed about Bottini's article published in Italian. However, it is strange that he should not have known Semmelweis's work, because Lister was able to read German, and because a summary of Semmelweis's work was published in the leading British medical journal *The Lancet* in 1848. Lister was born on 5 April 1827 in West Hampton, Essex (UK). He initially studied botany at University College, London, but after obtaining a bachelor of arts degree in 1847 he began to study medicine and graduated with honors as a bachelor of medicine. In 1854 he became assistant and friend of the professor of surgery James Syme at the University of Edinburgh. A few years later he was appointed a professorship in Glasgow, but returned to Edinburgh in 1969 as successor of Professor Syme.

During his time in Glasgow, Lister became aware of Pasteur's publications showing that spoilage of food or fermentation can only occur when microorganisms are present. Furthermore, he had heard that spoilage in the sewerage systems of Paris and Carlisle and their bad smell was suppressed by spraying creosote, an extract of tar containing various phenols.

This information stimulated him to study the usefulness of carbolic acid solutions for direct application to wounds. The immediate success prompted him to develop aseptic procedures for all medical activities in the hospital. Lister instructed all surgeons and staff members under his responsibility to wash their hands before and after operations in a 5 % aqueous solution of carbolic acid. Furthermore, all instruments, gloves, and clothes used during an operation had to be cleaned with the same solution, and assistants had to spray it in the operating theatre. His antiseptic methods were soon accepted by more and more colleagues. His fame grew steadily after his return to Edinburgh, and when he gave lectures there were frequently audiences of 400 or more. Lister died on 10 February 1912 at his home in Kent. In the Anglo-Saxon world he became famous as "the father of modern surgery."

Ignaz Phillip Semmelweis

Ignac Fülöp Semmelweis was born on 1 July 1818 in Tabán, Hungary, a village near Buda (today part of Budapest). He was the fifth out of ten children of a wealthy grocer family. His parents, Josef and Theresia (born Müller) Semmelweis, ran a wholesale business dealing with spices and various consumer goods in Taban and in Buda. He went to Vienna and began to study law in the fall of 1837. For unknown reasons he switched to medicine in the following year, and was awarded a doctoral degree in medicine in 1844. When his attempt to obtain a position in a clinic for internal medicine failed, he decided to specialize in obstetrics taught by Carl von Rokitansky, Josef Skoda, and Ferdinand von Hebra.

After a first position in a K.u.K. hospital, he was appointed assistant to Professor Johann Klein in the First Division of the obstetrical department of the Vienna General Hospital on 1 July 1846. He had to examine patients each morning in preparation for the professor's rounds, teach students in obstetrics, supervise difficult deliveries, and was responsible for the records of his ward. As described above, he found that the origin of childbed fever was a lack of hygiene. Students, like doctors performing in the morning autopsies of women who had died from childbed fever the day before, did not wash their hands, gloves, or instruments. Semmelweis was a militant and pugnacious person who in speeches, letters and article blamed his colleagues for being responsible for the death of hundreds and thousands of women. This behavior had two consequences. A minority of colleagues gradually accepted his arguments and felt depressed for being accessory to the numerous maternal deaths. One obstetrician, Gustav A. Michaelis (maternity clinic in Kiel) even committed suicide because his own cousin died from childbed fever after he had examined her after she gave birth.

However, Semmelweis's findings and theory were in conflict with the established theory of Galen, which explained the origin of any disease as an imbalance of the four bodily humors (see Sect.7.1), and in conflict with Aristotle's theory of spontaneous generation of microorganisms. Therefore, the vast majority of his colleagues felt insulted and became aggressive towards him. They tried to undermine his reputation and to terminate his career. Fortunately, the small number of colleagues supporting his concept slowly increased. Semmelweis himself and some of his students sent letters to directors of maternity clinics explaining their new observations and conclusions. Ferdinand von Hebra, editor of an Austrian medical journal, presented his results in two issues (December 1847 and April 1848). At the end of 1848, a former student of Semmelweis wrote a lecture explaining Semmelweis's theory, which was presented before the Royal Medical and Surgical Society in London. A review of this lecture was then published in the British medical journal *The Lancet*.

In 1848, revolutionary activities spread across Europe, including an Hungarian uprising against the Habsburgian monarchy. Professor J. Klein, supervisor of Semmelweis in the Vienna General Hospital, was a patriotic, conservative Austrian and apparently mistrusted Semmelweis, a born Hungarian, although Semmelweis was himself not engaged in revolutionary activities. Whether political considerations or the dispute about childbed fever was more decisive is unclear, but when the term of Semmelweis's employment ended, Professor Klein preferred a competitor, Carl Braun, for the open position and Semmelweis lost his job.

Immediately thereafter, Semmelweis petitioned the Viennese ministry to be made docent of obstetrics. Yet, because of Professor Klein's opposition the decision on his repeated petition was delayed and only 18 months later was he was finally appointed docent of theoretical obstetrics. This title allowed him to teach students, but he did not have access to cadavers or hospital facilities. Immediately after this decision of the Viennese administration he left Vienna abruptly and returned to Pest saying later that "I was unable to endure further frustrations in dealing with the Viennese medical establishment."

With help of supporters, on 20 May 1851 Semmelweis was appointed as the unpaid, honorary head-physician of the obstetric ward of the Szent Rokus Hospital, a position he held for 7 years. Under Semmelweis's influence the high mortality rate of childbed fever dropped to only 0.85 %. Despite this success, neither Ede F. Birly, professor of obstetrics at the University of Pest, nor other physicians in Budapest accepted his doctrine. Birly continued to believe that the uncleanliness of the bowels was responsible for childbed fever. Among the numerous contemporaries who attacked Semmelweis's theory in letters, articles, or books were the editor of the *Wiener Medizinische Wochenschrift* (Viennese Medical Weekly); Carl Braun, his successor in Vienna; August Breisky, an obstetrician in Prague; Joseph H. Schmidt, professor of obstetrics in Berlin; and Carl M.E. Levy, head of the Copenhagen maternity hospital. The most influential scientific enemy in central Europe was certainly the physician and anthropologist Rudolf Virchow (1821–1902) in Berlin, who was one of the highest authorities in medical circles and a member of the Reichstag (German parliament) with many political connections. Semmelweis's results and theories were more favorably received in England, but they were mainly considered as experimental evidence for the unproven hypotheses that Thomas Watson and Oliver W. Holmes had published in 1842 and 1843, respectively.

With the death of Professor Birly in 1854 most Hungarian physicians voted against Semmelweis as successor, but with support of authorities in Vienna he was appointed professor of obstetrics in 1855. On the basis of this position he dared to marry Maria Weidenhoffer, the daughter of a successful merchant in Pest. Together they had five children, but one daughter died at the age of 4 months. In 1858, Semmelweis began to publish his entire work in three stages. First, he wrote an essay entitled "The Etiology of Childbed Fever." Two years later, his second essay appeared with the title "The Difference in Origin between Myself and the English Physicians Regarding Childbed Fever." A complete book including all observations followed in 1861 entitled *The Etiology, Concept and Prophylaxis of Childbed Fever*, in which he lamented the slow adoption of his concept.

In 1861, after publication of his book, Semmelweis began to show signs of mental disorder. He became severely depressed and absentminded and turned any conversation to the problem of childbed fever. After several negative reviews of his book had appeared, Semmelweis began to attack his critics in a series of open letters addressed to various European obstetricians. These letters were full of desperation and fury and were extremely offensive. In a couple of letters he called his critics ignoramuses and murderers. Furthermore, he began to drink immoderately and showed absolutely inappropriate behavior, even in the presence of friends or members of his family. After his death, various speculations about the origin of this mental disorder were published, such as extreme frustration and depression, Alzheimer's disease, or syphilis, which was a common disease for obstetricians, who examined thousands of women including prostitutes.

In 1865, Janos Balassa (1815–1868), professor of surgery and leading authority in the Hungarian Medical Society, wrote a document recommending that Semmelweis should be treated in a mental institution. On 30 July of the same year, Ferdinand Hebra lured him to a Viennese insane asylum (Landes-Irrenanstalt) pretending to visit a new institute. When Semmelweis recognized the situation he tried to escape, but was severely beaten by guards and secured in a strait jacket. He was in fact tortured in this asylum; he was injured when fighting with guards and died from sepsis on 13 August 1865 at the age of only 47 years.

Initially, Semmelweis was buried in Vienna, but in October 1891 his remains were transferred to Budapest. In 1964 they were eventually transferred to the house in which he had been born. This house is today a historical museum and library in honor of Ignaz Semmelweis. His death was briefly mentioned in two medical journals, but no colleague wrote a commemorative article honoring his work. Under Janos Diescher, his successor at the maternal clinic of the University of Pest, the mortality rate increased to 6 %, seven times the rate Semmelweis had realized, but no obstetrician or physician in Budapest or Vienna uttered complaints or protest.

Semmelweis's work gradually gained acceptance only after Pasteur's publications proving the germ theory of infectious diseases. After the turn of the century his reputation and fame steadily increased. In 1904, a statute of him was unveiled in front of the Szent Rokus Hospital in Budapest. Later, a hospital in Miskolc, Hungary, and a hospital for women in Vienna were named after him. Later, a university for medicine and health-related disciplines was named Semmelweis University. In 1956, a stamp of the Bundespost (West Germany) and in 1958 a stamp of the DDR were issued showing a portrait of Semmelweis. Austria issued in 2000 a commemorative gold coin (50 €). A novel by Martin Thompson, *The Cry and the Covenant* (1949), is based on his life, and a film, a drama, and a play related to his life have been produced. Robert A. Wilson coined the notion "Semmelweis Reflex" as a label for a human behavior characterized by a reflex-like rejection of new theories because of their contradiction to existing paradigms. Perhaps the greatest honor is his later nickname "Saviour (or angel) of mothers."

The work and life of Ignaz Semmelweis is also a striking example of the fact that an earlier revision of a misleading paradigm in medicine (contrary to science) might have saved the lives of thousands of patients.

Bibliography

Bottini E. http://www.trecani.it/enciclopedia/enrico-botini

Carter KC, Carter Carter BR (2005) Childbed fever: a scientific biography of Ignaz Semmelweis. Transaction Publishers

Cartwright F (1963) Joseph Lister: the man who made surgery safe. Weidenfeld & Nicholson, London

Fischer RB (1977) Joseph Lister, 1827–1912. Stern & Day, NY

Hannimen O, Farago M, Monos E (1983) Ignaz Semmelweis, the prophet of bacteriology. Infect. Control 4(5):367

London I (2000) The tragedy of childbed fever. Oxford University Press, USA

Semmelweis IP. http://www.semmelweissociety.org/Biography.aspx

Semmelweis Museum. http://www.semmel.weismuseum.hu

Von Györy T (ed.) Semmelwei's Gesammelte Werke (collected works). VDM Dr. Müller, Saarbrücken

Wertz RW, Wertz DC (1989) A history of childbirth in America. Yale University Press

7.3 History of Local Anesthetics

> A mistake is instructive once it is understood and revised.
> (Manfred Rommel)

The term "local anesthetic" is the combination of a Latin and a Greek word and describes a drug that prevents or stops pain in a limited area of the human body without affecting the activities of the brain. In other words, the effect of a local anesthetic does not involve any narcotic effect. However, an appropriate injection may also have the purpose of blocking a nerve pathway or achieving paralysis, meaning loss of muscle power in a larger area of the body. From the viewpoint of chemical structure, clinical local anesthetics can be subdivided into two classes, esters and amides (see Fig. 7.1). Yet, together with their natural ancestor, cocaine,

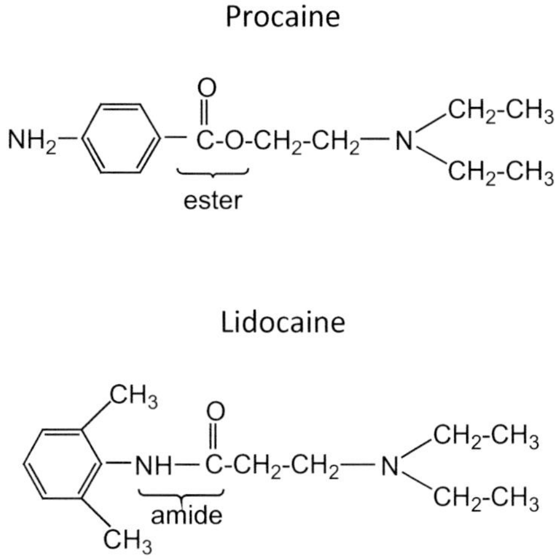

Fig. 7.1 Chemical structures of two widely used local anesthetics

Procaine

Lidocaine

all synthetic local anesthetics have a tertiary amine group in common that is decisive for the mode of application and for their physiological reaction. They are chemically stable and soluble in water only when the amino group is protonated, usually as a salt of hydrochloric acid. Therefore, almost all clinical applications are based on aqueous solutions of the salt form, but in blood or other bodily liquids free amine and salt form an equilibrium. It is the free amine that passes rapidly through the membranes of nerves and reacts with the inner end of certain tubuli that are responsible for the transportation of sodium ions across the membrane. The rapid influx of sodium ions is connected with a change in the local electric potential of the membrane. The rapid propagation of this combined chemical and electrical change along the axis of a nerve represents the propagation of a nerve signal. Hence, inactivation of the sodium-transporting tubuli by a local anesthetic blocks generation or propagation of nerve impulses.

In principle, all types of nerve fibers react to local anesthetics, their sensitivity towards anesthetics varies depending on the diameter and myelin content. The most sensitive nerves are the type B fibers (responsible for sympathetic tone), followed by type C (pain), type A delta (temperature), type A gamma (proprioception), type A beta (sensory touch and pressure), and type A alpha (motor). Myelination considerably enhances sensitivity and, thus, the myelinated type C fibers are more sensitive than the thin type C.

Local anesthetics can be used to prevent or alleviate acute pain resulting, for instance, from trauma, surgery, dentistry, infection, or other tissue injuries. They are also applied to treat chronic pain, usually in combination with other drugs such as opioids or anticonvulsants. The effect of local anesthetics rapidly vanishes when they diffuse away. Therefore, they are frequently combined with vasoconstrictor drugs to reduce local blood circulation and their rapid diffusion. All local anesthetics are chemically modified and inactivated in the liver. However, the ester-type anesthetics are also cleaved (hydrolyzed) by the enzyme cholinesterase, which is concentrated in the neighborhood of synapses. Another difference between ester and amide types of anesthetics is the higher tendency of ester anesthetics to provoke allergies.

The most common standard techniques used in clinics are:

1. Surface anesthesia. This technique applies solutions, sprays, or creams to skin. A short duration of the painkilling effect is usually intended.
2. Infiltration. This technique usually combines several nearly simultaneous injections of local anesthetics into a tissue.
3. Peripheral nerve block may be understood as an alternative or variant of infiltration. The local anesthetic is injected in the vicinity of a peripheral nerve to anesthetize that nerve's area of innervation.
4. Plexus anesthesia. In this case, the anesthetic is injected close to a nerve plexus (compact bundle of nerve fibers) with the intention of blocking a large tissue area innerved by all the fibers stemming from the plexus.

5. Spinal anesthesia. The injection is applied to the spinal fluid in the lower part of the back where it blocks the spinal nerve roots. The result is an anesthesia extending from the legs to abdomen or lower chest.
6. Intravenous anesthesia (Bier's block). First the blood circulation in a limb is interrupted (for a limited time), then a large volume of a local anesthetic is injected into a peripheral vein from where it rapidly migrates into other veins. Diffusion into the neighboring tissue results in anesthesia of those parts of the limb excluded from blood circulation.
7. Transwound catheter anesthesia. This technique is based on insertion of a catheter into a wound or incision. The catheter is aligned to the wound inside and allow continuous administration of a local anesthetic.

The history of local anesthetics begins with cocaine. In 1859/1860 the German chemist Alfred Niemann studied the extraction of imported coca leaves and isolated a drug that he called cocaine after the name of the plant. He also discovered the anesthetic effect of cocaine, but this discovery was not effectively published. The Viennese oculist Carl Koller (1857–1944) reported several years later to have learned from a self-experiment that when tasting cocaine, this drug anesthetized his tongue. After successful tests with animals, in1884 Koller used cocaine for the first time as a local anesthetic in eye surgery on a human. He used cocaine in the form of an aqueous solution that was applied by dropwise addition to the eye until a satisfactory anesthetic effect was achieved. Since then, Koller has been considered the father of local anesthesia, in as much as he invented and introduced this notion.

Whereas Koller used surface anesthesia, the German surgeon Maximilian Oberst (1849–1925, professor in Halle) developed in 1888 the plexus anesthesia for surgery of fingers (later called Oberst's block). Apparently without knowing the results of Koller, Oberst, and Carl L. Schleich (see below), in 1895 the British dentist William S. Halsted (1852–1925) began to introduce local anesthetics into dentistry.

In June 1892, during a conference of physicians and surgeons, the German surgeon Carl Schleich (1859–1922) presented his concept of a broad application of local anesthetics for a variety of surgeries . He recommended that general anesthesia, usually performed by inhalation of chloroform or diethyl ether, be substituted whenever possible by local anesthesia, because long-lasting anesthesia with chloroform and diethyl ether or its repeated application damages the liver. In view of the published results of Koller and Oberst, one might have expected that the audience would welcome and applaud Schleich's lecture demonstrating and explaining the advantages of the infiltration technique. Yet, apparently nobody was informed about the properties of cocaine. The still-existing protocol of that conference contains the last sentences of Schleich's lecture (translated by the author):

> I think it is not justified to apply general anesthesia with chloroform or similar inhalation techniques in the future, if the usefulness of local anesthesia has not been tested in advance. Only when local anesthesia is not satisfactory or applicable in individual cases, should general anesthesia be applied. However, I think, performing surgeries under general

> anesthesia that are feasible under the influence of local anesthetics is not justified anymore with the current state of the infiltration technique or similar methods of local anesthesia, considering the aspect of humanity and the consequences of criminal law.

The majority of the audience felt provoked or offended and reacted with an outcry and tumultuous scenes. The chairman of the conference, the physician von Bardesleben, had great difficulties in calming down the atmosphere and eventually posed the following question:

> Dear colleagues, when we are confronted with such a provocation and accusation as presented in the last sentences, we are allowed to deviate from our habit of avoiding criticism at the end of a lecture. Therefore, I now ask the audience, whoever is convinced that what he has heard is true should lift his hand.

Yet, nobody lifted a hand. Schleich tried to add a comment to his last words, but the chairman stopped him and Schleich had to leave the lecture hall.

At the beginning of the next conference of surgeons, two years later, the surgeon Ernst von Bergmann (1836–1907), professor in Berlin, pioneer of aseptic surgery, and convinced by Schleich's concept, invited colleagues to attend a surgery performed by Schleich in a nearby hospital of the university. Immediately thereafter, von Bergmann informed the audience about the successful course of the operation. In the same year (1894), Schleich's book *Schmerzfreie Operationen* (Painless Operations) was published by Springer in Berlin. From that year on, Schleich's infiltration technique gained rapid acceptance, at first in German-speaking countries and later all over Europe. In 1898, the German surgeon August Bier (1861–1944) introduced spinal anesthesia into surgical practice and 10 years later the technique of intravenous regional anesthesia. From that year on, all the major techniques of local anesthesia were established.

Whereas the first steps in successful application of local anesthetics were made by courageous and long-sighted surgeons, chemists continued the success story. Soon after the first applications of cocaine as a local anesthetic, it had become evident that this drug had several disadvantages, such as irritation of tissues, rapid decay of the anesthetic effect, and a high risk of abuse. It was the German chemist Alfred Einhorn (1856–1917) who invented the first synthetic local anesthetic. Einhorn was born in Hamburg, but because of the early death of his parents he was educated by relatives in Leipzig. He began to study chemistry in Leipzig and continued his studies in Tübingen, where he received the doctoral degree in 1878. In 1882, he was appointed assistant in the group of Professor Adolf von Bayer (1835–1917, Nobel Prize for Chemistry 1905) at the University of Munich. He left this group twice for short periods of time for his habilitation at the University of Darmstadt. He was eventually appointed professor at the University of Munich, where he stayed from 1891 until his death in 1917. In the course of his career, Einhorn synthesized more than 100 compounds that were tested with regard to their usefulness as local anesthetics. His first and biggest success was procaine (see Fig. 7.1), which after patenting in 1905 was commercialized by the Farbwerke Hoechst under the trademark Novocaine. Procaine is less irritant to tissue than cocaine, its effect lasts longer (up to 1 h), and its abuse potential is lower. Within a few years, procaine became the standard local anesthetic. Although it was largely replaced by Lidocaine (xylocaine, see Fig. 7.1) after 1943, procaine is still used in the twenty-first century in dentistry. Lidocaine became the standard local anesthetic after 1943, because its effect sets in more rapidly and lasts longer (up to 2 h). Since 1970, several local anesthetics have been developed that maintain the anesthetic effect for up to 6 h. In other words, in the twenty-first century, local anesthetics, mixtures of anesthetics, or combinations with other drugs can be optimized for any medical application.

Carl Ludwig Schleich

Carl L. Schleich was born on 19 July 1859 in Stettin (today Szczecin, Poland) as son of the oculist Carl Ludwig Schleich Sr. His father was interested in biology and the leading theories of his epoch, and invited not only physicians but also biologists to his home for intensive discussions about Darwin's theory of evolution. Because his father was an atheist, whereas his uncle Herrmann Frederik was a pastor, the young Carl Ludwig was exposed to vigorous debates about the existence of God and other aspects of the Christian religion. He was deeply impressed by these debates and mentioned later in his autobiography:

> It is strange to note how deeply many of the arguments for and against the existence of God and immortality impressed themselves on my mind. . . . Such arguments would occupy me for days, and I recalled them during divinity lessons, and even onto this day they have their repercussion in my philosophic reflections.

Apparently, his uncle had the better arguments; Schleich Jr. was confirmed and wished to become a pastor. He attended the Sundische Gymnasium (a high school) in Stralsund and eventually decided to study medicine. His studies included three stages, the University of Zürich, the University of Greifswald (near Stralsund and Stettin), and, finally, as an assistant to Professor Rudolf Virchow in the famous Hospital Charité in Berlin. He received his doctoral degree in 1887 from the University of Greifswald under the supervision of the surgeon Heinrich Helferich. He was appointed assistant, but left Helferich's group in 1889 to open his own small clinic for surgery and gynecology in Berlin-Kreuzberg. Because of his increasing reputation and fame as inventor of the infiltration technique he was honored with the titles Professor (1891) and Geheimrat (1899) and became director of the department of surgery at the municipal hospital of Groß-Lichterfelde (Berlin) in 1900. During his activities in Berlin, Schleich became more and more interested in the function of the nervous system with a focus on glia cells, which surround nerve fibers. He became a pioneer of glial research and developed a still up-to-date theory of their function. He assumed that an interconnected and interactive network of glia cells was a substrate for most brain functions. In this connection, he published an influential work on hysteria research entitled *Gedankenmacht und Hysterie* (Power of Thoughts and Hysteria).

Carl L. Schleich Jr. was a multitalented person and became a famous poet and writer in German-speaking countries. At this point it is worth noting that several members of his family (which had its roots in Munich) were artists, for instance, the painters Robert Schleich and Eduard Schleich the Elder (1812–1874). He wrote numerous articles for journals such as *Natur, Zukunft, Neue Rundschau, Arena, Gartenlaube*, and *Über Land und Meer*. In several essays he described and explained his philosophy. A summary of his thoughts was presented in his book *Phantasien über den Sinn des Lebens* (Fantasy about the Meaning of Life). He criticized an overinterpretation of Darwin's theory of evolution saying that "Darwinism by no means overturns the concept of creation." He was himself criticized as an enemy of science, but he opposed this criticism with a counterattack, saying that science is much too dogmatic (translated): "It is no longer an indisputable fact that natural sciences can be just as dogmatic as the Church. The stubborn adherence to prejudices, traditions, and comfortable habits is but a general human obstacle to progress no matter whether it is expressed by the Church, the State, or in the laboratory. We have these pretentions of infallibility here and there and the popes of science have been no less intolerant than those of the church and still are."

This critique was certainly justified for the decades before World War I.

In 1920 he published his most successful book, his memoirs, under the title *Besonnte Vergangenheit*. This book was one of the most successful autobiographies in German-speaking countries, and more than one million copies were sold. It was translated into several foreign languages and appeared in England under the title *Those were Good Days*. Schleich Jr. died in 1923 in Bad Saarow and was buried in Berlin in the Südwestkirchhof–Stahnsdorf cemetery. A street was named after him in Stralsund.

Bibliography

Bankoff (1943) The practice of local anesthetics. Heinemans, London

Brownsfield AJ, Ellis PM, Poolig K (2007) Early pain-free days. http://www.rsc.org/Education/EiC/issues/2007Nov/EarlyPainFreeDays.asp

Dyson GM, May P (1959) May's chemistry of synthetic drugs, 5th edn. Longmans, London

Goering M, Schulte EJ (1993) Carl Ludwig Schleich—pioneer exclusively in infiltration anesthesia? Anesthesiol Intensivmed Notfallmed Schmerzther 28 (2):113

Münch R, Absolon KB (1976) Carl Ludwig Schleich and the development of local anesthetics

Ruetsch YA, Böni T, Borgeal A (2001) From cocaine to ropivacaine: the history of local drugs. Curr Top Med Chem 1(3):175

Schleich CL (1936) Those were good days. Norton

Schleich CL (2013) Besonnte Vergangenheit: Lebenserinnerungen, Reprint. Severus Verlag (ISBN 9783863475093)

7.4 Origin of Peptic (Gastric) Ulcers

The most may err as grossly as the few
(John Dryden)

A peptic ulcer is a lesion of the mucosa (and possibly submucosa) that can occur in the stomach (gastric ulcer), in the first part of the small intestine (duodenum ulcer), or in the lower part of the esophagus (esophageal ulcer). The appearance of a peptic ulcer can be of the classic, concave, crater-like type, or convex type quite often resembling a colonic polyp. The concave type is most frequently found in the stomach, whereas the convex variant is more typical for duodenum ulcer. Another difference results from the observation that a gastric ulcer may be combined with a malignant tumor, whereas a duodenum ulcer is usually benign. The slow growth of a convex ulcer is typically painless until it metamorphosizes into pathogenic growth. In contrast, the concave gastric ulcer may be painful even in an early stage when the lesion of the mucosa is still small, but deep enough to allow the acidic gastric juice to attack the submucosa and the stomach wall. The intact mucosa protects itself with a layer of mucus consisting of special polysaccharides that are sufficiently stable to withstand the stomach acid for several days. To maintain the protective layer in an optimum state, the mucosa renews the mucus at time intervals of 3–4 days. If the stomach acid can penetrate the mucus and mucosa, the stomach can, in principle, digest its own stomach wall, which is a muscular tissue mainly consisting of protein. The stomach acid mainly consists of hydrochloric acid and the enzyme pepsin, a combination that has the purpose of digesting proteins (meat, fish, eggs), . The direct evidence for this scenario is, for instance, offered by restaurants in South Germany specializing in regional food. Those restaurants offer meals based on the sliced stomach wall from calves. The author has himself tested that the digestion of such meals is as easy as digestion of any other kind of meat.

Because the entering of food into the stomach stimulates production of acid, the timing of the symptoms in relation to the beginning of a meal is usually different for gastric and duodenal ulcers. The pain from a gastric ulcer increases during the meal, whereas the symptoms of a duodenal ulcer may be reduced, because the pyloric sphincter closes during the meal to concentrate the stomach content. Yet, 2–3 hours later, when the stomach begins to release digested food and acid into the duodenum, pain from an ulcer located in the upper part of this intestine increases. Increasing pain during or after a meal is a first indication of the existence of an ulcer and of its location. However, reliable diagnosis requires various tests, above all endoscopic screening of the esophagus, stomach, and duodenum.

All kinds of peptic ulcers together had and have an enormous effect on the mortality rate in all countries, but mainly in certain areas of Africa and India. As a result of improved medication, the mortality rates have fallen significantly during the past three decades. However, even in 2010 about 250,000 people died worldwide from peptic ulcers. In the USA, 4 million people still suffer from peptic ulcers and approximately 350,000 new cases are diagnosed each year. The number of duodenum ulcers is four times higher than that of gastric ulcers, but the number of people who die from duodenal ulcers and gastric ulcers is almost identical, namely around 3000 per year. Before 1983, the established theory regarding peptic ulcers considered their origin to be the attack of stomach acid on the mucosa and submucosa, favored by the following factors:

1. Genetic disposition
2. Stress
3. Acidic food
4. Smoking
5. Coffee
6. Alcohol
7. Acidic, pain-killing drugs such as aspirin, diclofenac, and ibuprofen (and other NSAIDs, non-steroid anti-inflammatory drugs)

In more recent studies, the negative effect of coffee or moderate consumption of alcoholic beverages was not confirmed, but long term use of NSAIDs indeed involves a high risk of peptic ulcer. It was found that NSAIDs block mucosal secretion of mucus.

A ground-breaking new theory, a paradigm shift, was published in 1982 by two Australian scientists Barry J. Marshall and Robin Warren. They found that infection with the bacterium *Helicobacter pylori* was the causative factor for most peptic ulcers. At first, their theory was not well received, because most scientists were convinced that bacteria cannot survive in the presence of stomach acid. Therefore, Marshall dared to perform a self-experiment and drank the contents of a Petri dish containing a culture of *H. pylori* extracted from a patient. The progress of the infection was monitored over short time intervals by endoscopy and biopsy. After about 5 days, Marshall indeed developed gastritis, the first stage of a peptic ulcer, but after 2 weeks the symptoms vanished. Nonetheless, he took antibiotics to eradicate the remaining bacteria.

This experiment was described in 1995 in an article of the *Australian Medical Journal* that became one of the most cited papers of the journal. Meanwhile, it has become clear that most, but not all, peptic ulcers are caused or favored by infection with *H. pylori*. Hence, detection of this infection in a patient is decisive for identification of a peptic ulcer and for optimum treatment. In contrast to previous decades, the standard treatment today is based on a combination of at least two antibiotics, with a proton pump inhibitor that reduces the amount of stomach acid. Reliable identification of peptic ulcers still requires esophago-gastro-duodenoscopy (EGD), a special variant of endoscopy, but infection with *H. pylori* can be detected by simpler means, such as the urea breath test and stool antigen test. More reliable and more informative, but also more costly, is an EGD biopsy test. When Marshall and Warren published for the first time their theory of *H. pylori* infections as cause of peptic ulcer, they were confronted by skepticism and ridicule quite analogous to that suffered by Ignaz Semmelweis. Yet, Marshall and Warren were eventually luckier and were awarded the Nobel Prize in Physiology or Medicine in 2005.

Barry J. Marshall

Barry J. Marshall was born on 30 September 1951 in Kalgoorlie, Western Australia, as the first of four children. He spent the first 8 years partially in Kalgoorlie and partially in Carnarvon, North-West Australia, where his father worked as engineer on a whaling station.

His father awoke his interest in engines and electric devices, while the books of his mother, a nurse, stimulated his interest in biology and medicine. When he was eight, his family moved to Perth where he attended Newman College. Although this high school satisfied his interest in science, he felt that his mathematical ability was not strong enough for a career as scientist. Hence, he decided to enter the medical school of the University of Western Australia.

Here in 1972 he met his future wife, Adrienne, a psychology student. He graduated from the university as a bachelor of medicine and surgery and continued his studies at the Queen Elizabeth II Medical Center of the Sir Charles Gairdner Hospital. He became interested in a university career and returned in 1978 to the University of Western Australia as a specialist physician. In 1979 he joined the Royal Perth Hospital to become more experienced in cardiology and heart surgery. The internal rotation took him in 1981 to the gastroenterology division, where he met Robin Warren. It was quite normal for an ambitious physician trainee to formulate a clinical research project every year. Following the suggestion of his supervisor Tom Water, he contacted Robin Warren who showed him slides of curved bacteria unexpectedly living in contact with stomach acid on or around gastric ulcers. Marshall became interested and started a cooperation that continued even when he was absent for several months for training at the Port Hedland Hospital, 2000 km north of Perth.

Marshall's part in this cooperation was inspection of patients and delivery of biopsies. Both scientists were eventually convinced that the bacteria, identified as *H. pylori*, were responsible for gastritis and peptic ulcers. When in October 1982 Marshall presented their results and hypotheses for the first time at a local meeting of the College of Physicians, the response was mainly skeptical or critical. Nobody was willing to believe that bacteria were capable of growing in the presence of stomach acid. He was confronted with the same skepticism when he presented his results for the first time outside Australia at a European *Campylobacter* meeting in September 1983. Nonetheless, Marshall and Warren continued their cooperation and published their results in 1983 and 1984. In those years, Marshall had the position of a senior registrar at the Freemantle Hospital in Perth, because after 1982 no registrar position was available at the Royal Perth Hospital. Because Warren did not have an appointment at Freemantle, the cooperation between them was loose but efficient. It was at Fremantle Hospital that Marshall tried to infect piglets to overcome the skepticism of his colleagues.

When these experiments failed, he decided to perform a self-experiment with *H. pylori*, which was successful and brought the first breakthrough for the theory after publication in 1985. In an autobiography he wrote:

> Becoming increasingly frustrated with the negative response to my work I realized I had to have an animal model and decided to use myself. Much has been written about this episode and I certainly had no idea it would become as important as it has. I didn't discuss it within the ethics committee at the hospital. More significantly, I didn't discuss it in detail with Adrienne. She was already convinced about the risk of these bacteria and I knew that I would never get her approval. This was one of those occasions when it would be easier to get forgiveness than permission. I was taken by surprise by the severity of the infection. When I came home with my biopsy results showing colonization and classical damage of

my stomach, Adrienne suggested it was time to treat myself. I had a successful infection, I had proved my point.

The increasing number of travels to Europe, Japan, and the USA in the years 1984–1994 had the consequence that more and more research groups turned their interest towards gastro-intestinal infections and some of those research groups began to focus their studies on *H. pylori*

A characteristic example of this situation was the work of David Graham, Peter Peterson, and Martin Blazer. At first, they were critics of Marshall's and Warren's theory and set out to disprove it, but quickly became leaders of *Helicobacter* research in the USA.

The following events pushed Marshall's and Warren's work in the direction of a big success. In 1984, a journalist of the American newspaper *Star* called Warren and asked "How do you know it's a pathogen and not a harmless commensal?" Although Marshall's self-experiment had not yet been published, Warren answered. "I know Barry Marshall has infected himself and he damn near died." The next day, *Star* presented the story under the title "Guinea-pig doctor discovers new cure for ulcers . . . and the cause." This story prompted an increasing number of patients in the USA to ask for help. Marshall developed a novel treatment based on antibiotics, which proved to be helpful in all cases of infection with *H. pylori*. This early success in the USA brought him into contact with Procter & Gamble (P&G), a company that produced a bismuth drug against stomach pain. P&G patented most of Marshall's results and helped to finance a position and a laboratory at the University of Virginia, where he eventually spent 10 years. An increasing number of donations from healed patients supported the research activities. Further support came from TriMed, which patented the breath test and managed approval by the US Federal Drug Administration. Marshall and Warren reached official acceptance of their work during a consensus meeting of the NIH (National Institutes of Health) held in Washington DC in February 1994. The final statement declared that the basis for successful treatment of duodenal and gastric ulcers is detection and eradication of *H. pylori*.

In 1996, Marshall decided to return to Australia, where he obtained an appointment at the University of Western Australia in Perth. Here he continues to work as physician and as researcher leading the newly founded *Helicobacter pylori* Research Center. Since the early 1990s, the number of invitations and honors steadily increased and culminated in the Nobel Prize for Physiology or Medicine in 2005. In 2007, Marshall was appointed Companion of the Order of Australia and in 2009 he was awarded an honorary doctor of science degree by the University of Oxford.

Robin Warren

Robin Warren was born on 11 June 1937, in Adelaide as the first son of Roger Warren and Helen Warren (nee Verco). His father had studied viniculture and was one of Australia's leading wine makers. His mother was a nurse and several members of the Verco family were physicians, and this background perhaps stimulated him to study medicine. At first, Warren attended the local, public primary school, the Westbourne Park School. The secondary school he joined afterwards was the oldest school in Adelaide, St. Peters College, a school that two generations of Warrens had attended before him. Here he learned Latin, French, mathematics, chemistry, and physics. In 1955, he entered the medical school of the University of Adelaide on the basis of a Commonwealth scholarship. This scholarship was offered to all good students to provide free tertiary education.

In the first year he enjoyed learning botany and zoology, but the following 2 years were focused on pathology and anatomy. In this connection he learned to dissect and analyze the structures of joints, inner organs, and the brain. During the three clinical years he learned more about surgery, medicine, gynecology, and obstetrics. His attempts to find a position as registrar in psychiatry failed and he accepted a position as registrar in clinical pathology at the Institute of Medical and Veterinary Science of the Royal Adelaide Hospital. This was an all-round position that gave him a broad experience in pathology, the basis for his future career. The next step of his studies was a short-term position in 1963 as temporary lecturer in pathology at the University of Adelaide. He found soon a better position as clinical pathology registrar at the Royal Hospital of Melbourne and moved for 4 years to Melbourne. There, he completed his studies with exams in hematology and microbiology and became a fully fledged pathologist.

In1967, he was hired by Professor Rolf ten Seldam as senior pathologist at the Royal Perth Hospital in Western Australia, where he spent most of his career. In this position the timetable was more flexible and he could afford more time for his scientific interests. In the early 1970s he became interested in the relatively new method of gastric biopsy. In 1979, he detected bacteria growing on the surface of biopsy samples. From then on, he focused his studies on these bacteria. He collected biopsies from gastric ulcers and biopsies from seemingly healthy parts of the stomach. He found that bacteria grew only on tissue of the ulcer and its close neighborhood, but not on biopsies of the antrum. He began to write a paper describing his discovery, but prior to its publication he met Marshall (in 1981) and they decided to undertake together a more complete clinical-pathological study. Their results were published in 1983 and 1984, the first reports on *Helicobacter pylori* and its role as initiator of duodenal and gastric ulcers.

At this point it should be mentioned that R. Warren married Winfried Williams, a former student of medicine. Together they had four sons and one daughter. Despite five children his wife slowly continued her study of psychiatry and later became an accomplished psychiatrist. Warren reported in his biography that before he met Marshall, his wife was the only person who supported his work and believed in his concept of bacterial infection as the cause of peptic ulcers.

Warren's cooperation with Marshall extended over 7 years and eventually demonstrated that eradicating *H. pylori* resulted in healing of the ulcer, and after complete healing the recurrence rate was extremely low. Around 1990, the international medical community began to recognize Marshall's and Warren's theory with the consequence that both received increasing number of honors, invitations to lectures, and requests for attendance at meetings.

In 1996 Warren was invited to Japan for a lecture tour and in 1997 to central Europe for a three month tour. The following list of honors is not complete:

1994 Warren Alpert Foundation Prize
1995 Award of the Australian Medical Association
1995 Distinguished Fellow Award of the Royal College of Pathologists of Australia
1996 Inaugural Award of the First Western Pacific Helicobacter Congress
1996 Medal of the University of Hiroshima
1997 Paul Ehrlich und Ludwig Darmstädter Award
1998 Foulding Florey medal
2005 Nobel Prize in Physiology or Medicine

Bibliography

Kato I, Abraham M, Nomura Y, Grant N, Chyou P-H (1992) A prospective study of gastric and duodenal ulcer and its relation to smoking, alcohol and diet. Am J Epidemol 135:521–530

Marshall BJ (2002) Helicobacter pioneers: firsthand accounts from the scientist who discovered Heliobacters. Wiley, Hoboken, NJ

Marshall BJ (2005) Autobiography. http://nobelprize.org/nobel_prizes/medicine/laureates/2005/marshall-autobio,html

Marshall BJ, Warren JR (1983) Unidentified curved bacilli on gastric epithelium in active chronic gastritis. Lancet 321:1273–1275

Marshall BJ, Warren JR (1984) Unidentified curved bacilli in the stomach of patients with gastritis and peptic ulceration. Lancet 323:3111–1315

Peptic Ulcer. http://www.nlm.nih.gov/medlineplus/ency/article/000206.htm and http://www.ncbi.nlm.nih.gov/pubmedhealth/PMH0001255/

Van DerWeyden MB, Armstrong RM, Gregory AT (2005) The 2005 Nobel Prize in Physiology or Medicine. Med J Aust 183:612–614

Warren JR (2005) Autobiography. http://nobelprize.org/nobel_prizes/medicine/laureates/2005/warren-bio.html

Warren JR. http://www.vianet.net.au/-jwarren/

7.5 Crohn's Disease and Ulcerative Colitis

Nothing is more damaging to a new truth than an old error.
(Johann W. von Goethe)

Crohn's disease (Morbus Crohn, Crohn syndrome) is an inflammatory bowel disease that preferentially affects the large intestine and the lower part of the small intestine (terminal ileum). However, in rare cases Crohn's disease may also affect other parts of the gastrointestinal tract including mouth and anus. Characteristic of Crohn's disease is that small segments of the intestine are inflamed, separated from each other by healthy segments. Hence, Crohn's disease is also called regional enteritis. Crohn's disease is frequently called an autoimmune disease, but its causes are complex and most likely involve infection of the intestine walls by certain bacteria.

The first reports on inflammatory bowel diseases are known from the Italian physician Giovanni B. Morgagni (1682–1771). A first description of what was later called Crohn's disease was published by the Polish surgeon Antoni Lesniowski in 1904 under the title *Ileitis terminalis*.

In 1932, Burrill B. Crohn, gastroenterologist at the Mount Sinai Hospital in New York, wrote a letter to the American Medical Association describing 14 cases under the title "Terminal ileitis: A new clinical entity." Several months later, he and his colleagues Leon Ginzburg and Gordon Oppenheimer published a paper entitled "Regional ileitis: a pathological and clinical entity." In later years, the name Crohn–Lesniowski disease was shortened in the Anglo-Saxon literature to Crohn's disease, whereas it is still called Lesniowski–Crohn disease in Poland.

The incidence of Crohn's disease is apparently similar in Europe and in North America, but lower in Asia and Africa. It is particularly high in Scandinavian countries and in the USA. The percentage of people having this disease in the USA and European countries varies between 0.03 and 0.2 %, depending on the country and depending on the source of the data.

Only slightly more women than men have this disease, but close relatives of people having Crohn's disease are 3–20 times more prone to develop this disease. Studies of twins revealed that, if one twin has the disease, the chance of the other twin also having it is around 55 %. In other words, the development of Crohn's disease is certainly favored by genetic disposition. Meanwhile, more than 30 genes have been detected that are somehow associated with the development of this disease.

Typical, but nonspecific, symptoms of Crohn's disease are abdominal pain, fever, and weight loss. If inflammation is severe, bloody diarrhea may occur. Further symptoms are loss of appetite, nausea, and vomiting. A dangerous complication in a later state of this disease is bowel obstruction, which is combined with a higher risk of bowel cancer and may need surgery. Further complications that may involve organs and tissues outside the gastrointestinal tract include abscesses in the intestines, skin rashes, anemia, arthritis, osteoporosis, gallstones, venous thrombosis, inflammation of the eye, and tiredness. The following factors are known or at least supposed to favor Crohn's disease:

1. Genetic disposition.
2. Stress. Psychological stress is certainly not a decisive cause, but may favor progress of Cohn's disease as a result of interaction with the immune system.
3. Smoking has been proven to enhance the risk of recurrent attacks.
4. The introduction of hormonal contraception in the USA in the 1960s was followed by a dramatic increase in incidence.
5. BecauseCrohn's disease has a higher incidence in industrialized highly developed countries, exaggerated hygiene, somehow misleading the immune system, is thought to favor the disease.

6. An alternative hypothesis says that a higher consumption of sugar, or animal protein and milk protein in industrialized countries contributes to the higher incidence.

Detection and identification of Crohn's disease is based on several diagnostic methods, including blood tests, radiologic tests and, above all, colon endoscopy possibly in combination with biopsy. Unfortunately, no medication or surgical procedure can provide a complete cure for the disease. Because the cause and origin of the disease has not been elucidated over the past 100 years, all medical treatment was (and still is) directed towards symptoms. These symptoms result from a disorder of the immune system, frequently called autoimmune disease, resulting from a malfunction of the T cells. Therefore, the medication based on this hypothesis includes two groups of drugs:

– Drugs designed to reduce inflammation and acute attacks. These drugs usually contain 5-amino salicylic acid (5-ASA) and corticosteroids (e.g., cortisone), with hydrocortisone against severe attacks.
– Drugs designed to modulate or suppress the immune response, such as azathioprine, methotrexate, infliximab, and adalimumab. The most effective, but also most costly members of this group are monoclonal antibodies (e.g., humira or romira), which immobilize proteins that maintain the inflammatory process. Their side effects are negligible, but they need permanent storage in a refrigerator.

Both groups, corticosteroids and immunosuppressive drugs, have in common that they cause severe side effects upon prolonged application. Surgery is useful to eliminate acute problems with abscesses, carcinoid tumors, and intestinal obstructions, but it is far from being a cure for the disease.

Over the past 20 years a new hypothesis is gaining acceptance and promises more successful treatment of Crohn's disease. This hypothesis is based on two assumptions: First, the patient has a weakened mucosal layer, allowing pathogenic microbes to attack the bowel wall. Second, impaired innate immunity is not capable of eradicating the pathogenic microbes during the first stage of the infection. Hence, the infection becomes chronic and entails a permanently disordered immune response. Various types of bacteria are suspected to be involved in this process, but *Mycobacterium avium* subspecies *tuberculosis* (MAP) seems to play a predominant role. This bacterium is known to cause a similar disease, called Johne's disease, in cattle. This hypothesis resembles the now-established theory that infection with *H. pylori* causes peptic ulcers (see Sect. 7.5). If this hypothesis proves correct, medication with an optimized cocktail of antibiotics may allow complete cure, if the disease is detected at a very early stage. At a later stage it may at least have the advantage of reducing application of corticosteroids and immunosuppressive drugs with their severe side effects.

Ulcerative colitis (UC) is a chronic inflammatory disease of the bowels that resembles Crohn's disease at first glance. In addition to the similarities, several characteristic differences have been found, which should be discussed first. UC is usually confined to the large intestine, the main part of which is called the colon, whereas Crohn's disease can affect any part of gastrointestinal tract from mouth to anus. At a late stage, UC can damage tissue neighboring the colon, above all the oviducts of women. Therefore, removal of the large intestine (colectomy) may result in complete healing at an earlier stage of UC, whereas removal of damaged segments of the colon never cures Crohn's disease. UC rarely generates fistulae, whereas carcinoid tumors and open sores in the colon wall are more frequent. Characteristic of UC are symptoms outside the gastrointestinal tract, such as aphthous ulcers in the mouth and open sores on the surface of the body. Furthermore, arthritis of the joints or spine are frequently observed, and ophthalmic (eye) problems. Smoking is a lower risk for UC and special diets helping the majority of patients have not been found yet. The maximum incidence of UC occurs for people aged 25–35, whereas in the case of Crohn's disease, teenagers and people aged 50 plus have a similarly high risk of infection.

UC, like Crohn's disease, is more prevalent in northern Europe or North America than in southern Europe and the percentage of UC patients in Africa is still lower. Whether this trend is a consequence of more efficient diagnosis in northern countries or a consequence of different diets or hygiene is not clear. UC is an intermittent disease, so that periods that are relatively symptom free and periods of intensive symptoms alternate. Symptoms are abdominal pain and a relatively constant diarrhea, frequently mixed with blood. UC always was and still is considered an autoimmune disease. Hence, the medication is analogous to that of Crohn's disease with only slight differences. Over the past 50 years and still today 5-aminosalicylic acid, its derivatives, and combinations with other drugs are the most widely used medications in the therapy of mild or moderate UC. In severe cases the same immunosuppressive drugs mentioned above for the treatment of Crohn's disease are applied. Yet, complete cure is never achieved in this way.

Genetic disposition seems to play a decisive role in the incidence of UC, as evidenced by aggregation of this disease in families and by studies of twins. No specific cause has been detected so far. Yet, the conspicuous similarity to Crohn's disease suggests that here again a weakened mucosa of the colon allowing penetration of pathogenic bacteria is the starting point of the disease. An insufficient immune response has, in turn, the consequence that an acute infection becomes a chronic disease. If this hypothesis proves correct, a new therapeutic concept based on application of antibiotics may be more successful than exclusive treatment of symptoms.

Bibliography

Baumgart DC, Carding SR (2007) Inflammatory bowel disease: cause and immunology. The Lancet 369:1627–1640

Casanova JL, Abel L (2009) Revisiting Crohn's disease as a primary immune deficiency of macrophages. J Exp Med 206:1839–1843

Cho JH, Brant SR (2011) Recent insights into the genetics of inflammatory bowel disease. Gastroenterology 140:1704–1712

Danese S, Fiocci C (2011) Ulcerative colitis. N Engl J Med 365:1713–1725

Dessein R, Chammaillard M, Danese S (2008) Innate immunity in Crohn's disease. J Clin Enterol 42:144–147

Kirsner JB (1988) Historical aspects of inflammatory bowel diseases. J Clin Gastroenterol 10:286–297

Lalande JD, Behr (2010) Mycobacteria in Crohn's disease: how immune deficiency may result in chronic inflammation. Expert Rev Clin Immunol 6: 633–641

Langan RC, Gotsch PB, Krafczyk MA, Skilling DD (2007) Ulcerative colitis: diagnosis and treatment. Am Fam Physician 76:1323–1330

Lichtarowicz AM, Mayberry JP (1988) Antoni Lesniowski and his contribution to regional enteritis. J R Soc Med 81:468–470

Marks DJ, Rahman FZ, Sewall GJ, Segal AW (2010) Crohn's disease: an immune deficiency state. Clin Rev Allergy Immunol 38:20–31

Sandborb WJ (2012) Crohn's disease. The Lancet 380:1590–1605

Stefanelli T, Malesel A, Repici A, Vetrano S, Danese S (2008) New insights into inflammatory bowel disease pathophysiology.: Paving the way for new targets. Curr Drug Targets 9:413–418

Tysk C, Lindberg E, Järnerot G, Floderes-Myrhed B (1988) Ulcerative colitis and Crohn's disease in an unselected population of monozygotic and dizygotic twins. A study of heritability and the influence of smoking. Gut 29:990–996

Yamamoto-Furusho JK, Korzenik JR (2006) Crohn's disease: innate immunodeficiency? World J Gastroenterol 12:6751–6755

7.6 What Is the Vermiform Appendix Good For?

Only those who think can err.
(Horst Friedrich)

The term "vermiform appendix" comes from Latin and means worm-shaped organ appending to the colon or more precisely, to the cecum. The cecum is the upper part of the colon, the junction of the small and large intestines. On average, the appendix has a length of 10–12 cm and a diameter of around 0.8 cm. The appendix is closed at the external end, so that no digestive liquid or stool can pass through it. Its opening to the cecum is guarded by the Gerlach valve, named after a German anatomist. The appendix is usually located on the right side of the abdomen close to the right hip bone.

In humans, the most common diseases are appendicitis and appendix cancer. Appendicitis is an inflammation of the appendix, usually resulting from a bacterial infection. In the early stage, pain is poorly localized, but as the inflammation progresses and the peritoneum becomes inflamed, the pain begins to localize to the right lower quadrant of the abdomen. Acute appendicitis requires rapid removal of the appendix, a surgery called appendectomy (the Greek word *ectomy* means cutting an organ or tissue out of the body). Antibiotics may delay the need for emergency surgery, but do not heal the appendicitis. If untreated, appendicitis and the ensuing peritonitis result in sepsis followed by death.

For nearly 150 years the human appendix was considered to be a vestigial structure, the remainder of an organ that lost its original function in digestion during the course of evolution. This hypothesis was proposed by Charles Darwin in his famous book *The Descent of Man and Selection in Relation to Sex* (1871, see Sect. 8.3). He assumed that the cecum was originally needed for digestion of leaves. The appendix of primates was assumed to result from a shrinking process of the cecum when apes began to eat more fruit. This hypothesis was seemingly supported by the following observations. Herbivorous animals, such as horses and even koalas have extremely long ceca. In the ceca of many extinct herbivores mutualistic bacteria were found, microorganisms that help animals to digest cellulose, a main component of all plants. The progress of human evolution had the consequence that human food began to include more and more fruit and berries and eventually even meat. Hence, the digestive activities of the appendix finally became obsolete. Darwin's hypothesis was also supported by the observation that absence of the appendix after an ectomy did not cause negative side effects.

However, in recent studies (2012, 2013) it was found that numerous animals, not only apes, possess an appendix with widely varying size and shape. The biologist Michel Laurin at the National Museum of Natural History in Paris together with several Australian and American colleagues compared 361 mammalian species with regard to the evolution of their intestines. Six species had lost the appendix, but for all other species 32 modifications of the appendix were found, suggesting a continuous evolution of this organ. Whenever a species possessed an appendix, it contained tissue connected to the immune system and the appendix walls were strong and muscular. All these findings are in contradiction to Darwin's hypothesis.

It has been known for a long time that various species of microorganisms in the large intestine play an important role in the efficient digestion of food. Yet, the usefulness of this symbiosis was only seen in improved utilization of food. Over the past two decades, the interaction between the digestive function of intestine walls, gut flora, and immune system has attracted increasing interest. The useful and helpful bacteria not only supersede useless and infectious bacteria, but it has also become apparent that that they actively support the intestine walls and the lymphoid system in fighting infections. It is already known that the appendix and other parts of the gut are surrounded by lymphatic tissue (gut-associated lymphatic tissue, GALT) that carries out a variety of important functions. Although hormone-secreting cells are also found in the appendix, no explanation exists for a distinctive function, in as much as appendectomy seemingly does not have negative consequences.

As early as 1999, Loren G. Martin, professor at Oklahoma State University, speculated that the appendix plays a role as a lymphatic organ. Aliya Zahid speculated in 2004 that it contributes to the training of the immune system of the large intestine. A more precise explanation was forwarded by William Parker, Randy Bollinger, and other colleagues of Duke University in 2007. They proposed that the role of the appendix is that of safe haven for useful bacteria when diarrhea or a similar infection flushes all bacteria from the large intestine. This interpretation is in perfect agreement with other well-known features of the appendix, such as its location at the beginning of the large intestine, its pouch-like architecture, and its association with large amounts of lymphatic tissue. The reservoir of useful gut flora in the appendix may help to repopulate the intestines following a bout of diarrhea, cholera, or other dysenteries. It fits this theory that physicians of the Winthrop University Hospital reported in 2012 that patients without an appendix are four times more likely to have a recurrent infection of *Clostridium difficile*. The connection between the appendix and gut immune tissue with the evolution of mammals and humans was explained by Heather F. Smith of Arizona State University as follows:

> Recently . . . improved understanding of gut immunity has emerged with current thinking in biological and medical science, pointing to an apparent function of the mammalian cecal appendix as a safe-house for symbiotic gut microbes, preserving the flora during time of gastrointestinal infection in societies without modern medicine. This function is potentially a selective force for the evolution and maintenance of the appendix. Three morphotypes of cecal-appendices may be described among mammals, based primarily on the shape of the cecum: a distinct appendix branching from a rounded or sac-like cecum, an appendix located at the apex of a long and voluminous cecum, and an appendix in the absence of a pronounced cecum. In addition, long narrow appendix-like structures are found in mammals that either lack an apparent cecum, or lack a distinct junction between the cecum and appendix-like structure. A cecal appendix has independently evolved at least twice, and represents yet another example in morphology between Australian marsupials and placentals in the rest of the world. Although the appendix has apparently been lost by numerous species, it has also been maintained for more than 80 million years in at least one clade.

In summary, it is somehow instructive and amusing to see that Charles Darwin not only initiated a mega paradigm shift, but also introduced an incorrect paradigm.

Bibliography

Bollinger RR, Barbas AS, Bush EL, Lini SS, Parker W (2007) Biofilms in the large bowel suggest an apparent function of the human vermiform appendix. J Theoret Biol 249:826

Everett ML, Polertrant D, Miller SE, Bollinger RR, Parker W (2004) Immune exclusion and immune inclusion: a new model of host bacteria interaction in the gut. Clin Appl Immunol Rev 5:321

Laurin M, Everett MI, Parker (2011) The cecal appendix: one more immune component with a function disturbed by post-industrial culture. Anat Rec 294 (4):567

Martin LG (1999) What is the function of the human appendix? Did it once have a purpose that has since been lost? Sci Am. http://www.sciam.com/article.cfm?id=what-is-the-function-of-t

Smith HF, Fisher RE, Everett ML, Thomas AD, Bollinger RR, Parker W (2009) Comparative anatomy and phylogenetic distribution of the mammalian cecal appendix. J Evol Biol 22:1984

Smith HF, Parker W, Koke SA, Laurin M (2012) Multiple independent appearance of the cecal appendix in mammalian evolution and an investigation of related ecological and anatomic factors. Compt Rend Paleovol 12(6):339
Sommerburg JL, Angement LT, Gordon JL (2004) Getting a grip on things: how do communities of bacterial symbionts become established in our intestine? Nat Immunol 5(6):569
Zahid A (2004) The vermiform appendix: not a useless organ? J Coll Phys Surg Pak 14(4):256

7.7 Homosexuality: Sin or Disease?

To err is human, but the consequences may be inhuman.
(Hans R. Kricheldorf)

The words "heterosexual" and "homosexual" are Greek–Latin hybrids based on the Latin word *sexus* meaning sex. The element hetero- is derived from the Greek *heteros* meaning different, otherwise, the other (or second) of two. Homo- is derived from *homoios*, which means equal, the same, quite similar. Hence, homosexuality summarizes sexual emotions, affections, and sexual acts between members of the same sex. An interpretation of homosexuals as a term derived from the Latin *homo* meaning man and exclusively denoting gay men is incorrect. The term homosexual has a negative connotation in English-speaking countries, and the abbreviation homo is usually understood as an offense. Therefore, the US organization Gay & Lesbian Alliance Against Defamation (GLAAD) has advised the media to avoid terms such as homosexual and homosexuality. In modern literature, homosexual is frequently replaced by the term same-sex, and in colloquial English gay denotes same-sex men while same-sex women are called lesbians. However, more recently, lesbians are also called gay women.

Hetero- and homosexuality are words that were first used in 1868 by the Austrian writer Karl M. Benkert (pseudonym Karl M. Kertbeny, 1814–1882) in pamphlets against a Prussian antisodomy law. The German physician Richard von Krafft-Ebing (1840–1902) used the terms hetero- and homosexual throughout his book *Psychopathia sexualis* published in 1886. This book became very popular among laymen and experts, so that hetero- and homosexual became widely accepted for the characterization of sexual orientation. This late invention of the term homosexual also indicates that that the understanding of same-sex behavior and same-sex activities in past centuries was so different from the modern view that such a term (or similar words) was not needed.

Same-sex behavior has been documented over the past 3000 years in all major civilizations from Japan across China, India, near Orient to ancient Greece and the Roman empire. In the case of ancient Greece, written information on same-sex activities is scattered in numerous texts and fragments, but is plentiful. Famous authors, such as Perikles, Plato, Xenophon, and Athenaeus have contributed to our knowledge of the sexual life in ancient Greece. Furthermore, same-sex activities were illustrated on vases, bowls, and cups. Such illustrations are relatively rare considering that many thousands of pieces of Greek pottery have survived, but they are taboo-free and unmistakable. Characteristic of the ancient world in Europe was the almost total absence of information on same-sex love for women. This observation is a direct consequence of the fact that women were so-to-say second-class humans, who did not pay any role in public social life, politics, or warfare. However, there is one noteworthy exception, Sappho.

Sappho was a lyric poet who lived between 630 and 570 B.C. on the island of Lesbos (the exact dates are unknown). Little is known for certain about her entire life, but she was apparently married to a rich merchant and had one daughter. She was certainly born into an aristocratic family and, regardless of whether she had a wealthy husband or not, she certainly had a luxurious life and plenty of free time. Because of political troubles in Lesbos she was exiled to Sicily between 604 and 594 B.C. Nonetheless, she is supposed to have written 12,000 lines of poetry, but only around 600 lines have survived. Sappho was also the head of a *thiasos*, a kind of school for girls. In such a community, young Greek women were taught a variety of skills. It was quite normal that same-sex relationships existed between individual girls and between girls and their mistress. Hence, the majority of Sappho's poems praise and admire the grace, charm, loveliness, and beauty of the girls and women surrounding her. The fame of Sappho's work did not diminish with time, and the library of Alexandria collected all her poems and hymns. Furthermore, she was later counted among the ten most famous poets of Greece. An epigram in the *Anthologia Palatina* attributed to Plato declares: "Some say the Muses are nine: how careless! Look, there is Sappho too, from Lesbos, the tenth." Her fame has survived many centuries, and the term lesbian, invented in the nineteenth century, was derived from her homeland, Lesbos.

The ancient Greeks did not understand homosexuality as a natural property of a certain fraction of each gender, as do modern societies in the twenty-first century. They believed that, in principle, every man is bisexual with an interest in same-sex activities and classified any kind of homosexual behavior according to the roles active and dominant or passive and submissive.

The active role was connected with terms such as adulthood, virility, and higher social status, while the passive role corresponded to terms such as youth, femininity, and lower social status. Hence, the passive role was typical for women in general, for boys of any social class, and for male slaves. The same-sex relationship of a free adult Greek and a male slave was, thus, the normal case of homosexual (and bisexual) behavior and was fully accepted in society.

However, ancient Greek society is also known for a special version of same-sex relationships between men, *pederastia* (boy love), which is unknown for other ancient and modern societies. Pederasty means an intense relationship between an older man and a boy. In this context, boy means a youth up to the age of 17–18, when the beard begins to grow. The term *pederastia* is composed of the words *pais* (child, boy) and *erastes*, the name of the adult in this relationship. The origin of the pederasty lies in the tribal past of the Greeks. The social organization of the tribal communities was based on age groups. The passage from early boyhood to the world of adults required the company of an older, experienced man. This *erastes* had the duty to educate his young partner in various skills, to supervise his military training, and to play the role of model and idol in political affairs and in warfare.

The most famous example of the role pederasty played in wars is the fate of the Sacred Band of Thebes. This band was group of several hundred elite soldiers, consisting exclusively of homosexual pairs. In the battle of Chaeronea (338 B.C.), the Greek fought against Philip II of Macedon (382–336 B.C.) and were defeated. After his victory, Philip II found all the pairs of the Sacred Band lying dead on the battlefield. He was so deeply impressed that he erected a monument in memory of their bravery. Pederasty was characteristic of the upper social classes, but it was well accepted by all social classes and sometimes even admired. However, same-sex relationships between two adult men of similar social status were not well received and were associated with stigmatization of the passive partner. Male prostitutes offering the passive role had the lowest reputation.

Numerous sources provide written information on same-sex orientation in ancient Rome. Illustration on pottery is rare, but graffiti on the walls of Pompeii and Herculaneum provide excellent insight into all kinds of sexual activities in everyday life. Analogous to the situation in ancient Greece, almost nothing is known about the homosexual behavior of Roman women, and a Roman equivalent to Sappho does not exist. The attitude towards same-sex love changed in the course of 800 years from a more puritanical, conservative attitude in the early and middle period of the republican period to a quite liberal and hedonistic attitude during the imperial period. The close contact with Greek culture after the conquest of Greece (143 B.C.) and the import of Greek teachers influenced the presentation and reception of homosexual relationships in literature and in public life, but a direct influence on Roman mentality is unlikely and unproven. Regardless of the period, the same-sex attitude of the Romans may be summarized in three points, already mentioned for the Greeks:

1. The active role was associated with properties such as masculinity and dominance. It was part of a cult of virility.
2. The passive role was considered submissive and feminine and, thus, was despicable for an adult man.
3. It was normal behavior for an upper class Roman to be bisexual, having sex with his wife and generating children and also enjoying sex with a young male slave.

Another important aspect that the world of Greeks and Romans had in common was the absence of any conflicts with religious rules or laws. In other words, homosexual activities were never a sin. In this respect, the emergence of the three monotheistic religions, Judaism, Christianity, and Islam, entailed a dramatic paradigm shift.

The three monotheistic religions and related smaller religious groups (e.g., Druzism or the Bahaí faith) have the same roots in the history and mentality of the near Orient. Their mental background was quite different from that of the contemporary Greek world and condemned same-sex relationships between men, regardless of age, social status, and circumstances. In the Bible chapter Genesis, God advised Adam to generate numerous descendants with his wife, whereas any other kind of sexual activity was considered sinful. In the chapter Leviticus several grave sins were enumerated, including same-sex acts by men and the sexual relationships of men and women with animals. The condemnation of same-sex boy love is repeated in another line of the same book, in Genesis (19.5–11), and in Paul's letter to the Romans (19.22–25).

Same-sex rape is commented on in connection with an episode in Lot's life. Citizens of the city of Sodom came to Lot's house to rape his guests. Lot offers his daughters as substitute, but the aggressors decline. However, Lot's guests were angels and blinded the aggressors.

This episode formed the basis for the later "invention" of the term sodomy as a pejorative label for the same-sex relationships of men. In his comments on the Old Testament, Augustine (354–430), archbishop of Hippo (North Africa), explicitly justified the extermination of Sodom by fire because of the extreme same-sex activities (including rape and prostitution) of its citizens. A text that has shaped the attitude of Christianity towards the homosexuality of men for all the centuries up to the twenty-first is the so-called canon (Kanones) of John Nesteutes (522–595), Patriarch John II of Constantinople. This compilator and author wrote a list of recommendations for father confessors. The homosexual act of men was classified as a grave sin, but other same-sex activities (e.g., oral sex) were not mentioned in the Bible, in the canons, nor in later comments on same-sex behavior.

The term sodomy appeared for the first time in a treatise of the monk Petrus Damianus addressed to Pope Leo IX (1002–1054). Damianus exclusively used this term for anal sex of men and did not include sexual contact with animals. This understanding of sodomy prevailed until the middle of the nineteenth century when new insights and new terms became public. It was the famous theologian Thomas of Aquinas (1224–1274) who declared sexual contact with animals to be the gravest sin.

Until the end of the twelfth century, homosexual behavior or sex with animals were sins without any legal consequences in those European countries controlled by the Catholic church. Yet, during the crusades, the Muslims were blamed for systematic same-sex rape of Christian youth, and in the course of the thirteenth century, the attitude towards same-sex activities changed dramatically in all Christian countries. Same-sex relationship became a crime, which in severe cases was punishable by death. A particularly negative consequence of this change was the misuse of the accusation of sodomy as a political weapon, frequently in combination with an accusation of heresy. As early as 390 B.C., Emperor Theodosius I. of Constantinople decreed a law condemning passive males to be burned at the stake. Justinian enacted a second law that prohibited any kind of unnatural sex, with the death penalty as punishment. Justinian applied this law not only against critics and political opponents, but also against rich people just for the sake of profit.

The most conspicuous drama in the history of sodomy was eradication of the Knights Templar. The Templars were not only a pugnacious order, but also a wealthy one. During the crusades they had organized the first international banking system in Europe. They gave credits and enabled money transfer across Europe, for instance, from England via France to Sicily and Palestine. Philip IV (1268–1314), King of France, was angry about this strong and independent power in his kingdom. He had borrowed an enormous sum of money from the Templars and felt offended because his application for membership had been declined. On 13 October 1307 (a Friday), a combined force of policeman and soldiers arrested almost all Templars in France. The imprisoned knights were accused of sodomy and heresy, tortured, and sentenced to death. Pope Clement V, imprisoned in Avignon, was forced by Philip IV to agree.

In the following centuries, prosecution and execution of men suspected of sodomy took place in numerous cities of Europe. Emperor Charles V (1500–1558, Charles I in Spain) initiated a penal code for his empire that regulated, in article 116, the punishment for any kind of unnatural sexual activities. However, in practice, the handling of this penal code varied from country to country and from city to city. For example, extensive prosecution of homosexual men is reported for Firenze and other cities in northern Italy in the decades following the plague epidemic of 1432. In Amsterdam and London prosecution campaigns were known even in the eighteenth century. The last execution for sodomy in the Netherlands is reported for 1803 and in England for 1832.

In the early nineteenth century the situation began to change slowly but irreversibly. In the eighteenth century two British writers, Thomas Cannon and Jeremy Bentheim, tried to publish books defending homosexuality, but their books were suppressed. About 100 years later the writers Havelock Ellis, Edward Carpenter, and Addington Symonds were more successful and published numerous pro-homosexual articles, pamphlets, and books. Several decades earlier and more influential were the activities of the German jurist Karl H. Ulrich. He published after 1864 a series of twelve articles under the collective title "Research on the Riddle of Man-Manly Love." In 1867, at a Congress of German Jurists in Munich, he came out himself and tried to launch a campaign against anti-homosexual laws in Prussia. Yet, his speech ended with an outcry of the conservative, narrow minded audience.

A real paradigm shift was induced by the book *Psychopathia sexualis* (1886). The author, R. von Krafft-Ebing, was forensic physician and psychiatrist. He had access to prisons and psychiatric clinics and, thus, he had a broad experience with gays. He interpreted homosexuality as an innate psychic disease and inheritable disorder of the nervous system. This new understanding of same-sex behavior had several long-term consequences. A first consequence of minor importance concerned the definition of sodomy, the meaning of which became focused on sex with animals. Particularly important was a change in the fundamental understanding of homosexuality. In the ancient world every man was assumed to be able and, under favorable circumstances, willing to act as a homosexual. For the monotheistic religions, gay men were sinful regardless of age, social class, or circumstances of the relationship. R. von Krafft-Ebing viewed homosexuality as a hereditary property confined to a small group of the population. The third consequence concerned the penal codes of European countries.

After 1805, the *Code Napoleon* became effective in central Europe, and in the French penal code execution for sodomy was replaced by sentence to prison. In the following decades, other European countries changed their penal codes accordingly. In the early twentieth century a stay in prison was gradually replaced by treatment in a closed psychiatric clinic. Whether this change was substantial progress for a homosexual under the conditions before World War II is at least questionable. It took until the end of the twentieth century to remove any kind of punishment from the penal codes of European countries (e.g., 1994 in Germany, 2002 in Austria).

Sigmund Freud, the inventor of psychoanalysis, considered same-sex orientation to be an anomaly of heterosexuality, but he publicly pleaded that homosexuality should not be considered a disease or crime. The beginning of an organized movement supporting emancipation of homosexuals in German-speaking countries can be traced back to the activities of the gay physician Magnus Hirschfeld, who founded in 1897 in Berlin the Wissenschaftliche Humanitäre Komitee. However, this organization only comprised about 500 members and focused on the elimination of punishment from the German penal code. Far more important and influential were two new organizations founded in 1919 (Deutscher Freundschaftsverband) and in 1920 (Bund für Menschenrechte). This second organization, fighting for all kinds of human rights, comprised 50,000 members after 1930, and associated organizations existed in Austria, Switzerland, Czechoslovakia, and even in New York, Argentina, and Brazil. Despite Freud and those organizations, a fourth, particularly negative consequence of Krafft-Ebing's theory became evident during the Nazi period in Germany. To protect the Aryan race from any inheritable weakness or disease, an estimated number of 10,000–15,000 gay men were arrested and transferred to concentration camps, where most of them died.

After World War II the emancipation of gays and lesbians was mainly fueled by an increasing number of organizations and protest movements in the USA (e.g., Christopher Street riot, 1969). Harry Hay, Bob Hull, Chuck Rowland, Dale Jennings, and Rudi Gernreich founded the Mattachine Society in 1951, and Del Martin and Phyllis Lyon founded the lesbian organization Daughters of Bilitis in 1955. These organizations called themselves homophilic to avoid too much emphasis on the sexual aspect. Nonetheless, both organizations were more or less suppressed in the McCarthy period. In the middle of the 1960s, Dick Heitsch (New York) and Frank Kumery (Washington, DC) began to reorganize and unify the numerous local same-sex movements, and a new organization, Radicalesbians, was founded. These activities and the scientific studies of Evelyn Hooker had the consequence that in 1973 the American Psychiatric Association eliminated homosexuality from its *Diagnostic and Statistic Manual of Mental Disorders* (DSM-II). The American Psychoanalytic Association originally opposed this modification of DSM II, but published one generation later (in 1991) an excuse entitled

"Declaration on Homosexuality." As late as 1992 the World Health Organization removed homosexuality from the *International Classification of Disease* (ICD 10).

The third and, hopefully, last paradigm shift away from homosexuality as a disease to a natural property of mentally and psychically healthy humans was and is supported by an increasing number of biologists. In the last quarter of the twentieth century more and more zoologist began to study the same-sex behavior of animals in their natural habitat. The same-sex activities of various mammals and birds in zoos have been known for decades, but were ascribed to their unnatural environment. However, up to the year 2006, same-sex activities have also been observed for more than 1500 animal species in their natural environment and this number is rapidly increasing. To avoid misunderstandings, the term "same-sex behavior" needs a comment in this context. First, same-sex behavior in animals refers to a variety of activities and is not confined to the sexual act. Second, the same-sex behavior of almost all animals studied so far is embedded in their bisexuality. An exclusively homosexual orientation is extremely rare, but has been observed for penguins and domesticated animals such as male sheep.

Amongst the most intensively studied animals are penguins and giraffes. The same-sex behavior of penguins was first observed by George Murray Levick in 1911 at Cape Adare. His report was considered to be too shocking and was suppressed. Private copies were written in Greek to prevent them from becoming widely known. A Greek copy was discovered about 100 years later and published in *Polar Record* in 2012. Broader studies after World War II revealed that, on average, 10 % of penguin nests are occupied by male couples. It was also observed that a male couple "hired" a female to produce an egg, and when the female began to incubate the egg it was kicked off; the male couple continued with incubation and hatched the chick. Many more interesting and sometimes amusing observations have been made in zoos. For instance, zoos in Japan have documented that male pairs of penguins built nests together and used stones as substitutes for eggs. The zoo in Bremerhaven (Germany) tried to improve the reproduction of the endangered Humboldt penguins by importing females from another zoo and separating their own three male pairs. Yet this experiment was unsuccessful, because the relationships of the male pairs were too strong. When German gay groups protested against breaking-up the male pairs, the director of the zoo answered: "We don't know whether the three male pairs are really homosexual or whether they have just bonded because of a shortage of females. ... Nobody here wants to forcibly separate homosexual couples."

Giraffes have also attracted a lot of interest because their sexual acts are easy to observe and to document with photographs and videos. Furthermore, comparison of various populations has revealed that the numbers of same-sex acts amounts to 30–75 % of all sex acts. Despite this high percentage, giraffes have survived for millions of years, demonstrating that their same-sex activities are part of their bisexuality and not in contradiction to their heterosexuality. Another group of mammals with a high percentage of same-sex activities are dolphins and whales, and in the case of Amazon river dolphins even group sex has been documented.

Concerning the extent of same-sex behavior among animals, the following comments and statements should be cited. In early studies of animals the extent of homosexual activities was underestimated because of prejudices caused by social attitudes to same-sex behavior, because of innocent confusion, or because of fear of being ridiculed by colleagues.

Bruce Bagemihl, author of the book *Biological Exuberance...*, remarked that initially the homosexual behavior of giraffes was overlooked: "Every male that sniffed a female was reported as sex, while anal intercourse with orgasm was only revolving around dominance, competition or greetings." Peter Bockman, scientific advisor of an exhibition in Oslo in 2007, speculated: "No species has been found, in which homosexuality has not been shown to exist with the exception of species that never have sex at all, such as urchins or aphis. Moreover, a part of the animal kingdom is hermaphroditic, truly bisexual. For them, homosexuality is not an issue." The exhibition in Oslo's Museum of Natural History was entitled "Against Nature?" and presented for the first time a comprehensive documentation of the same-sex behavior of animals. The director of the museum, Geir Sölli, also presented the final conclusion (adopted from the book of B. Bagemihl) that animals enjoy their same-sex activities.

Numerous hypotheses about the origin of homosexuality in animals have been proposed. Bruce Bagemihl, Joan Roughgarden, Thierry Lodé, and Paul Vasey suggested that hetero- and homosexuality have a social function and serve to strengthen alliances and social ties within a clan or herd. Some researchers assume that the same-sex behavior of males originates from their social organization and fight for dominance, in analogy to prison sexuality. Other researchers object that the social organization hypothesis is incorrect, because it cannot explain certain observations, such as the life-long stability of same-sex relationships between penguins, where the males refuse to pair with females.

Concerning human homosexuality, the various research results and hypotheses are summarized by Allen Brooky in a critical review entitled "Reinventing the Male Homosexuality. The Power and Rhetorics of the Gay Gene." The two most important and most widely discussed theories are:

- Sexual orientation is fixed before birth
- Sexual orientation is mainly influenced by events and social contacts during childhood and puberty

The origin of homosexuality in animals and humans and its role in evolution is a particularly challenging problem for classical Darwinism. Homosexuality neither contributes to an increasing number of individuals nor to an increasing diversification of races and species. For individual species it was found that the existence of homosexuals in a clan or herd may improve the rearing of descendants. Yet, a universal theory explaining why homosexuality is an advantage for evolution of most or all species is still lacking. Possibly, the role of homosexuality in evolution may contribute to a new paradigm shift in the theory of evolution (see Sects. 8.2 and 8.3).

Studies of the homosexuality of humans and animals also present excellent examples of the different methodologies of sociology and psychology, on the one hand, and the natural sciences, on the other. The most important sources of information in sociology and psychology are interviews and the polling of individuals and groups of humans. However, the answers depend on several factors, such as:

1. Structure of the questions
2. Emotional situation of the interviewed person
3. Economic situation and social status
4. Life in a liberal society or in a society suppressing homosexuals

Furthermore, humans can purposely lie. All these problems do not exist in scientific studies. The sexual behavior of animals is open to direct observation and can be combined with experiments in laboratories, such as genetic and physiological investigations.

The results obtained from scientific studies of animals and humans allow the unambiguous conclusion that homosexuality is a natural property of healthy animals and humans and is certainly not a disease. When the faithful of the monotheistic religions are convinced that there is a creator of the world, it is a simple and logical conclusion that homosexuality was intentionally implanted in creation and, thus, homosexuality cannot be a sin. Unfortunately, there is still a large number of people (religious persons or not) who are reluctant to learn this lesson, although globally the percentage of people considering homosexuality to be a disease is steadily decreasing.

Bibliography

Adam B (1987) The raise of a gay and lesbian movement. G.K. Hall & Co

Answers to your questions. For a better understanding of sexual orientation & homosexuality. http://www.apa.org/topics/sexuality/sorientation.pdf

Bagemihl B (1999) Biological exuberance: animal homosexuality and natural diversity. St. Martin's Press

Bailey NW, Zhu N (2009) Same-sex sexual behavior and evolution. Trends Ecol Evol 24(8):439–446

Bailey JM, Zucker KJ (1995) Childhood sex-typed behavior and sexual orientation: a conceptual analysis and quantitative review. Develop Psychol 31(3):43

Boswell J (1980) Christianity, social tolerance and homosexuality: Gay people in Western Europe from the beginning of the Christian Era to the fourteenth century. University of Chicago Press

Bullough VL (2002) Before Stonewall: activists for gay and lesbian rights in historical context. Routledge

Calimach A (2002) Lover's legends: the gay Greek myths. Haiduk Press, New Rochelle

Dictionary of sexual terms. http://www.sex-lexis.com/

Dynes WR, Johansson W, Percy WA, William A, Donaldson S (1990) Encyclopedia of homosexuality. Garland Pub

Ellis H, Symonds JA (1975) Sexual inversion. Arno Press (reprint)
Feray JC, Herzer M (1990) Homosexual studies and politics in the 19th century: Karl M Kertbeny. J Homosexuality 19:1
Foucault M (1986) The history of sexuality. Pantheon Books
Gordon M (1984) Slavery and homosexuality in Athens. Phoenix XXXVIII:308–324
Hubbard TK (2003) Homosexuality in Greece and Rome. University of California Press
Iemmola F, Ciani AC (2009) New evidence of genetic factors influencing sexual orientation in men: female fecundity increases in maternal line. Arch Sexual Behav (Springer) 38:393–399
Kinsey's heterosexual-homosexual rating scale. http://www.kinseyinstitute.org/research/ak-hhscale.html
Langström N, Rahman Q, Carlström E, Lichtenstein P (2008) Genetic and environmental effects on same-sexual behavior: a population study of twins in Sweden. Arch Sexual Behav 39:75–80
Laumann EO, Gagnon JH, Michael RT (1994) The social organization of sexuality: sexual practices in the USA. Chicago Univ. Press
McGinn TAJ (1998) Prostitution, sexuality and the law in Rome. Oxford University Press
Nussbaum MC (1999) Sex and social justice. Oxford Press
Perin EC (2002) Sexual orientation in child and adolescent health care. Kluwer Academic/Plenum Pubs
Shere H (2004) The Hite report: a nationwide study of female sexuality. Seven Stories Press, New York
Sommer V, Vasey P (eds) (2006) Homosexual behavior in animals. Cambridge University Press
Wellings K, Field J, Johnson A, Wadsworth J (1994) Sexual behavior in Britain: the national survey of sexual attitudes and lifestyles. Penguin Books, London
Williams CA (1999) Roman homosexuality. Oxford University Press

Chapter 8
Biology

8.1 Is Spontaneous Generation Possible?

> The great enemy of knowledge is not error but inertness.
> (Henry T. Buckle)

First of all, the term "spontaneous generation" needs definition. In this book, spontaneous generation means the spontaneous formation of primitive animals, such as microbes, worms, and even insects, from dead matter such as dirt, mud, and garbage. The question that has been discussed for more than 2000 years is whether such spontaneous generation (*Spontanzeugung*) occurs every day and everywhere. A quite different scenario is the first emergence of living cells on the early earth about 3.5–4 billion years ago (*Urzeugung*). This event (if it happened on earth at all) was a unique event and the starting point of the entire evolutionary process (an alternative hypothesis is an influx of living cells from space via meteorites or asteroids).

The theory of permanent spontaneous generation of primitive animals has been known since the days of the pre-Socratic philosophers. Depending on the philosopher, this theory was adorned with different details. For example, the cosmology of Empedokles (c. 490–430 B.C.) included long cosmic cycles involving a complete change of the surface of the earth (or even the disappearance of the earth) with extinction of all living organisms. This point was followed by a regeneration (or reappearance) of the earth, including spontaneous generation of primitive organisms. Emergence of the first primitive organisms was, in turn, followed by an evolutionary process generating complex organisms and finally humans. More precise information from the viewpoint of modern biology is available from Aristotle (384–322 B.C.). He mainly described his own observations and concluded that the proliferation of most plants and animals proceeds via seeds or sperms and sexual reproduction. However, for certain plants and insects he postulated spontaneous generation from dirt and rotting organic material. His view survived the Middle Ages and was effective until the second half of the nineteenth century.

H.R. Kricheldorf, *Getting It Right in Science and Medicine*,
DOI 10.1007/978-3-319-30388-8_8

A paradigm shift was induced around 1674 by the Dutch hobby scientist Antonie van Leuwenhoek (1632–1723, see Biography). He did not have any scientific education, but was greatly interested in the construction of microscopes and their application. At that time, microscopes were of poor quality, reaching a magnification of 50 times in the optimum case. He developed a special technique for the production of better lenses that enabled magnification of at least up to 270 times. He never published any description of his procedure and never allowed anybody to assist with the preparation of lenses and microscopes. In the middle of the twentieth century, English and Russian scientists tried to reconstruct his technique, which apparently consisted of two steps. First, he drew glass fibers from molten glass and then heated one end of a fiber in a flame until a glass pearl was formed. These pearls served as lenses in small microscopes, which had to be kept close to the eye. The entire microscope was not longer than 5 cm. He sold many microscopes to other scientists and laics, but held the best instruments back for his own research. In this way, he maintained a monopoly on new microscopic discoveries and more than 100 years elapsed before better microscopes became available.

Van Leuwenhoek did not understand Latin, German, or English. He noted down all his observations in a Dutch dialect and never dared to publish them. However, he disclosed his first observations to a friend, the prominent physician Reinier de Graf, who was a member of the Royal Society in London. This friend praised and recommended van Leuwenhoek's work and in 1673 The Royal Society published a first letter from van Leuwenhoek. Until the end of his life, he submitted around 190 letters to the Royal Society. Among his first observations were tiny rapidly moving organisms, called "animalcules," in dirty water or salvia. In the years 1674–1776 he reported in several letters the first observations of single-cell organisms, and he was able to distinguish different morphologies, such as sphere-shaped bacteria (*cocci*), rod-shaped bacteria (*bacilli*), and spiral-shaped bacteria (*spirillae*). At first, these observations received a skeptical or critical response or were even ridiculed. Yet, when other scientists (some were invited to his laboratory) confirmed his results his reputation grew immensely and he was elected member of the Royal Society.

His most important findings concerned spermatozoa and their role in sexual reproduction. He identified the sperm of insects and men. Furthermore, he observed that certain beetles, shellfish, and flee as grew from eggs and not by spontaneous generation from dirt, mud, or rotting materials. These results were again met with skepticism or ignored, and finally they were forgotten until Louis Pasteur reinvestigated this problem using better microscopes.

After the Napoleonic Wars, all kinds of science began to flourish in Europe and a new discipline appeared on the scene, "organic chemistry" (see Sect. 9.3). Several scientists began to study alcoholic fermentation (see Chap. 9) and in 1854 Louis Pasteur, new full professor at the University of Lille, joined this group. At first, he concentrated his efforts on the fermentation of glucose under conditions yielding lactic acid. This kind of fermentation was, for instance, responsible for the formation of sour milk or sour wine. At that time he did not yet know that the microbes producing lactic acid were quite different from the yeast responsible for alcoholic fermentation. In 1857, he presented to the Société des Sciences de Lille his conclusion that the formation of lactic acid is caused by microbes (see Sect. 8.2). He applied his knowledge of lactate fermentation to alcoholic fermentation and postulated that certain microbes were responsible for both processes He expanded the meaning of the term "fermentation" to all other kinds of rottenness and putrefaction of organic materials. He stated that rottenness and putrefaction involving sulfur compounds (later identified as the amino acids cysteine and cystine) were responsible for the bad smell.

Beginning in 1860, Pasteur published five articles dealing with the fermentation and rotting of organic materials and he eventually concluded that spontaneous generation does not exist. This conclusion was heavily criticized by numerous experts, in particular by Felix A. Fouchet, director of the Museum of Natural History in Rouen. To settle this debate, the French Academy of Sciences offered the Alhumbert Prize of 2500 Francs to any scientist who could convincingly demonstrate by experiments that the hypothesis of spontaneous generation was correct or not. At a meeting of the Academy in 1862 Pasteur presented a summary of his experimental results, convinced the audience, and won the prize. His oral presentation described the following experiments:

1. An aqueous solution of glucose containing yeast and other germs was boiled for a while to kill the germs and was then stored in a closed flask. The content remained sterile even after several weeks.
2. Urine, milk, or a yeast-containing sugar solution were added to several glass flasks, the opening of which was transformed into a swan-neck duct having a small opening to air. In one series of experiments the content was stored at room temperature without any modification, whereas in a parallel series the content was boiled for a while. In the first series, fermentation was detectable within a few days, whereas the liquids of the second series remained sterile. When the swan-neck ducts were removed, so that the sterile flasks had a wide opening, fermentation started within a short time.
3. Twenty glass flasks were filled with yeast-containing sugar solution, boiled, and closed immediately thereafter. These sterile glass flasks were transported to the Jura mountains and opened for a few minutes at an altitude of 250 m. After closing, generation of germs was observed in eight flasks. This experiment was repeated with another series of 20 flasks at an altitude of 850 m, and growing of germs was detected in five flasks. In a third series, the flasks were opened at an altitude of 2000 m and only one flask lost its sterility.

Pasteur concluded: "Never will the doctrine of spontaneous generation recover from the mortal blow of this simple experiment. There are no known circumstances in which it can be confirmed that microscopic beings came into the world without germs, without parents similar to themselves." In 1863, Pasteur proved that samples of urine and blood taken under sterile conditions from the bladder and veins, respectively, of an animal remained sterile for many weeks when transferred to a sterile flask. In 1877, the English scientist Henry Charlton reported to have observed spontaneous generation of microbes in a sample of urine sterilized by heating. This report prompted coworkers of Pasteur to reexamine the heat treatment of various liquids and solutions. They found that some bacteria and fungi generated spores that have an extraordinarily high thermostability, so that they could survive temperatures of around 100 °C.

The results of Pasteur and his coworkers had several far-reaching consequences in various directions

(i) The paradigm of spontaneous generation was definitely buried.
(ii) Beverages prone to fermentation, such as beer, milk, or sugar-containing wine can be stabilized for days or weeks without loss of taste or nutritive quality by heating to temperatures of around 55–60 °C. However, this procedure (today called "pasteurization") does not kill all germs.
(iii) When liquids and solutions containing organic substances are heated to temperatures above 120 °C, all germs, including spores, are destroyed. For medicine, this is an important procedure and is called sterilization. Because not all useful materials survive heating above 120 °C, UV light or special chemicals (e.g., oxirane) are today frequently used for sterilization.
(iv) Whereas the Romans liked intensive bathing and washing, the Europeans of later centuries believed that intensive contact with water is unhealthy. They "varnished" their own bad smell by means of perfume and toilet powder. After Pasteur's discoveries, the concept of hygiene was reestablished in western civilization. However, this concept was now not limited to cleaning the human body, it also encompassed the washing of dishes and clothes and the cleaning of rooms and apartments. The hygiene concept also included a reorganization of the water and sewage systems in European cities. Furthermore, the chemical industry was stimulated to produce enormous amounts of soaps, detergents, and chemicals for disinfection.

Even in the twenty-first century, the consequences of Pasteur's work play a tremendous role in the health and lifespan of Europeans and their descendants everywhere in the world.

Antonie van Leuwenhoek

Antonie van Leuwenhoek was born on 24 October 1632 in Delft (the Netherlands) as the youngest of five children of Philips A. Leuwenhoek and his wife Margaretha (nee Bel van den Berch). His father was a basket maker, who died 5 years after the birth of A. van Leuwenhoek. His mother came from a wealthy brewer's family and married a painter sometime after the death of her first husband. Little is known about Antonie's education. First, he attended a school in or near the city of Leyden, but soon moved to Benthuizen, where he lived in the house of an uncle, an attorney and town clerk. Apparently he learned physics and mathematics from his uncle. At age 16, his mother took him to Amsterdam, where he became an apprentice in a linen-drapers shop. In July 1654 he married Barbara de Mey and returned to Delft, where he spent the rest of his life. He and his wife had five children, but only one daughter survived for more than 5 years.

In Delft he opened a draper's shop and became a successful businessman. His profit and perhaps some financial support from his wife or his mother allowed him to buy a house and install a laboratory there. His wife died in 1666 and 5 years later he married Cornelia Swalmius.

None of their children survived early childhood. As a result of his increasing reputation as a successful businessman and trustworthy person, he also made a political career in his hometown. In 1660, he became chamberlain of the Delft sheriff's assembly chamber, a position he held for nearly 40 years. Later van Leuwenhoek also became a municipal "wine gauger," responsible for the city's wine import. In 1669 he was elected surveyor at the Court of Holland.

Delft was a small city at that time with approximately 24000 inhabitants and, thus, it is highly likely that he was at least an acquaintance, if not a friend, of the famous painter Johan Vermeer (1632–1675). When the painter died, he was executor of his last will. Throughout his life, van Leuwenhoek wrote many letters to the Royal Society and other scientific institutions reporting his observations and conclusions. In the last few letters he also described his illness. Therefore, in the twentieth century it has been possible to identify his disease. He suffered from a rare disease, today called "van Leuwenhoek's disease." He died on 26 August 1723, at the age of 90 and was buried in the Oude Kerk of Delft.

Soon after having established his draper business in Delft van Leuwenhoek became interested in microscopy. His high income allowed him to buy the materials necessary for the construction of microscopes and he was able to spend a lot of time on the construction of microscopes and on microscopic observations. He is said to have constructed more than 200 microscopes, most of which were sold to other scientists or interested amateurs, but he kept the best instruments for his own studies. Nine microscopes of van Leuwenhoek's collection have survived, demonstrating a magnification of up to 275 times. However, he is supposed to have reached a magnification of up to 500 times. In addition to the detection of single-cell organisms and spermatozoa van Leuwenhoek made other important observations and discoveries, such as the vacuole of the cell and the banded pattern of muscular fibers. Furthermore, he made drawings of the cross-sections of plants, together with a careful description of details, and also drew and described numerous tiny insects. His discoveries made him a famous man who was visited by numerous VIPs of his time, for example Queen Anne, William III of Orange and his wife, Tsar Peter the Great, and the famous scientist Gottfried W. Leibnitz. In the aftermath he was nicknamed "Father of microbiology."

Louis Pasteur

Louis Pasteur was born on 27 December 1822, in the French village of Dole, Jura. He was the third child of Jean-Joseph Pasteur and his wife Jeanne-Etienne (nee Roqui), a poor tanner family. When he was 5 years old, the family moved to Arbois, where he joined the primary school in 1831. During his studies he showed a lot of interest in fishing and sketching and not so much in science. The pastels and portraits he made at the age of 15 have survived and are stored in the museum of the Pasteur Institute. In the first half of 1838, he went to Paris to enter the Institution

Barbet, but he returned homesick in November of the same year. In the following year he entered the Collège Royal de Besancon, where he received a bachelor's degree in 1841. Afterwards, he continued his studies in Dijon, where he earned a bachelor of science, but with a low grade in chemistry. In 1842, a first attempt at the entrance test for the highly reputed École Normale Supérieure in Paris failed, but he succeeded in 1844.

In 1845 he graduated from the École Normale Supérieure and in 1846 he was appointed professor of physics at the Collège de Tournon a Ardèche, but his former supervisor Antoine J. Ballard wanted him back at the École Normale as an assistant for teaching chemistry.

After his return, Pasteur immediately began research in crystallography and in 1847 submitted two theses, one in physics and one in chemistry. After a short stay in Dijon, in 1848 he accepted a position as full professor of chemistry at the University of Strasbourg in the faculty of sciences. There he became acquainted with Marie Laurent, daughter of the rector, and married her on 26 May 1849. They had five children, but three of them died of typhoid in childhood. This tragedy stimulated him later to study methods for curing infectious diseases.

In 1854 he moved to Lille where he was appointed to the Chair of Chemistry and after a few months he was named dean of the new faculty of sciences. The new faculty was financially supported by local and regional industry and, thus, Pasteur had to concern himself with the chemical problems of that industry. The contacts with breweries motivated him to begin studies of fermentation. Furthermore, he reorganized the curriculum and the courses in chemistry so that students had to perform experimental work. This is today a triviality for any student of chemistry, but it was revolutionary around 1860, which reflected a new understanding of scientific research.

In 1857, Pasteur became director of scientific studies at the École Normale Supérieur and in this new position he began again to reform the curriculum and the quality of scientific work. The next step in his career came in 1862, when he was appointed professor of chemistry, geology, and physics at the École National Supérieur des Beaux-Arts. He held this position until his retirement in 1867. Financed by the government, Pasteur focused his work in the following two decades on the fight against infectious diseases. His great success in this working field allowed him to collect a considerable sum of money from industry and from private donations, which enabled the foundation and building of the Pasteur Institute in Paris. This institute became fully operational in 1888, with Pasteur as its first director. Unfortunately his health was affected by several strokes since 1868, and he was later partially paralyzed and unable to perform experiments in his new institute. When he died on 28 September 1895, he was given a state funeral. He was first buried in the Cathedral of Notre Dame, but his remains were later moved to a crypt of the Pasteur Institute.

The work and success of Pasteur go far beyond the topic discussed above. He made an immensely important contribution to the evolution of modern chemistry in general, and to organic chemistry and pharmacy in particular. This contribution concerned the symmetry of compounds containing tetravalent carbon atoms connected to four different groups (substituents). Tartaric acid contains two such carbons. Pasteur observed that a solution of tartaric acid isolated from plants (e.g., wine lees) rotated the plane of linearly polarized light passing through it. In contrast, synthetic tartaric acid did not, although all other properties were identical. Pasteur examined numerous salts of synthetic tartaric acid and discovered (with a great deal of luck) that certain salts yielded two forms of nonsymmetrical crystals, which were mirror images of each other. He was able to pick up individual crystals with tweezers and found that one sort of crystals rotated polarized light to the left, whereas the mirror image crystals rotated it to the right. A 1:1 mixture of the two types of crystals did not cause any rotation.

This finding had two important consequences. First, it caused a paradigm shift in the understanding of organic molecules. Previously it was believed, in connection with the theory of vitalism (see Sect. 9.3), that only living organisms can produce nonsymmetrical, optically active substances. Pasteur proved that they can also be synthesized in a laboratory, a second and final blow against vitalism in chemistry. Second, Pasteur's finding enabled the Dutch chemist Hendrik van't Hoff (first Nobel Prize laureate in chemistry, 1801) to complete the formula language of chemistry in such a way that it allowed a perfect description of all aspects of chemical structure. His pertinent publication in 1874 is considered to be the final step in the emancipation of chemistry from alchemy, and many chemists think that Pasteur's discovery and interpretation of synthetic nonsymmetrical substances was his greatest and most original contribution to science.

However, in his later work on the fight against infectious diseases he also made an immensely important (but less original) contribution to medicine. When he worked on chicken cholera he found by chance that the cholera microbes, after long exposure to air, made the infected chicken immune to cholera not sick. Pasteur believed that the infectious bacteria were somehow damaged or weakened by oxygen and, thus, were not aggressive enough to kill their hosts, but still active enough to stimulate an immune response. He called damaged microbes "vaccines" and developed the concept of vaccination by means of chemical modification of bacteria and viruses (the difference was still unknown at that time). He was successful in stimulating immunity against anthrax infection of cattle and against rabies in rabbits and humans.

Stimulation of immunity by inoculation with a weaker form of infectious microbes was not new. It was developed before 1800 by the British physician Edward Jenner (1745–1823), who used cowpox, which is less aggressive in humans, to stimulate immunity against the deadly smallpox. However, only the chemical modification of microbes or viruses enabled technical production of vaccines. This new development was Pasteur's contribution, which allowed him to collect enough money for his institute and helped to save the lives of millions of infected people.

However, when Pasteur's notebooks became available after 1995, it became evident in a review by Gerald Geison that Pasteur had published and patented misleading accounts and been deceptive about this most important discovery. For instance, the patented vaccination of anthrax was not achieved by oxygen (as claimed) but by potassium dichromate solution, a procedure stolen from his rival Lean-J.H. Toussaint, a veterinarian in Toulouse. In a recent biography, the author Patrice Debré admitted that Pasteur was sometimes unfair, combative, arrogant, inflexible, dogmatic, and unattractive in attitude. Nonetheless, he remains for France a scientific hero and national monument.

Pasteur received numerous honors during his lifetime, for example, a prize of the French Pharmaceutical Society in 1853, the Rumford Medal of the Royal Society in 1856, the Copley Medal in 1874, the Motion Prize of the French Academy of Sciences in 1859, the Jecker Prize in 1861, and the Alhumbert Prize in 1862. He became Knight of the Legion of Honor in 1863, Commander in 1868, and Grand Officer in 1878. He was elected to the Academie Nationale de Medecine in 1881 and to the French Academy of Sciences in 1887. He became Commander of the Brazilian Order of the Rose in 1873, and he won the Leuwenhoek Medal in 1895. The Ottoman Sultan Abdul Hamid II awarded him the Order of Medjidie in 1886. However, he refused to accept any honor or order from a German institution, because he was an ardent patriot full of hatred for Germans. After his death, streets and buildings in numerous countries were named after him and the asteroid 4804 carries his name. A statue was erected at the St. Rafael High School in California. UNESCO created an "Institut Pasteur" medal 100 years after his death, which is given every second year "in recognition of outstanding research contributing to the beneficial impact on human health."

Bibliography

Aristotle (1908-1952) The works of Aristotle translated into English under the editorship of D. Ross, 12 volumes. Clarendon Press, Oxford

Caddedu A (2005) Les vérités de la science. Pratique, récit, histoire: Le cas Louis Pasteur. Leo Olschki, Florence

Carter KC (1988) The Koch-Pasteur dispute on establishing the cause of anthrax. Bull Hist Med 62(1):42

Carter KC (1991) The development of Pasteur's concept of disease causation and the emergence of specific causes in nineteen-century medicine. Bull Hist Med 65 (4):528

Debré P (1994) Louis Pasteur. Flammarion, Paris

Ford BJ (1991) The Leuwenhoek legacy. Biopress and Farrand Press, Bristol and London

Geison GL (1995) The private science of Louis Pasteur. Princeton University Press, Princeton

Nestle W (1934) Aristoteles—Hauptwerke. Kröner, Stuttgart

Payne AS (1970) The cleere observer: a biography of Antonie van Leuwenhoek. Macmillan, London

Ruestow EG (1996) The microscope in the Dutch republic: the shaping of discovery. Cambridge University Press, New York

Williams E (2010) The forgotten giants behind Louis Pasteur: contributions by the veterinarians Toussaint and Galtier. Vet Herit 33(2):33

8.2 Alcoholic Fermentation

Experience is the fruit of revised past errors, hence one has to err from time to time.
(Johannes Nestroy)

Alcoholic fermentation, also called ethanol fermentation, is a biochemical process that transforms sugars, such as glucose, fructose, or sucrose, into ethanol (alcohol) and gaseous carbon dioxide. Beer is probably the oldest alcohol-containing beverage of humankind, possibly dating back to the beginning of cereal farming in the early Neolithic around 9000–10,000 B.C. The earliest chemical evidence for production of beer by fermentation of barley was found during excavation of Godin Tepe in the Zagros Mountains in western Iran. The Sumerian *Hymn to Ninkasi* as well as the *Epic of Gilgamesh* refer to the consumption of beer. The Ebla Tablets discovered in Ebla, Syria in 1974 indicate that beer was brewed in that city around 2500 B.C. From China, an alcoholic beverage resulting from fermentation of rice and fruit has been known since 7000 B.C. As documented in the Bible, the production of wine from grape juice was known in the Mediterranean area even before its consumption was described in the manifold literature of the ancient Greeks. Tacitus reported in his book *Germania* around 100 A.D. that the Teutons produced mead by fermentation of honey. In the following centuries, the Europeans learned to produce alcoholic beverages from any kind of sugar- or starch-containing fruit or cereal.

When in the early nineteenth century organic chemistry began to emerge as a new discipline of science, it was obvious that chemists would become interested in detailed study of the fermentation process. In previous centuries, the alchemists had preferentially studied the properties and reactions of inorganic materials such as metals, minerals, and various salts, because they were mainly keen on finding an inexpensive method for the production of gold. Extracts of plants were prepared for medical purposes and not for chemical studies, in as much as the necessary analytic methods did not yet exist. In the first half of the nineteenth century, more and more biologists and organic chemist became interested in the isolation, purification, and characterization of soluble substances in plants or animals, and in this connection an increasing number of research groups concentrated on detailed analyses of food and beverages. A protagonist of this new working field was the German chemist Justus von Liebig, Professor in Gießen and Marburg who, together with his coworkers, dissected and analyzed numerous plants and a few animals.

In the year 1815, the French chemist Joseph Louis Gay-Lussac (1778–1850) formulated for the first time an overall equation for the transformation of glucose into alcohol and carbon dioxide. Two of the most important chemists of that time, Jöns J. Berzelius (1779–1848) and Justus von Liebig held that the presence of a catalytically active chemical was responsible for the fermentation. In this connection, Berzelius coined the terms "catalysis" and "catalyst." Berzelius and von Liebig were not the only chemists adhering to this hypothesis, but they were not able to retrieve any experimental evidence.

In 1837, three scientists reported simultaneously but independently of each other that alcoholic fermentation was caused by living organisms, namely by yeast. These three researchers were the French chemist Charles C. de la Tour (1777–1859) and the Germans Theodor Schwann (1810–1882) and Friedrich Kützing. After 1854, Louis Pasteur entered this scientific battlefield. In 1854, he was appointed professor and dean of the new faculty of sciences at the University of Lille. Because the faculty was financially supported by regional industry, Pasteur became involved in the chemical problems of that industry, which included breweries and factories producing alcohol from beet sugar (sucrose, a combination of glucose and fructose). In both cases, better understanding and optimization of the fermentation process were desired.

At first, Pasteur concentrated on the fermentation yielding lactic acid, a process responsible for the formation of sour milk (and sour wine). The clear solution of lactic acid obtained after filtration of the fermentation broth showed the unexpected property of rotating the plane of linearly polarized light, a phenomenon called optical activity. This phenomenon was not absolutely new, because the optical activity of glucose, fructose, and several α-amino acids (the building blocks of proteins) was known. Pasteur adhered for several years to the vitalistic theory that only living organisms were endowed with a special, nonsymmetrical cosmic power, part of the *vis vitalis* (vital spark or vital energy) that, according to the doctrine of vitalism, was characteristic of any kind of living organism (see Sect. 9.3). In 1857, Pasteur presented a summary of his results in a lecture at a meeting of the *Société des Sciences de Lille*. These results included the following points:

1. Fermentation is caused by living organisms
2. Different kinds of fermentation are caused by different types of microbes
3. The microbes need nutrients from the fermentation broth
4. Different types of microbes compete for the nutrients
5. The relevant microbes exist everywhere in normal air

Pasteur knew that the microbes responsible for the formation of lactic acid (*lactobacilli*) were quite different from the yeast responsible for alcoholic fermentation and applied his concept to alcoholic fermentation. However, he discovered an unexpected property of yeast, namely their activity under anaerobic conditions, later called the "Pasteur effect." All animals including microbes can be subdivided into two categories, organisms that need oxygen to maintain their metabolism (aerobic organisms) and those that do not need it (anaerobic organisms). Yeast belongs to the relatively rare species that can live under either aerobic or anaerobic conditions. In the latter case, the rate of reproduction is slow and the end products of metabolism are different from those generated under aerobic conditions. When enough oxygen is available, glucose is completely oxidized, so that water and carbon dioxide are the only reaction products, whereas a limited supply of oxygen results in formation of alcohol. However, increasing the concentration of alcohol reduces and finally stops this fermentation. Hence, beverages containing a high percentage of alcohol ($>$20 %) require enrichment of alcohol by distillation.

Pasteur was also aware that alcoholic fermentation was accompanied by formation of byproducts such as glycerol, succinic acid, and fat. He believed that this complex mixture of products is characteristic of the metabolism of living organisms, whereas a simple, clean reaction pathway was expected according to the theory of Berzelius and von Liebig. He was also able to demonstrate that alcoholic fermentation only proceeded when certain nutrients, such as ammonium tartrate and various inorganic salts, were added to the fermentation broth. Nonetheless, the vast majority of experts rejected Pasteur's theory. This dispute was particularly fierce because von Liebig was German and Pasteur was an extreme French patriot hating all Germans.

It was an irony of fate that all further research confirmed a theory that included both of the apparently contradictory concepts that had been so intensively disputed in previous decades.

The German chemist Moritz Traube (1826–1907) proposed in 1858 a new hypothesis that was a compromise between the embattled doctrines of von Liebig and Pasteur, and in 1860 Traube's hypothesis was supported in a publication by Marcelin Berthelot (1827–1907). Both scientists postulated that fermentation was catalyzed by enzymes (biocatalysts) that were produced by microbes but might be active when the microbes were killed. J. von Liebig declared his consent in 1868, but Pasteur was not willing to accept the new hypothesis, in as much as those authors had not presented sufficient experimental evidence. A publication of Eduard Buchner (see Biography) in January 1897 brought the decision in favor of the Traube–Berthelot hypothesis. Buchner had isolated a cell-free complex of proteins called "zymase," which was capable of catalyzing the entire fermentation process. For this discovery Buchner was awarded the Nobel Prize in 1907.

In the following decades, various research groups contributed details to the full elucidation of the fermentation process. It turned out that zymase was a complex of several enzymes, because fermentation is a multistep reaction sequence that requires a specific biocatalyst for each individual step. In 1929, the English scientist Arthur Harder (1865–1940) and the German chemist Hans von Euler-Chelpin (1873–1964) were awarded the Nobel Prize for further important contributions to the elucidation of alcoholic fermentation.

In the twenty-first century, all reaction steps and enzymes are well known and the ongoing research activities concentrate on genetic aspects of fermentation. The background of this research are the questions: "How and why has evolution invented alcoholic fermentation?" and "Is the creator of the world himself, like Zeus, a fan of alcoholic beverages?"

Eduard Buchner

Eduard Buchner was born on 20 May 1860, in München (Munich) as the younger of two brothers. His father was physician and associated professor of forensic medicine. After graduating from the high school and finishing basic training in the army he entered the University of München in 1878 and began to study chemistry and medicine. Simultaneously, he attended lectures on inorganic chemistry at the Technical University. In 1879 he began in his spare time to research the stability of canned food in the tinning factory (cannery) of Walter Nägeli (apparently an acquaintance of the family). It was at this time that he first came in contact with fermentation processes, a topic that interested him until the end of his life. Under the influence and supervision of his older brother Buchner worked from 1882 through 1884 in the Botanic Institute of the University of München, where he studied the properties of fission fungi (schizomycetes). In parallel with this experimental work he studied botany under the guidance of Prof. Carl W. von Nägeli.

In 1884, Buchner became more interested in chemistry and focused his further studies on chemistry under the guidance of Prof. Adolf von Bayer (1835–1917, Nobel Prize in 1805). Von Bayer and his assistant Theodor Curtius (1857–1928) soon recognized the extraordinary talent of Buchner and hired him for their working group. Buchner and Curtius became friends and when Curtius accepted a position as assistant professor in Erlangen (1887/1888) Buchner followed him to Erlangen for short time to complete his thesis, but he received the doctoral degree from the University of München. Despite a severe dissonance in the relationship between von Bayer and Curtius, Buchner maintained an excellent relationship with von Bayer and worked as assistant in his laboratory. He was allowed to perform independent research and on the basis of this work he completed his habilitation in 1891. In 1893, Buchner moved to Erlangen as assistant professor to join his friend Curtius. In 1896, he was appointed associate professor for analytical and pharmaceutical chemistry at the University of Tübingen. In this position he began to study alcoholic fermentation in the absence of yeast. Only 2 years later he moved to Berlin, where he was appointed full professor at a technical university specializing in agriculture. He held this position for 11 years, and for his successful work on alcoholic fermentation was awarded the Nobel Prize in 1907.

In1909, Buchner accepted a full professor position at the University of Breslau and in the same year was elected member of the Bavarian Academy of Sciences ("Leopoldina"). Because he was not satisfied with the working conditions and with his private situation in Breslau he accepted a full professor position at the University of Würzburg in 1911.

Despite his age of 55 and despite his reputation as Nobel Prize laureate, the government called him for war service in the first half of 1915. He served as leader of a transportation unit in the rank of Major. The University of Würzburg, which badly needed professors for teaching purposes during the war, strived hard to get him back and succeeded in 1916.

However, in April 1917 when the USA entered the war, Buchner deliberately returned to war service and organized the transport of ammunition to the eastern front. In August 1917 he was severely wounded near the village of Focsani in Romania and died nine days later from his infected wounds.

In 1900, Buchner had married Lotte Stahl, daughter of Hermann Stahl, professor at the University of Tübingen. With his death he left a widow and three daughters.

Bibliography

Barnett JA, Barnett L (2011) Yeast research: a historical overview. ASM Press, Washington, DC

Buchner E. http://nobelprize.org/chemistry/laureates/1907

Cornish-Bowden A (1999) The origins of enzymology. Biochemist 19(2):26
Kohler R (1971) The background to Eduard Buchner's discovery of cell-free fermentation. J Hist Biol 4(1):35
Kohler R (1972) The reception of Eduard Buchner's discovery of cell-free fermentation. J Hist Biol 5(2):327
Pasteur L (1857) Memoire sur la fermentation appelée lactique. Comptes Rendus Chimie 45:913
Pasteur L (1858) Nouveaux faits concernant l´histoire de la fermentation alcoholique. Comptes Rendus Chimie 47:1011
Stryer L (1975) Biochemistry. W.H. Freeman

8.3 Theory of Evolution and Darwinism

Nothing in biology makes sense except in the light of evolution.
(Theodosius Dobzhansky)

There is presumably no other topic in science that was and still is the origin of so many (frequently incorrect) reviews, treatises, and reports (partly contributed by scientists, but primarily by journalists) as the theory of evolution. Therefore, another short review may look obsolete at first glance, and a detailed analysis or description of all aspects of this topic is beyond the scope of this book. However, the theory of evolution is too important for the history of science and for the self-esteem and world view of humankind to be ignored. Furthermore, this topic offers numerous examples of errors or fallacies and their revisions. Probably, there is no other research area in science where discoveries, insights, and theories have been so frequently misunderstood, distorted, criticized, rejected, or defamed. For a proper understanding of this working field it is essential to differentiate between the fundamental theory of evolution and evolutionary mechanisms that explain how evolution proceeds. Charles Darwin discovered and described two evolutionary mechanisms, but in contrast to numerous popular articles and reviews, he is not the inventor of the theory of evolution.

As may be expected from the history of science in Europe, the roots of the theory of evolution can be traced back to the pre-Socratic philosophers in ancient Greece. It is handed down that Anaximander of Miletus (c. 611–c. 546 B.C.) was the first to describe an evolution of animals. He believed that all life came from the sea and that land animals were an adaption to life on dry earth. He regarded certain viviparous sharks and dogfish as intermediates between fish and animals. He also speculated that humankind descended from animals. Empedokles of Akragas (Agrigenti, Sicily, c. 495–c. 435 B.C.) was the second to advocate evolutionary processes for all living organisms. He held that the entire world with all its dead objects, plants, and animals was involved in cosmic cycles under the influence of two driving forces, "love" ($\phi\iota\lambda\iota\alpha$) and "strife" ($\nu\varepsilon\iota\kappa o\sigma$). The beginning of a cosmic cycle included the birth of primitive organisms that gradually generated plants and animals of greater complexity as the cosmic cycle proceeded.

The Romans had little interest in evolutionary theories and for 1500 years following the end of the Roman Empire (counted from the demolition of the limes) nobody showed any interest in this topic, in as much as the evolution of living organisms was in contradiction to the teaching of all three monotheistic religions.

Within the frame of modern science, evolutionary processes on the surface of the earth were first postulated and described by geologists (see Sect. 10.6). The discovery of shells and other remains of primitive organisms in various sediments fueled a new debate at the beginning of the nineteenth century on the extent and causes of evolution of life. The leaders of this debate were three biologists working in Paris, Jean-Baptiste Lamarck (1744–1829), Etienne Geoffroy Saint Hilaire (1772–1844), and George L. C. D. (Baron) de Cuvier (1769–1832).

Lamarck was born in the Picardie and served as soldier in the Pomeranian war against Prussia. He had become interested in botany as a boy and took advantage of the frequent changes of his garrison for botanic studies. With financial support from the government, in 1778 he published a three volume work, *Flore Francaise*. In 1779, he became member of the French Academy of Sciences. In 1793 he was appointed professor of zoology at the new Museum of National History in Paris. There he was responsible for the collections of insects and invertebrates (a term he coined himself) and published numerous articles on the classification and properties of invertebrates. His articles dealing with chemistry, geology, and meteorology found little interest, but his work *Philosophie Zoologique* (1809) found international recognition. In this work he described the first cohesive theory of evolution. This theory included the following points:

1. Primitive organisms resulted and still result from spontaneous generation by a natural life force (see Sect. 8.1).
2. Complex organisms evolve from more primitive ones under the influence of a complexing force. He formulated a "ladder of progress," an idea borrowed from Aristotle.
3. An adaptive force favored the adaption of new organisms to their specific environment.
4. In his own words: "All the acquisitions or losses wrought by nature on individuals through the influence of the environment in which their race has long been placed and hence through the influence of the permanent use or disuse of any organ, all these are preserved by reproduction of the new individuals which arise, provided that the acquired modifications are common to both sexes, or at least to the individuals which produce the young."

Point (4) represents Lamarck's second law, which is today called self-inheritance, and has developed into the research area of epigenetics. Lamarck's theory of the inheritance of acquired properties was well accepted in the nineteenth century before the Darwin–Wallace theory of evolution became popular, but it was rejected in the twentieth century. However, modern genetics (see Chap. 9) has discovered that Lamarck's vision is partly correct.

Geoffroy St. Hilaire was born the son of a lawyer in Étampes near Paris. First, he studied natural philosophy and theology and, after graduation, became canon in his hometown. He began to attend lectures in various disciplines of science and eventually accepted a position at the Cabinet of Natural History. In 1793, he was appointed professor of zoology at the new Museum of Natural History as a colleague of Lamarck. In 1798, he accompanied Napoleon on his great scientific expedition to Egypt. His theory of evolution included the following points:

1. The anatomies of vertebrates and invertebrates have many features in common, suggesting that both groups of animals have the same origin.
2. The environment has a direct influence on the evolution of species.
3. Evolution may involve saltational transformations of one species into another, in his own words: "monstrosities could become the founding fathers (or mothers) of new species by instantaneous transformation from one form to the next."
4. Birds could have arisen from reptiles via an epigenetic saltation.

George Cuvier was born on 23 August 1764 in Montbéliard (southern Alsace), a town belonging to the Duchy of Württemberg at that time. He grew up and studied in Stuttgart. He spoke German and French and learned Latin and ancient Greek at school. After graduation, he pursued a career in France and his first position was that of a teacher in Normandy. With support of St. Hilaire he was appointed professor in the College de France in Paris. In 1803, he was elected permanent secretary of the French Academy of Sciences and in 1806 became foreign member of the Royal Society (London). He also had a rising political career. His most famous book, *La Regne Animal* (The Animal Kingdom), appeared in 1817. His interest and work included:

- Structure and classification of mollusca
- Systematic order and comparative anatomy of fish
- Anatomy of fossil mammals and reptiles and the osteology of living forms belonging to the same group
- Paleontology and stratigraphy (his most important sources of information)

Cuvier's view of natural history differed from that of Lamarck and St. Hilaire in the following points:

I. Cuvier adhered to the geological theory of catastrophism (see Sect. 10.6) and held that all previous animals were extinct during the last catastrophe and all extant classes and species were created after the last catastrophe
II. Permanent evolution did not and does not exist
III. Extant animals belong to four large groups that were created simultaneously after the last catastrophe

The fierce debates between Cuvier and Lamarck or St. Hilaire not only found the interest of French biologists and geologists, they were also followed by British, Italian, and German scientists (e.g., Johann W. von Goethe). The harshness of

Cuvier's criticism and his high social rank and reputation discouraged naturalists from speculating about an evolutionary process that proceeded by gradual transmutation of species, until Darwin and Wallace (see Biographies) published the results of their expeditions.

Following the advice of his father, Charles Darwin had studied theology, but his main interest was scientific studies. Attending lectures in botany and geology, collecting beetles, and accompanying John Henslow (1796–1861), professor of botany, and Adam Sedgwick (1785–1873), professor of geology, on excursions in England allowed him to acquire a broad knowledge of biology and geology. The recommendation of Henslow and financial support from his wealthy father enabled him to participate in an expedition on board the HMS Beagle under the command of Captain Robert FitzRoy (1805–1865). Originally, this was intended to be a 2-year expedition along the coasts of South America, but the expedition ended after 5 years as a circumnavigation of the globe. Darwin's observations of plants and animals, his paleontological finds, his collection of various specimens, and the reading of Charles Lyell's book *Principles of Geology* yielded the nutrient broth for his theory of evolution.

During the first years after his return in October 1834, Darwin spent most of his time writing his travel book (Journal), ordering his huge collection, and organizing the evaluation of his collection by other experts. In 1837, he began on the evaluation of his evolutionary theory. Two events stimulated the progress of his work. The first stimulus came from the ornithologist John Gould (1804–1881) who had examined the (dead and preserved) birds Darwin had collected on the Galapagos Islands. Gould informed him that the finches from different islands were different species not different varieties. Darwin then elaborated a scheme of how their distribution changed from the northern to the southern islands. The second stimulus was a letter from the naturalist Alfred R. Wallace including an article entitled "On the Law Which has Regulated the Introduction of New Species." On two expeditions, one to South America and the other to the Malayan Archipelago, Wallace had acquired a broad knowledge of geology, zoology (notably on entomology), and comparative anatomy. Wallace knew that Darwin was a friend of Lyell, the most famous geologist in England at that time, but he was not informed on Darwin's elaboration of an evolutionary theory. Wallace hoped to obtain from Lyell a positive comment or review of his article.

At that time Darwin had not yet published anything about his theory of evolution. Letters or short articles delivered to friends were the only written documents of his work. Lyell and another friend, the botanist Joseph D. Hooker (1817–1911), pushed Darwin to finish a short summary of the intended book on evolution and they arranged a joined presentation at the Linnean Society on 1 July 1848, followed by publication. However, neither article attracted much attention. On 22 November 1849, Darwin published an improved and expanded version of his treatise under the title *On the Origin of Species by Means of Natural Selection of the Preservation of Favored Races in the Struggle for Life*. The main constituents of his theory can be summarized in five points:

1. Evolution exists and proceeds by transmutation of species
2. All living organisms have the same origin in common
3. All alterations and modifications of races and species proceed slowly and gradually
4. Natural selection is the most important, but not the only effective, mechanism of evolution
5. New species arise and propagate in large populations

The first edition of Darwin's book (1250 copies) was sold out within a few weeks and the book became popular in the following decades. At least among naturalists, the evolution of living organisms (with the exception of humans) was widely accepted, but natural selection in the struggle for live was a revolutionary concept that was rejected by many scientists, even by some friends of Darwin.

Darwin formulated two selection mechanisms. First, slow gradual alterations of races under the influence of environmental changes. Those races that were fittest for survival could propagate more rapidly and effectively and eventually generate new species, whereas other races died out. Second, Darwin was aware that certain animals (notably birds and hoofed animals) had developed features that were not helpful in the daily struggle for life. Yet, these features (e.g., colored feathers, big horns) symbolized health and power and enhanced the sexual attractiveness of the males. As discussed in Sect. 8.4, modern genetics has proved the first selection mechanism to be largely wrong, but in the nineteenth century this concept looked convincing. In an introduction to a modern reprint of Darwin's book, Joseph Carol, professor at the University of Missouri, St. Louis, wrote: "The Origin of Species has special claims on our attention. It is one of the two or three most significant works of all time—one of those works that fundamentally and permanently alter our vision of the world. It is argued with a singularly rigorous consistency, but it is also eloquent, imaginatively evocative, and theoretically compelling."

This hymn is acceptable for the book, but Darwin himself and his work need a more objective and critical comment. First of all, it should be emphasized that Wallace elaborated an almost identical theory at the same time and Darwin profited from Wallace's results and thoughts. Furthermore, Darwin was a careful thinker but not a highly original one. All the main constituents of his theory were picked up from other sources. For instance, a theory of evolution, although less detailed and consistent, had been published by Lamarck five decades before. Lamarck's theory also included a "ladder of complexity" of all organisms, resembling the tree formulated by Darwin. Transmutation of races or species under the influence of the environment was speculatively mentioned in the work of Pierre L. Maupertuis (1698–1759) and later in the work of St. Hilaire. The slow and gradual progress of all alterations and changes was a concept advocated by Lyell for the history of earth's surface. The struggle of life with the survival of the fittest was a concept published by the English economist Thomas R. Malthus (1766–1834) in his book *An Essay on the Perspective of Population*. Darwin confessed in his autobiography: "In October 1838 that is 15 months after I had begun my systematic enquiry, I happened to read for amusement Malthus on population, and being well prepared to appreciate the struggle for existence which everywhere goes on from long-continued observation of the habits of animals and plants, it at once struck me under these circumstances favorable variations would tend to be preserved, and unfavorable ones to be destroyed. The result of this would be the formation of new species. Here then I had at least got a theory by which to work. . ..”

Neither Darwin nor Wallace dared to discuss in their first publications the inevitable consequence that evolution of all living organisms included the descent of men from apes.

Other scientists, notably the German physician and professor of comparative anatomy Ernst Haeckel (1834–1919) drew and published this conclusion before Darwin. He advocated the Darwin–Wallace theory of evolution and in his book *Natürliche Schöpfungsgeschichte* (1868; the English edition, *History of Creation* was published in1876) gave a detailed description and discussion of men’s descent from apes. He used morphology to reconstruct the evolutionary history of life. He also mapped a genealogical tree of humankind beginning with invertebrates and even more primitive organisms. An improved and expanded treatise of this aspect was published in 1894 under the title *Systematische Phylogenie* (Systematic Phylogeny). He coined the term *Pithecanthropus alalus* as the name for the missing link between humans and ancient apes. Haeckel was a highly talented draftsman and painter and his books included impressive illustrations. Therefore, in continental Europe his books were more popular than Darwin’s for several decades.

Darwin was a religious man who had studied theology and this background hindered him for many years in publishing his final conclusion of the theory that was in contradiction to his religious view of humankind. Hence, his book *The Descent of Man and Selection in Relation to Sex* appeared as late as 1871. This book fueled intensive debates between laics and between scientists. His conclusions were rejected by almost all the faithful of the three monotheistic religions, and even most colleagues and friends of Darwin were critical. The crucial point was the driving force of evolution. Was humankind nothing but the result of natural selection and chance events such as mutations that optimized the survival of species? This view, the core of classical Darwinism, was maintained by Darwinians for many decades. The widely known book by the French Nobel Prize laureate Jacques Monod (1916–1976) *Le hasard et la nécessité* (Chance and Necessity), which appeared in 1970, is a typical example.

The opposite view was and is characteristic of religious people who accept evolution in general and evolution of humankind only if evolution has the purpose of generating a humankind that is the crown of creation and God’s image. However, in the Darwin–Wallace theory neither Aristotle’s entelechy nor God’s image play a role. In England, this controversy culminated on 30 June 1860, in a fierce debate between Samuel Wilberforce, Bishop of Oxford, and Captain FitzRoy, on the one hand, and the scientists Hooker and Thomas H. Huxley (called Darwin’s bulldog) on the other. This controversy has continued over decades and still continues in the twenty-first century.

The term “Darwinism,” coined by Huxley, has also survived until the twenty-first century and the evolutionary biologist Ernst Mayr (1904–2005) wrote about it:

"The term Darwinism ... has numerous meanings depending on who has used the term and at what period. A better understanding of the meaning of this term is only one reason to call attention to the composite nature of Darwin's evolutionary thought." Nonetheless, most dictionaries and encyclopedias define Darwinism in a quite similar manner, as demonstrated by the following representative examples:

Webster's College Dictionary: "The Darwinian theory that species originate by descent with slight variation from parent forms through the natural selection of individuals best adapted for survival and reproduction"
Collin's English Dictionary: "The theory of the origin of animals and plant species by evolution through a process of natural selection"
The Gale Group: -ologs & -isms: "The theory of evolution by natural selection of those species best adapted to survive the struggle for life"

These definitions need to be complemented by definitions of the terms "neo-Darwinism" and "Social Darwinism." According to the *Encyclopedia Britannica* the term neo-Darwinism was first used after 1896 to describe the theories of Alfred. R. Wallace and August Weisman (1834–1904) who, in contrast to Darwin, denied the inheritance of acquired characters. In the twentieth century, neo-Darwinism is also used as a label for what is correctly called "synthetic theory," a combination of classical Darwinism and certain results of genetics (see Chap. 9). Yet, not all biologists accept this deviation from the original definition. Social Darwinism may be defined as follows:

Webster's College Dictionary: "Social Darwinism is a nineteenth century doctrine that the social order is a product of natural selection of those persons best suited to existing living conditions."
The American Heritage New Dictionary of Cultural Literacy: "A theory arising in the late nineteenth century that the laws of evolution, which Charles Darwin had observed in nature, also apply to society. Social Darwinism argued that social progress resulted from conflicts in which the fittest or best adapted individuals, or entire societies, would prevail. It gave rise to the slogan "survival of the fittest."

The second definition is partly misleading, because it ignores the fact that Darwin borrowed his concept of "struggle for life and survival of the fittest" from the "social Darwinism" of Malthus. Decisive contributions to the progress of the theory of evolution have come from modern genetics, including a paradigm shift of classical Darwinism, in general, and neo-Darwinism, in particular (see Chap. 9).

Charles Robert Darwin
Charles Darwin was born on 12 February 1809, in Shrewsbury, Shropshire, England, in the house of his parents. His father, Robert Darwin, was a wealthy physician and his mother Susannah belonged to the influential and wealthy Wedgwood family. Although his father was a free thinker, he agreed that Charles Darwin and his five siblings attend the Unitarian Chapel together with their mother. At first he had a private teacher, but he began to attend a day school in Shrewsbury in 1817.

After the death of his mother in the same year, he and his older brother Erasmus entered the Anglican Shrewsbury School as boarders. At the end of 1825, Darwin joined the medical school of the University of Edinburgh, known as the one of the best medical schools in the UK at that time. However, he found the lectures and lessons in surgery distressing rather than stimulating and spent more time learning taxidermy from John Edmonstone, who had accompanied Charles Waterton in the rainforests of South America. Furthermore, he assisted Robert E. Grant in his study of the anatomy and lifecycle of invertebrates in the Firth of Forth. One day, Grant discussed Lamarck's theory of evolution and Darwin was surprised about his audacity, but he had read similar theories in his grandfather's journals. Darwin also learned the classification of plants and helped to improve the collections of the University Museum.

Darwin's father was angry about his neglect of medical studies and sent his son to Cambridge to attend Christ's College to study theology. The bachelor of arts degree should have become the first step in his career as an Anglican parson and theologian. However, Darwin was more interested in riding and shooting, and his cousin William Darwin Fox stimulated his interest in the popular hobby of collecting beetles. He became a friend of Henslow and accompanied him on many excursions. Eventually Darwin focused more and more on his studies and passed the examination in January 1831 with remarkable success. Before and after his examination, he studied William Paley's book *Evidences of Christianity* and *Natural Theology or Evidences of the Existence and Attributes of the Deity* that advocate divine design in nature and God acting through laws of nature. He also read the new book by the astronomer William Herschel (1804–1881), which advised the reader to find laws of nature by inductive conclusions based on observation. Furthermore, he was impressed by Alexander v. Humboldt's *Personal Narrative* about his expeditions. All these books inspired Darwin to make a significant contribution to the progress of science. As preparation for a voyage to Tenerife, he attended geology courses given by Sedgwick, travelled with him to Wales, and learned how to map strata.

When he returned to home on 21 August 1831, he found a letter from Henslow who had proposed him a suitable naturalist and companion of Captain Robert FitzRoy in a self-financed position on the HMS Beagle. Originally, Robert Darwin opposed his son's plan and his opinion was decisive, because he had to finance his son's participation, but he was persuaded to agree by his brother-in-law Joshua Wedgwood. The private funding had the additional benefit that Charles Darwin retained full control over his collection after his return.

A first meeting with Captain FitzRoy was satisfactory for both sides and Darwin was accepted as companion. The expedition lasted almost 5 years and Darwin spent most of the time on land, building up a collection of beetles and other specimens and studying the geology of his environment. At certain time intervals Darwin sent parts of his collection and copies of his notebook to Cambridge to avoid total loss in the case of shipwreck. Henslow conveyed summaries of Darwin's letters to selected naturalists, and thus Darwin had already acquired a high reputation by the time he returned on 2 December 1836. Darwin and Henslow selected the scientists who would have the privilege of studying and analyzing Darwin's collection.

After visiting relatives and friends, Darwin took lodgings in Cambridge in December 1836 to rewrite his journal. He also wrote his first publication on the geology of South America and presented it in January 1837 (supported by Lyell) to the Geological Society of London. On the same day he presented examples of the collected birds and mammals to the Zoological Society. In March, he moved to London and began to work on the transmutation of species, the first step towards his famous book *On the Origin of Species*. Furthermore, he edited and published expert reports on his collection. Henslow helped him to obtain a grant of £1000 (£80,000 in 2014) for writing and editing a multivolume work, *Zoology of the Voyage of the HMS Beagle* and he planned a second book on geology (the third edition was completed in 1846). Furthermore, he accepted the duties as Secretary of the Geological Society in March 1838.

Darwin's health began to deteriorate under this burden of duties and commitments. His doctors advised him to abandon his work for a while and to spend several weeks in the countryside. Visiting his relatives, he came into closer contact with his 9 month older cousin Emma Wedgwood. Emma was an experienced nurse and Darwin felt the need of a female hand. He proposed marriage and purchased a house in London (called Macaw House). They married in January 1839 and had ten children, but not all of them reached adulthood.

However, three of his sons, George, Francis, and Horace, became Fellows of the Royal Society as astronomer, botanist, and engineer. Also in January 1853, Darwin was elected Fellow of the Royal Society and was awarded the Royal Medal. In 1854 he became Fellow of the Linnean Society of London. All his duties and activities caused periods of intensive stress and this stress resulted in bouts of sickness, which continued throughout his life. These bouts of illness included stomach pain, vomiting, severe boils, trembling, headaches, and heart problems. No specific origin of Darwin's illness was discovered and no successful treatment was found.

Darwin decided to spend the rest of his life living in the countryside. In 1842, he purchased a house near the village of Down and avoided any further travel. He became a friend of Hooker (1817–1911) and of the biologist Thomas H. Huxley (1825–1895), who became the most influential supporters of his theory of evolution. Darwin developed in the following years his theory on the descent of man, published under this title in 1871. However, he also worked on a broad variety of other topics, mainly focused on sexual selection and botany. Books covering this part of his activities had the following titles:

Fertilization of Orchids (1862), *The Variation of Animals and Plants under Domestication* (1868), *Insectivorous Plants: The Effects of Cross and Self-Fertilization in the Vegetable Kingdom*, and *The Formation of Vegetable Mould through the Action of Worms*.

In 1882, physician diagnosed angina pectoris, a heart disease based on coronary thrombosis. At that time no promising treatment was known and when Darwin died on 19 April 1882 in Down House, the official diagnose was heart failure. His family intended to bury him in St. Mary's churchyard at Down, but at the request of his colleagues the President of the Royal Society arranged burial in Westminster Abbey. The funeral on 26 April was attended by thousands of people and Alfred R. Wallace accompanied his coffin.

Alfred Russel Wallace

Alfred R. Wallace was born on 8 January 1823, in Llanbadoc near Usk, Monmouthshire (Wales) as the seventh of the nine children of Thomas Wallace and his wife Marie Anne (nee Greenell). His father had graduated in law, but tried to generate his income from inherited property. Yet, due to bad investments the financial situation of the family steadily deteriorated. In 1828, the Wallace family moved to Hartford, the hometown of his grandmother. Here he attended the Grammar School until 1836, when the family withdrew him for financial reasons.

Wallace moved to London, where he lived for a short time together with his brother John until his oldest brother William could take him as an apprentice surveyor. During his stay in London he attended lectures and read books at the Mechanics Institute. He lived and worked with William for 6 years. In 1839, they moved to Kingston, Hereford, and eventually to Neath in Glamorgan, Wales. He worked as land surveyor until 1843, when William ran out of money. After a short time of unemployment he was hired by the College School in Leicester as a teacher of drawing, mapmaking, and land surveying. Wallace spent many evenings in the library and had the opportunity he met the young entomologist Henry Bates. They became friends and Bates stimulated Wallace to collect insects. When his brother William died in March 1845, Wallace left his teaching position to continue William's work, together with his brother John. However, they were unable to overcome the financial difficulties. Wallace then found a position as an engineer in a firm that was surveying for a proposed railroad in the Vale of Neath. This work required spending much time in the countryside, so that he had many opportunities to collect insects. Nonetheless, he persuaded John to found a new architecture and engineering firm, which developed numerous successful projects. From their income they purchased a cottage, where they accommodated their mother and one sister (the father had died in 1843).

Wallace had continued to read scientific books concerning biology and expeditions to foreign countries, such as those of Charles Darwin, William H. Edwards, and Alexander von Humboldt. He exchanged comments on all important books with Henry Bates, and finally both friends decided to undertake a trip to Brazil in 1848. They intended to collect insects and other small animals for their collection and to sell duplicates to collectors in Britain.

During this travel they met the botanist Richard Spruce and Alfred's younger brother Herbert. Although Herbert died 3 years later from yellow fever, Spruce and Bates spent 10 years in South America. Wallace returned after 4 years on the brig Helene. After 26 days at sea, the Helene caught fire and the crew abandoned ship. After 10 days in an open boat they were picked up by another brig and returned to London on 1 October 1852.

Wallace had lost almost all specimens collected during the past 2 years, but he had saved some of his notebooks and drawings. For 18 months Wallace lived from insurance payments and from selling the few specimens he had sent to London before his departure from Brazil. During this time he wrote six scientific articles and two books. He made connections with other British scientists and came into contact with Darwin.

In 1854, Wallace started his second and particularly successful expedition to Asia. He spent 6 years in Indonesia, Malaysia, and Singapore. He collected more than 125,000 specimens including more than 80,000 beetles. Several thousand specimens were unknown at that time. He also observed that animals south and north of a narrow strait were quite different and proposed a zoogeographical boundary, later called the Wallace Line. During this expedition he also gained new insights into the principles of natural selection and in 1858 he sent an article to Darwin describing his theory of evolution. It was published together with an article on Darwin's new theory in the same year. A book describing the most important observations and results of his expedition was published in 1869 under the title *The Malay Archipelago*. It became one of the most popular books of scientific expeditions in the nineteenth century and was praised by Darwin and Lyell.

After his return, Wallace arranged his collection and gave numerous lectures about his expedition to scientific societies. In these lectures and further articles he explained and defended his theory of evolution. He also visited Darwin in Down House and became acquainted with Lyell and Spencer. Furthermore, he met his friend, the botanist R. Spruce, after his return from South America. Spruce introduced him to William Mitten, an expert on mosses, and his family. Wallace made the acquaintance of the daughter Anne Mitten and married her in 1860. They had three children, but the youngest son, Herbert, died at the age of 7. Wallace built a house in Essex where the family lived for 10 years. During his expedition to the Malay Archipelago, Wallace had sent beetles and other objects to England, which had been sold by an agent for a considerable amount of money. Yet, after his return, Wallace made several bad investments in railways and mines and lost most of his money.

Although assisted by friends Wallace was unable to find an acceptable permanent position such as a curatorship in a museum. He generated an insecure income from various resources, such as grading government examinations, royalties from published articles in scientific journals, and helping Lyell and Darwin edit their own papers. Darwin, who knew Wallace's poor financial situation, eventually achieved a government pension for him. When this annual pension was awarded in 1881, it formed a small but reliable basis for the rest of his life. Around 1903, Wallace built a country house called "Old Orchard" near Broadstone, Dorset. There he spent the rest of his life until his death on 7 November 1913 (aged 90). Some of his friends tried to arrange a burial in Westminster Abbey, but his wife insisted on obeying his wish and had him buried in the cemetery of Broadstone. However, his friends founded a committee and had a medallion of Wallace placed in Westminster Abbey close to the burial site of Darwin.

Wallace was not only active as a biologist but after his return from the second expedition he also became an active socialist. In 1918, he was elected the first president of the Land Nationalization Society. He wrote political articles on social issues and published books such as *Land Nationalism: Its Necessity and Aims* (1882), *The Wonderful Century: Its Success and its Failures* (1898), and *The Revolt of Democracy* (1913).

His main merit was, of course, the elaboration of a theory of evolution, which agreed largely with that of Darwin but had a superior understanding of the selection mechanisms. However, Wallace was also recognized as one of the first, and perhaps most important, founders of a new biological discipline, the biogeography of animals. He drew maps and wrote comments on environmental factors influencing the distribution of animals. His studies and theories also include paleontology, considering the migration of species over the past 1000 years. He summarized his results and insights 1876 in a book entitled *The Geographical Distribution of Animals*. In this work he wrote: "We live in a zoological impoverished world from which all the hugest, and fiercest and strangest forms have recently disappeared." A supplementary book entitled *Island Life* followed in 1880 and his final work *World of Life* in 1911. Wallace also became famous for his hypotheses and comments on astrobiology. His book *Man's Place in the Universe* was the first serious discussion of life on other planets. Even more famous was his book *Is Mars Habitable?* In this treatise he collected and explained arguments against the theory of the British astronomer Percy Lowell, who had postulated that intelligent organisms had created a network of canals on Mars.

Wallace received many honors before and after his death. Examples include the Royal Medal of the Royal Society (1868), the Darwin Medal (1870), the Linnean's Society Gold Medal (1892), the Geographical Society's Founders Medal (1892), the Order of Merit (1908), and the Darwin–Wallace Medal (1908). He was elected head of the anthropology section of the British Association in 1886, president of the entomological Society of London in 1870, head of the biology section of the British Association in 1876, and Fellow of the Royal Society in 1893.

Several buildings in the UK, a crater on the Moon, and one on Mars are named after him.

Bibliography

Bannister RC (1989) Social Darwinism: science and myth in Anglo-American social thought. Temple University Press, Philadelphia

Barlow N (ed) (1958) The autobiography of Charles Darwin 1809-1882 with the original omissions restored; edited with appendix and notes by his granddaughter Nora Barlow. Collins, London

Benton T (2013) Alfred Russel Wallace—explorer, revolutionist, public intellectual: a thinker for our own times. Siri Scientific Press, Manchester

Bowler PJ (2003) Evolution: The history of an idea, 3rd edn. University of California Press

Browne JE (1995) Charles Darwin: vol. 1. Voyages. Jonathan Cape, London

Browne JE (2002) Charles Darwin: vol. 2. The power of place. Jonathan Cape, London

Burnet J (1964) Greek philosophy: Thales to Plato. McMillan, London, Reprint of Forgotten Books 2012

Costa J (2014) Wallace, Darwin and the origin of species. Harvard University Press, Cambridge MA
Coyne JA (2009) Why evolution is true. Oxford University Press
Fichman M (2004) An elusive Victorian: the evolution of Alfred R. Wallace. University of Chicago Press, Chicago
Glass B (1959) Forerunners of Darwin. John Hopkins University Press, Baltimore
Hodge C. What is Darwinism?: http://www.gutenberg.org/files/19192/19192-8.txt
Joel H. What is Darwinism?: http://www.talkorigins.org/faqs/darwinism.html
Johnson PE. What is Darwinism?: http://www.arn.org/docs/johnson/id.htm
van Wyhe J (ed) (2002) The complete works of Charles Darwin online. http://darwin-online.org.uk/biography.html
van Wyhe J (ed) (2012) Wallace online. The first complete online edition of the writings: http://wallace-online.org/

8.4 Genetics and Darwinism

How often it is the error which has to bring the breakthrough to new knowledge.
(K. H. Bauer)

The term "genetics" is derived from the Greek *γενεσισ* (genesis) meaning origin. The adjective *genetisch* (genetic) was already used by the German writer and naturalist Johann W. von Goethe (1749–1832) in the beginning of the nineteenth century. In the following decades, this adjective was applied by scientists to studies of the ontology of living organisms, so that its meaning was different from what it means today. In a publication of 1906, the British botanist William Bateson (1861–1926) coined the term genetics as a label for a new research area in biology. The roots of this new discipline go back to the work of the Austrian scientist Johann Mendel (1822–1884), who had studied mathematics and natural sciences in Vienna. After becoming an Augustinian monk in Brünn (today Brno in Czech Republic) he was given the name Gregor. In addition to his teaching duties, he worked in the garden of the monastery on the crossbreeding of plant hybrids, notably of peas. He studied thousands of plants over a period of 8 years. He found hereditary rules that contradicted the doctrines of that time that purported hybrids cannot create stable new forms (species) and individuals inherit a smooth blend of traits from their parents.

In 1866, Mendel published his results in the *Proceedings of the National Science Society of Brünn* under the title "Experiments on Plant Hybrids." Although this treatise was mentioned in the *Encyclopedia Britannica*, his work was ignored or misunderstood in the following decades. After 1900, the biologists Hugo de Vries (1848–1935), Carl Correns (1864–1933), and Erich Tschermak (1871–1962) published new studies that confirmed Mendel's results and made his work public. There followed rapid progress of the new discipline genetics, a detailed description of which is outside the scope of this chapter. However, a number of highlights and landmarks should be mentioned.

Modern genetics can be subdivided into four directions:

1. Classical genetics, which follows the footprints of Mendel, studying the transmission of certain traits in crossbreeding experiments.
2. Population genetics, which concerns the changes and distribution of allele frequencies in a population in relation to the four evolutionary processes: natural selection, gene flow, mutation, and gene drift. It also includes studies of the influence of recombination, population subdivision, and population structure.
3. Molecular genetics, part of molecular biology, which is concerned with the molecular (chemical) basis of inheritance. The structure, chemical reactions, and biological functions of deoxyribonucleic acids (DNAs) are the focus of this research area.
4. Epigenetics, which explores environmental influences on the regulation of gene activities. It refers to heritable changes in gene expression that do not modify the DNA, in other words, changes in the phenotype without changes in the genotype.

Research on the chemical structure and three-dimensional order of genes and chromosomes has a tradition of almost 150 years, with contributions from numerous biologists, chemists, and physicians. In 1869, the German physician Friedrich Miescher (1844–1895) isolated from the nuclei of (eukaryotic) cells a component that he called "nuclein." Friedrich Altmann demonstrated in 1884 that this nuclein consists of an acidic component and a basic protein, a histone. One year before, the German professor of anatomy Heinrich W. Waldeyer (1836–1921) observed under the microscope thread-like components of cell nuclei that became detectable after staining and called them "chromosomes" (colored bodies). In the years 1903 and 1904, the American physician Walter Sutton (1877–1916) and the German biologist Theodor Boveri (1862–1915) published, independently, results that founded the chromosome theory, which states that chromosomes are the material basis of heredity.

In 1907, the American biologist Thomas H. Morgan (1866–1944, Nobel Prize 1933) began genetic studies of the fruit fly (*Drosophila melanogaster*), which became the most intensively studied organism in genetics over the following decades. Morgan was able to demonstrate that chromosomes are an array of hundreds or even of thousands of genes that carry information on protein syntheses. In 1928, the British bacteriologist Frederik Griffith (1877–1953) discovered that both harmless and infectious bacteria (e. g., *Streptomyces pneumonia*) can exchange genetic information, a process called "transformation." The Canadian physician Oswald T. Avery (1877–1953) and his coworkers proved in 1944 that in higher organisms DNA is the only class of biopolymers which carries genetic information.

The DNA strands are made up of sequences of building blocks, called nucleotides, which consist of phosphate groups, the saccharide (sugar) deoxyribose, and a small nitrogen-containing cyclic molecule, the nucleobase. The nucleobase is laterally attached to the saccharide unit. DNA contains four different nucleobases called thymine (T), cytosine (C), adenine (A) and guanine (G). In 1950, the Austro-

Hungarian biochemist Erwin Chargaff (1905–2002) discovered that the numbers of T and A or G and C are identical, whereas the T/C and A/G ratios may vary. T and A or G and C form pairs between two neighboring DNA chains via hydrogen bridges that stabilize an antiparallel arrangement of complementary DNAs. The three-dimensional structure, called a double helix, resembles that of an extremely long corkscrew. The American biologist James Watson (born 1928) and the British biologist Francis Crick (1916–2004) were awarded the Nobel Prize in Medicine in 1953 for the elucidation of this unusual molecular architecture. The sequence of the four different nucleobases along the DNA chain carries the information for the sequence of α-amino acids in proteins. A sequence of three nucleobases (codon) represents the code for one amino acid.

The vast majority of the organic materials in the human body consist almost exclusively of peptides and proteins, which are composed of 20 different α-amino acids. Hair and muscles, toenails and blood vessels, tendons and lungs, heart and nose, and all the enzymes (the biocatalysts that enable all biochemical reactions) consist of proteins. Hence, growth and survival of the human body requires that the syntheses of all proteins occur at the right time in an optimum quantity. The information needed for their syntheses is provided by the DNA of a gene when the cell stimulates its delivery. The synthesis of a protein is performed in a complex molecular machine called the ribosome, which consists of proteins (e.g., enzymes) and special ribonucleic acids (ribosomal RNAs). RNAs differs from DNAs in the structure of the saccharide, because the ribose of RNA contains one more -OH group than the deoxyribose of DNA. One sort of RNA (messenger RNA) copies the information of the DNA and another type of RNA (transfer RNA) carries activated amino acids to the ribosomes. After the discovery of the double helix structure of DNA, Crick said "We have discovered the secret of life." This statement is an extreme exaggeration because the production of proteins, RNA, and enzymes (i.e., proteins) are of equal importance and the secret of life is contained in the organization of the gene regulation of the entire metabolism. This secret still remains unveiled and elucidation of the chemical structures and reactivities of individual components of metabolism has not solved this problem (see Sect. 4.4).

Until the 1960s many biologists adhered to a paradigm of genetics that was based on the following three rules or dogmas:

1. One gene codes the information for one type of protein. This dogma was established by the American geneticists George Beadle (1903–1989) and Edward Tatum (1909–1979) in the year 1940. For this dogma, which was later revised, they were awarded the Nobel Prize in Medicine in 1958.
2. The flow of information proceeds from DNA to (messenger) RNA and from RNA to proteins, but never in the opposite direction. This so-called central dogma of genetics was coined by Crick in 1958.
3. Genes never change the position they have in a chromosome.

Modern Darwinism is called "evolutionary synthesis," "modern synthesis," "new synthesis" or (incorrectly) "neo-Darwinism" and includes those dogmas. One of those terms was invented by the British zoologist Julian Huxley

(1887–1975) in the title of his book *Evolution: The Modern Synthesis*. Modern synthesis arose before 1950 from contributions by several prominent biologists. It combines the classical theory of Darwin, based on natural selection and gradual changes of race favoring survival of the fittest for life, with newer results from classical genetics, molecular genetics, population genetics, ecology, and paleontology. Evolutionary synthesis, which is still the prevailing paradigm at the beginning of the twenty-first century, is again based on three dogmas:

- Genetic changes of species proceed slowly and gradually in a quasi-linear mode.
- Modifications of extant species having the potential to create new species exclusively proceed under the influence of accidental events with regard to both quality and timing.
- The role of selection is interpreted in terms of maximum propagation, which is considered to be decisive for survival of the species. Hence, successful selection exclusively favors those changes and modifications or organisms that enhance the efficiency of propagation.

Various research activities, notably genetic research, during the past 60 years have falsified the above-mentioned dogmas of genetics and Darwinism. The paradigm shift was induced by the American biologist Barbara McClintock (see biography). At a conference in Cold Spring Harbor (Long Island, USA) in 1951 she presented revolutionary results concerning the reactivity of maize genes. She had discovered that genes can be transferred from their original chromosome onto another chromosome or from one location to another on the same chromosome. McClintock called those molecules initiating and supporting the transfer of genes "controlling elements" (later the term "transposons" was introduced). As expected for a revolutionary paradigm shift (see Chap. 6), her presentation and publications provoked skeptical, critical, and even hostile comments. In 1953, she published a summary of all her new results in the journal *Genetics*, but afterwards she stopped publishing (but not her experimental work) on gene regulation to avoid further confrontation with colleagues. After 1960, more and more papers by other scientists appeared that dealt with genetic regulation mechanisms in other organisms, such as yeast or bacteria. Now, McClintock again began to publish on gene transposition and gained increasing international recognition. Eventually, she received the Nobel Prize in Medicine in 1983.

The paradigm shift was accelerated and completed by sequence analyses of single DNA strands and complete genomes. Sequencing of DNA and RNA (or proteins) is based on a combination of chemical, enzymatic, and chromatographic methods that allow determination of the sequence of building blocks (monomer units) from one chain end to the other. For the sequencing of DNA, two different methods were developed. The first approach was elaborated by Allan Maxam (born 1942) and Walter Gilbert (born 1932) and a second method in 1975 by Frederik Sanger (born 1918). Sanger received two Nobel prizes, in 1957 for sequence analyses of proteins and in 1980 for the sequencing of DNA. After several improvements, the "Sanger methods" proved suitable for computer-aided automation.

Based on the automated Sanger method it was possible to analyze the complete genome of a primitive bacterium (*Haemophilus influenzae*) in 1995. The sequence analysis of the yeast genome followed in 1997 and the human genome was deciphered in the subsequent 10 years. All the genetic data of the human genome elaborated by numerous scientists in an international cooperation (Human Genome Project) have been published on the internet since 2003. Comparisons of the human genome with those of various animals will enable scientists in the future to delineate the course of evolution and to establish a reliable phylogenetic tree.

Analysis of the human genome reveled that less than 2 % of all DNA located in the cellular nucleus carries information for protein syntheses. However, this small percentage corresponds to approximately 23,000 different genes that, depending on the circumstances, provide the information for 34,000 different proteins. The vast majority of DNA chains do not in fact act as genes in the original sense and were considered to be "junk DNA." This junk DNA was understood to be either a useless remainder of evolution or selfish genes that had no purpose other than their reproduction, as advocated by the British Darwinian Richard Dawkins (born 1941).

Quite recently, it was discovered that at least 50 % of human DNA plays a role as transposons that are active in the regulation of gene activities and in the reorganization of the genome when adaption of the cell to environmental changes becomes necessary. The percentage of transposons in the entire DNA of a genome may vary largely for different species. In frogs (*Rana esculenta*), a high percentage of transposons (75 %) is known whereas in the case of yeast (*Saccharomyces cerevisiae*) the percentage is a low as 3–5 %. The transposons play a particularly important role in the modification of the genome under environmental stress. Their activities include:

(a) Transfer of genes or copies of a gene from one position to another (as discovered by McClintock)
(b) Changing the position of parts of a chromosome (containing several genes) in the genome
(c) Fusion of two genes or gene segments to create a new gene
(d) Combination of genes with new promoters, which decide when and how often a gene is activated for protein syntheses

The transposons of various animals agree in all basic features. The transposons of human and mammalian cells can be subdivided into three groups having different functions. However, all transposons serve the purpose of enabling, if necessary, the transformation of an entire genome in such a way that new races and eventually new species emerge. The trend of these transformations (the direction of the evolution) is not influenced by accidental events such as mutations (see below). It is the entire cell that stimulates the transposons to start their activities via oligopeptides or proteins that play the role of signal molecules. Those genes needed for the basic functions of a cell, the so-called hox-genes, are protected against an attack of transposons. Furthermore, when a transposon is not needed for a while, its activity is blocked by the cell.

An important aspect of this concept of organized cooperation is the necessity that information flows in all directions, for instance, from protein to RNA and from RNA to DNA, and not only in one direction. This is in contrast to the aforementioned central dogma. The cooperative modification of genomes enables the cell to respond to slow or rapid changes in the environment. Modification and transformation of genes obey three fundamental principles, which are characteristic for the biosphere: cooperativity, communication, and creativity. In contradiction to the statement by Crick, the secret of life is not a certain class of molecules, it is the coordination of all metabolic activities, including modification of the genome. In the course of her Nobel Prize lecture, McClintock summarized this insight in the following statement: "The cell makes wise decisions and acts upon them."

This view is also in contradiction to any kind of Darwinism and, thus, in contradiction to Dawkins concept of the "selfish gene" (title of his book first published in 1971). He held that selfishness is a fundamental principle of life and he locates selfishness in the genes. Genomes and cells are machines constructed for an optimized reproduction of genes. In his own words (1989 edition, p. 4): "If you look at the way natural selection works, it seems to follow that anything that has evolved by natural selection should be selfish." On p. 11 he states that "I shall argue that the fundamental unit of selection, and therefore of self-interest, is not the species, nor the group, not even strictly the individual, it is the gene, the unit of heredity." On p. 21, he writes the following:

> Different sorts of survival machines appear very varied on the outside and in their internal organs. An octopus is nothing like a mouse, and both are quite different from an oak tree. Yet, in their fundamental chemistry they are rather uniform, and, in particular, the replicators that they bear, the genes, are basically the same kind of molecule in all of us—from bacteria to elephants. We are all survival machines for the same kind of replicators—molecules called DNA—but there are many different ways of making a living in the world, and the replicators have built a vast range of machines to exploit them. A monkey is a machine that preserves genes up trees, a fish is a machine that preserves genes in water; there is even a small worm that preserves genes in German beer mats.

In Darwinian theory, mutations play a decisive role in the course of evolution and, thus, for the evolution of humankind. This view was exemplarily presented in another widely known book, published in 1971 by the French biologist and Nobel Prize laureate Jacques Monod (1910–1976) under the title *Le hasard et la nécessité* (Chance and Necessity). He wrote (Vintage Books 1972) on p. 21: "The cornerstone of the scientific method is the postulate that nature is objective. In other words, the systematic denial that true knowledge can be got by interpreting phenomena in terms . . . of purpose.." On p. 145/146, he writes: "The universe was not pregnant with life nor the biosphere with man. Our number came up in the Monte Carlo game." He continues on page 172/173: "If he [man] accepts this message—accepts all it contains—then man must at least wake out of his millenary dream; and in doing so, wake up to his total solitude, his fundamental isolation. Now does he at least realize that, like a gypsy he lives on the boundary of an alien world." Then, on p. 180 Monod states: "The ancient covenant is in pieces; man knows at last that he is alone in the universe's unfeeling immensity, out of which he emerged only by chance." With these statements and with the book title Monod followed the footsteps of Demokritos (460–370 B.C.) who said: "Everything existing in the universe is the fruit of chance and necessity."

The term "mutation" was introduced into the language of genetics by the biologist Hugo de Vries shortly after 1900. A punctual mutation is today defined as the destruction of a nucleotide or pair of nucleotides (forming hydrogen bonds) in a DNA sequence. Mutations can arise from two different sources:

I. Endogenic processes: this term means that errors in the formation of base pairs can occur when two complementary (antiparallel) strands of DNA recombine. Recombination is an extremely frequent process, because copying of DNA information requires temporary separation of the complementary DNA strands of a double helix.
II. Exogenic processes: This means damage of DNA caused by external events such as:

    (a) UV radiation, which, depending on the frequency (see Sect. 10.4), can destroy any kind of chemical bond
    (b) Radioactive radiation, which originates either from space or from the decay of labile elements in the earth's crust
    (c) Aggressive chemicals, some of which are called cancerogenic

A cell cannot hinder mutations from occurring, but it can control their effect. Those DNA sequences that, like the hox-genes, should remain unchanged are subject to rapid and effective repair.

In those DNA segments that are subject to transformation under the influence of transposons, mutations are welcome to accelerate modifications of the gene and to enhance diversification. Cells frequently utilize mutated DNA sequences via combination with analogous unchanged gene sequences.

This process is usually combined with doubling of the gene. The doubling of intensively used genes is typically the first response of a cell to environmental stress. Gene doubling in combination with mutations was also characteristic for the first stage of the evolution from unicellular organisms to multicellular ones about 600 million years ago. The same process was again responsible for the so-called Cambrian Explosion 540–560 million years ago, which yielded a broad variety of species within a short period of time. In contrast to modern genetics, Darwinism does not yield a satisfactory explanation for these dramatic events in the history of evolution.

In summary, modern genetics combines two paradigm shifts in the Darwinian theory of evolution: First, it demonstrates that the mechanism yielding new species is based on cooperative reorganization of genomes, and second, it changes the role of accidental events (e.g., mutations) from a God-like position, where they decide if and in what direction evolution should proceed, to that of an engine accelerating the modification of the genome.

Barbara McClintock

Barbara (Eleanor) McClintock was born on 16 June 1902, in Hartford, Connecticut. She was the third of four children born to the physician Thomas H. McClintock and his wife Sara Handy (nee Grace). When she was a young girl, her parents decided that her name Eleanor was too feminine for her and they began to call her Barbara. She proved to be a solitary and independent child, and she later described this trait in her character as her "capacity to be alone." She had a difficult relationship with her mother, and because four children were a heavy financial burden for the young physician Thomas McClintock, her parents accepted the offer of an aunt and uncle to care for her education in Brooklyn, New York. The McClintock family followed her in 1908.

Barbara McClintock attended the Erasmus Hall High School in Brooklyn and graduated in 1919. Supported by her father, but in conflict with her mother, she decided to study botany at the College of Agriculture of Cornell University in Ithaca, New York. In addition to her studies of botany she became interested in student government and in jazz music. Her interest in genetics was stimulated in 1921 during a first course in that relatively new working field. This course and a similar course at Harvard University were taught by C. B. Hutchison, a geneticist and plant breeder. He was aware of McClintock's extraordinary interest and talent and invited her to participate in a graduate genetics course at Cornell in 1922. McClintock reported much later:"Obviously this telephone call cast the die for my future. I remained with genetics thereafter." She received a bachelor's degree in 1923.

During her graduate and postgraduate studies McClintock worked in a research group that was focused on the new field of cytogenetics in maize. This group included Marcus Rhoads, Harriet Creighton, and George Beadle (1903–1989, Nobel Prize in Physiology or Medicine 1958, shared with E. L. Tatum and J. Lederberg) and was supported by Rollins A. Emerson, head of the Plant Breeding Department.

McClintock's studies of maize chromosomes resulted in several important discoveries. The pertinent publications enhanced her recognition as an expert in genetics and she was awarded several postdoctoral fellowships from the National Research Council. On the basis of these fellowships she continued to work at Cornell in the years 1929–1934. A fellowship of the Guggenheim Foundation allowed her in 1934 to work for 6 months in Berlin under the supervision of the geneticist Richard B. Goldschmidt, head of the genetics department of the Kaiser-Wilhelm-Institute. After her return to the USA, a grant of the Rockefeller Foundation, obtained with the help of Emerson, allowed her to continue with cytogenetic research on maize at Cornell.

In 1936, the geneticist Lewis Stadler, full professor at the University of Missouri, Columbia, offered her an assistant professor position he had created for her. McClintock accepted this offer and worked in this position until spring 1941. During this time she worked with Stadler who taught her to use X-rays as mutagen. Despite further progress and recognition of her work (she was elected vice-president of the Genetics Society of America) she felt her position was in jeopardy when Stadler left for a new position in California. Therefore, she decided to leave the University of Missouri in spring 1941 and to accept a visiting professorship at Columbia University offered by her former Cornell colleague M. Rhoades, who had meanwhile become full professor at Columbia University.

Decisive progress in her career resulted from an invitation from Milislaw Demerec, the new director of the Cold Spring Harbor Laboratories of Genetics, a foundation of the Carnegie Institution in Washington DC. After 1 year on a temporary position, she was appointed a full-time research position, the first permanent position in her career. Because of her continuing success, she became in 1944 the first female president of the Genetic Society of America and was elected to the National Academy of Sciences. Following a suggestion of Beadle, who invited her for several months to Stanford University, she undertook cytogenetic studies of the fungus *Neurospora crassa*. Based on her work, *N. crassa* has since become a model species for classical genetics.

In the second half of 1944, McClintock began systematic studies of the mosaic color patterns of maize seeds. In the years between 1945 and 1950 she discovered that the message of genes is not stable over many generations, in contrast to the established doctrine. She discovered so-called controlling elements that were mobile and capable of inhibiting or modulating the activities of genes. She developed the hypothesis that the regulation of genes by transposons could explain why complex multicellular organisms made of cells with identical genomes can have cells of different function. When she presented her revolutionary results in publications or lectures, she stirred skeptical or critical reactions from colleagues. Therefore, in 1953 she stopped publishing on this topic, but continued her research activities.

In 1967, McClintock officially retired from her duties at the Cold Spring Harbor laboratory and became a Distinguished Service Member of the Carnegie Institution This position allowed her to continue research as "scientist emerita" together with colleagues and postdoctoral students.

McClintock's work on transposons was ignored for almost 10 years until other scientists, such as Francois Jacob and Jacques Monod (1910–1976), described similar observations for other organisms. With every new publication on transposons McClintock's reputation increased and culminated in the Nobel Prize for Physiology or Medicine in 1983. She was the first woman to earn the prize unshared. Even before the Nobel Prize, she had received numerous honors:

In 1959 she was elected Fellow of the American Academy of Arts and Sciences
In 1967 she received the Kimber Genetics Award
In 1970 she was given the National Medal of Science by President Richard Nixon
In 1973 the Cold Spring Harbor laboratory named a building after her
In 1978 she received the Louis and Bert Freedman Foundation Award and the Louis S. Rosenstiel Award
In 1981 she was awarded the Albert Lasker Award for Basic Medical Research, the Wolff Prize in Medicine, and the Thomas H. Morgan Medal
In 1982 she obtained the L. Gross Horwitz Prize from Columbia University.

In 1986 she was introduced into the National Woman's Hall of Fame.
In 1989 she was elected a Foreign Member of the Royal Society.
In 1993 McClintock received the Benjamin Franklin Medal.
She was awarded 14 honorary doctoral degrees.
In the final years McClintock led a more public life, in as much as Evelyn Fox Keller published a biography of her in 1983. She maintained a regular presence in the Cold Spring Harbor community and died on 2 September 1992, in Huntington, NY, at the age of 90.

Bibliography

Alberts B, Bray D, Hopkin K, Johnson A, Lewis J, Raff M, Roberts K, Walter P (2013) Essential cell biology, 4th edn. Garland Science
Bowler P (2003) Evolution: history of an idea. University of California Press, Berkeley
Coe E, Kass LB (2005) Proof of physical exchange of genes on the chromosomes. Proc Natl Acad Sci 102(19):6641
Comfort NC (1999) The real point is control: the reception of B. McClintock's controlling elements. J Hist Biol 32(1):133
Dawkins R (1976) The selfish gene. Oxford University Press, Oxford, New York
Fedoroff NV (1994) Barbara McClintock: 16 June 1902–2 September 1992. Biographical memoirs V68. The National Academic Press, National Research Council, Washington, DC
Futuyma DJ (1998) Evolutionary biology, 3rd edn. Sinauer Associates, Sunderland, MA
Kin RC, Mulligan PK, Stansfield WD (2013) A dictionary of genetics, 8th edn. Oxford University Press, Oxford.
Margulis L, Sagan D (2002) Acquiring genomes: a theory of the origin of species. Basic Books, New York
Monod J (1971) Chance and necessity: an essay on the natural philosophy of modern biology. Alfred A. Knopf, NY and Vintage Books, 1972
Stanley AR (2009) Encyclopedia of evolution. Infobase Publishing

8.5 The Piltdown Man

> In the history of mankind the most instructive chapter both for heart and soul are the annals of man's mistakes.
> (Friedrich Schiller)

The so-called Piltdown man consists of a few bone splinters and several fragments of a skull that were found between 1908 and 1912 in a gravel pit close to the village Piltdown near Uckfield, East Sussex (UK). The discoverer was Charles Dawson, a lawyer and hobby archeologist. He reported at a meeting of the Geological Society of London in December 1912 that a workman at the gravel pit had given him a fragment of a skull in 1908. Dawson conjectured that he had the skull of an early human in hand, revisited the gravel pit several times in the following years, and found more fragments of a skull. Dawson took this collection of fragments to Arthur Smith-Woodward (1864–1944), who was head of the geological department of the British Museum at that time. Woodward was highly interested in Dawson's finds and accompanied him several times in June and September 1912 to the gravel pit. Nonetheless, Dawson alone found further fragments including part of a jaw bone. With exception of the first skull fragment, which was apparently discovered in situ, all other finds were made in the spoil heaps of the pit.

At the same meeting on 18 December 1912, Woodward presented the conclusion he had drawn from the British Museum's reconstruction of the skull. The skull was in various aspects quite similar to that of a modern human, but with two significant exceptions. The part that sits on the spinal column was different and, particularly conspicuous, the size of the brain was only about two-thirds that of a modern human. Furthermore, both molar teeth resembled those of a modern human, and other teeth had not been found at that time. Surprisingly, the jaw bone looked rather like that of a chimpanzee. In summary, Woodward concluded that all skull fragments together belonged to an early human who played the role of a missing link in the evolution of *Homo sapiens* from early apes. This message provoked a worldwide echo, with numerous positive and few skeptic comments. The public opinion in England was particularly positive, not to say enthusiastic, for two reasons. First, a well-developed cranium in combination with an ape-like jaw supported the then prevailing hypothesis that the "evolution of man had begun with the evolution of the brain." Second, the British nation was proud to have an ancestor with a highly developed brain. This new ancestor obtained the scientific name *Eoanthropus dawsoni* (Dawson's dawn man) with the nickname "Piltdown Man."

The worldwide acceptance of the new ancestor of humankind was not only based on the positive comments of journalists, even experts from other countries signaled their agreement with Woodard's reconstruction and with his interpretation of the finds. For instance, the highly reputed German paleontologist Otto Schoetensack who had found the man of Maur (near Heidelberg) in 1907, hailed the Piltdown fragments as an excellent piece of evidence for an ape-like ancestor of modern humans. Furthermore, it must be taken into account that only few examples of bones of early human were known at that time, namely the Neanderthal man (1856), the Java man (1891), and the man of Maur (1907), so that bone or skull fragments that allowed a detailed comparison were scarce at that time. The optimistic resonance of European experts also had an emotional background. The Piltdown man supported the hypothesis that evolution was accompanied by three particularly important achievements: upright standing and walking, invention of tools, and development of language. The enthusiasm in England also had two negative consequences: British paleontologists ignored for almost 40 years new finds in other parts of the world, for instance the discovery of remains of *Australopithecus*. Second, the skeptical voices in England itself were ignored.

However, soon after Woodward's presentation of the Piltdown man in 1912, skeptical and even critical comments were uttered by several experts. For example, David Waterston, member of King's College in London, launched an article in

Nature claiming that the Piltdown head was composed of a human cranium and an ape mandible. In 1915, the French paleontologist Marcellin Boule published the same conclusion. The American zoologist Gerrit Smith Miller argued that the jaw originated from a fossil ape. However, the first and most serious critique came in 1913 from Prof. Arthur Keith (later Sir Arthur Keith).

He and his coworkers reconstructed at the Royal College of Surgeons a skull from exact copies of the fragments that Woodward had in hand, but they arrived at a quite different model. Because this model very much resembled a modern human, it was called *Homo piltdownensis* to emphasize the difference from *Eoanthropus dawsoni*.

A particularly crucial aspect of Woodward's reconstruction, in contrast to Keith's model, were the long canine teeth. To overcome the critique of this point, in August 1913 Dawson and Woodward began a systematic search for further teeth. The third member of this group was the French Jesuit Pierre Teilhard de Chardin, who spent a lot of time in the neighborhood of Piltdown. He was a trained paleontologist and geologist who had also accompanied Dawson several times before 1913. It was Teilhard who soon found a canine exactly fitting the previously found jaw. Interestingly, Teilhard returned to France immediately after his find and never cared about the "fate" of the Piltdown man. Dawson and Woodward were happy, hoping to end the discussion on their reconstruction. However, Keith criticized that the canine was too high. Evolution of the molars in a human jaw was the result of sideward motions upon chewing. Yet, high canines prevent this chewing motion unless they are not higher than the molars. At the next meeting of the Royal Society, Grafton Elliot Smith, a fellow anthropologist in Woodward's team, accused Keith of not being objective, but exclusively motivated by ambition. This affront ended the personal contact and disputes between the two groups.

In 1915, Dawson made public that he had found three fragments of a second scull at a site 2 miles away from the gravel pit. However, Dawson neither described the exact site nor the circumstances of his new find. Furthermore, he did not present these fragments to the Royal Society, but handed them over to Woodward. Woodward presented the Piltdown II fragments 5 months after Dawson's death in 1916 and mentioned that he knew where they had been found. This so-called Sheffield Park find dampened the criticism, at least in England, for a long time. The jaw of an ape and the skull of a human might have come together in a gravel pit by chance, but the same event 2 miles away was absolutely unlikely. In 1921, the President of the American Museum of History, Henry Fairfield Osborne, examined the Piltdown and Sheffield Park finds and declared that the jaws and skulls belonged together and that the Sheffield Park fragments "were exactly those which we should have selected to confirm the comparison with the original type."

Sir Arthur Keith was not completely convinced, but he did not have new arguments for his criticism. Therefore, he unveiled on 23 July 1938, a memorial at Barkham Manor, Piltdown, to mark the site where Dawson had discovered Piltdown man. Keith finished his speech with the following words:

> So long as man is interested in his long past history, in the vicissitudes which our early forerunners passed through, and the varying fare which overtook them, the name of Charles Dawson is certain of remembrance. We do well to link his name to this picturesque corner of Sussex—the same of his discovery. I have now the honor of unveiling this monolith to his memory.

The following inscription was carved on the memorial stone:

> Here in the old river gravel Mr. Charles Dawson, FSA, found the fossil skull of Piltdown Man, 1912–1913. The discovery was described by Charles Dawson and Sir Arthur Smith-Woodward, *Quarterly Journal of the Geographical Society*, 1913–15.

In connection with this event, the nearby public house changed its name to "Piltdown Man." Yet, even this event did not bring an end to the story. The interpretation of Dawson's finds remained a challenge to experts. In 1923, the German Franz Weidenreich studied the remains and reported that they were a combination of a modern human cranium with an orangutan jaw containing filed-down teeth. With the steadily increasing number of fossils belonging to various types of early men, more and more scientists began to interpret the Piltdown man as an enigmatic aberration from the normal path of human evolution.

The big bang came on 21 November 1953, in the form of an article published in *Time*. Kenneth P. Oakley, Sir Wilfrid E. Le Gros Clark, and Joseph Weiner presented unambiguous evidence of forgery. They proved that the Piltdown skull was composed of parts from three different species: a human cranium having a maximum age of 1000 years, the jaw of a 500-year-old orangutan, and teeth of a chimpanzee. These teeth had been filed to match the shape of human teeth and they were treated with solutions of iron ions and chromic acid to achieve an "antique" color. Keith Wahley, chief archeologist of the Natural History Museum, had made a decisive contribution to the elucidation of this forgery. A new analytical method allowing the determination of the age of antique objects via their fluoride content brought the success. This new method was invented in 1950, and its results were later confirmed by radiocarbon measurements. The oldest fragment of the Piltdown finds was not older than 1000 years. These results were an enormous disaster for all the experts who had accepted that the Piltdown man was a link in the evolution of *Homo sapiens*, but Woodward and other contemporaries of Dawson had already passed away.

Once again, the story of the Piltdown man had reached a new stage, but not its end. Now the public interest and that of experts concentrated on the question of who the forger was. Numerous people were suspected, in the first place Charles Dawson. In addition to him, Teilhard de Chardin, A. C. Hinton, Horace de Vere Cole, Arthur Conan Doyle (author of the Sherlock Holmes stories), and even Sir Arthur Keith were suspected. Teilhard had traveled in Africa, where one of the fragments originated, he resided frequently in the Weldon area since the date of the first find, he discovered the much desired canine, and disappeared immediately afterwards. A. C. Hinton had left a trunk in the Natural History Museum of London which, when opened in 1970, contained teeth and bones carved and stained in a way similar to the Piltdown bones. A. C. Doyle was suspected of the intention to fool experts, because he was heavily criticized by various experts for his engagement in spiritualistic sessions and theories, and he lived at a distance of only 10 miles from Piltdown.

The focus on Dawson as the only or main forger is strongly supported by increasing evidence that he perpetrated numerous other archeological hoaxes in the two decades before his Piltdown finds. When Miles Russell, archeologist at Bournemouth University, examined Dawson's collection of archeological objects he found that at least 38 items were fakes. Among these fakes were stained and filed teeth almost identical with those apparently found in the Piltdown pit. He wrote: "Piltdown was not a 'one-off' hoax, more the culmination of a life's work." An acquaintance of Dawson, Harry Morris, had the chance to study one of the flints obtained by Dawson at the gravel pit and suspected that it was artificially "aged" by staining with the intention to defraud, but he never accused Dawson publicly.

A new campaign intended to identify the forger unambiguously was launched in 2012 by Prof. Chris Stringer, expert on the evolution of humans at the Natural History Museum. This campaign included a detailed study of all documents and letters concerning the Piltdown and Sheffield finds. To the best knowledge of the author, the results of this campaign have not yet been published, although the manuscript was finished at the end of 2015.

The astonishing and amusing story of the Piltdown man is presented in this book as an example of the fact that progress in science is not only plagued by unintentional and undesired mistakes and fallacies, but also by black sheep publishing faked facts. Black sheep existed and exist in all branches of science, but fortunately their number is quite small. How fast mistakes, fallacies, and forgeries can be detected and revised depends mainly on three factors: chance, development of new analytical methods, and the number and intensity of studies devoted to a certain working field. The exposure of the Piltdown man is an example of the success of a new analytical method. Finally James Watson, Nobel Prize laureate of medicine in 1962 should be cited: "It does not suffice to fake research, it is also necessary to be awarded the Nobel Prize for it."

Bibliography

Craddock P (2012) Scientific investigation of copies, fakes and forgeries. CRC Press

Gee H (1996) Box of bones clinches identity of Piltdown paleontology hoaxer. Nature 381:261

Stringer C (2012) The 100 year mystery of Piltdown man. Nature 492:177

Walsh JE (1996) Unraveling Piltdown. The science fraud of the century and its solution. Random House, New York

Chapter 9
Chemistry

9.1 What Is an Element?

> A person's mind stretched by a new idea can never go back to its original dimension.
> (Oliver W. Holmes)

The meaning of the term "element" was a matter for debate and discussion for European philosophers and scientists over the whole period of written history, beginning with the first Greek philosophers. The grandfathers of Greek philosophy were Thales (c. 610–546 B.C.), Anaximander (c. 611–547 B.C.), and Anaximenes (c. 585–524/528 B.C.), when, as usual, Sokrates and Plato (427–347 B.C.) are considered fathers of philosophy. All three pre-Socratic philosophers, or rather cosmologists, lived in Miletos, the unofficial capital of the Ionians (today, the west coast of Turkey). A central problem in the thinking of all three cosmologists was the origin (*αρχη*, *Urgrund*) or material cause of all real things. Thales thought that water is the primary matter and the earth is a disk floating on water. More is not known about the cosmology and philosophy of Thales, because even Aristotle (384–322 B.C.), who handed down most of the information on the pre-Socratic philosophers, did not have an original text written by Thales.

Anaximenes also believed in a concrete element as the primary matter, but he favored air as the material basis of the cosmos. He held that dilution of air generates fire, condensation yields water, and high pressure yields earth. For him, the earth was again a flat disk floating on dense air. Anaximander was the most abstract thinker of those three Milesian cosmologists. He postulated the "boundless" as the *αρχη* and declared that all real things emerge from the boundless and return to it. In a handed-down fragment of his work he says that "things give satisfaction and reparation to one another for their injustice, as is appointed according to the ordering of time." Anaximander demonstrated for the first time a characteristic mental property of Greek natural philosophers, namely a dualistic way of thinking. The boundless was an abstract principle, on the one hand, but real matter, on the other.

H.R. Kricheldorf, *Getting It Right in Science and Medicine*,
DOI 10.1007/978-3-319-30388-8_9

Whereas the three Milesian cosmologists postulated one single *αρχη*, Empedokles (c. 490–430 B.C.) designed a new paradigm, the “Four element theory.” In combination with some modifications (see below) this theory survived in European philosophy and medicine for almost 2000 years. Empedokles was born in Akragas (Agrigento, Sicily) and spent most of his life in Sicily, but he was exiled to the Peloponnese for several years. Empedokles was a multitalented man, who was not only a philosopher, but also poet, physician, and politician and he was known as a gifted speaker. He formulated his insights and conclusions in the form of two long poems (each comprising around 2000 lines) entitled *On Nature* and *Purifications*. Fragments of these poems, amounting to approximately 10 % of the original texts, have survived. According to myth, this extraordinary man committed suicide in the flames of the volcano Etna.

Empedokles believed that the universe and all living organisms are subject to permanent change or evolution following a cosmic cycle. The driving forces were *φιλοτεσ* (love) and *νεικοσ* (strife). However, the stable material underlying any change was based on the four indestructible and indecomposable elements: earth, water, air, and fire. The different appearance of all things was the consequence of different mixtures of these elements. In other words, any change, any increase or decrease in properties and quality resulted from aggregation or segregation of different quantities of the four elements. In this respect, Empedokles was a forerunner of the atomists Leukippos and Demokritos (see Sect. 10.1).

Empedokles did not use the term element (*στοχειον*), which was later introduced by Plato, Empedokles called his elements *ραζιοματα* (rhizomata), the direct translation of which is “roots.” In analogy to Anaximander, Empedokles had a dualistic understanding of these roots, because, one the one hand, they were real matter and, on the other, represented abstract principles. Yet, in contrast to Anaximander his principles had a religious character and were associated with the four divinities Hera, Zeus, Hades, and Nestis. The dualistic thinking was later continued by Aristotle who postulated a fifth element, the ether. Ether was a sort of matter moving around the earth in an outer sphere and it was the principle of motion, imparting mobility and motion to the other elements and, thus, to any kind of object.

In later decades and centuries, the four element theory was modified and expanded in various ways, as outlined in Table 9.1. Plato combined the four elements with four symmetrical forms. Aristotle associated the elements with properties of real matter, such as hot and dry, cold and moist, etc. Furthermore, the signs of the zodiac were correlated with the elements. Particularly influential were the theories of the famous physicians Hippokrates of Kos (460–370 B.C.) and Galen (124–c. 200/216 A.D.). Hippokrates combined the elements with four bodily humors: blood, phlegm, black bile, and yellow bile. Galen, in turn, associated these humors with the four human temperaments: sanguine (blood), phlegmatic (phlegm), melancholic (black bile), and cholesteric (yellow bile). This concept had a great influence on medicine until the seventeenth century (see Sect. 7.1).

The next and most important paradigm shift was induced by the English/Irish scientist Robert Boyle (1626–1691). In his second book *The Sceptical Chymist* Boyle advocated, like Francis Bacon (1561–1626) before him, the primacy of

Table 9.1 The four elements of Empedocles and their later added attributes

Element	Space form	Properties	Direction	Signs of the Zodiac	Temperament
Fire	Tetrahedron	Hot and dry	South	Aries, Leo, Sagittarius	Cholesteric
Air	Octahedron	Hot and moist	East	Gemini, Libra, Aquarius	Sanguine
Water	Icosahedron	Cold and wet	West	Cancer, Scorpio, Pisces	Melancholic
Earth	Cube	Cold and dry	North	Taurus, Virgo, Capricorn	Phlegmatic

empiricism. Bacon established the scientific methodology, which consists of three steps:

1. Performing experiments
2. Testing their reproducibility
3. Drawing an inductive conclusion, resulting in a hypothesis that explains the experiments and suggests new (crucial) experiments

Boyle was mainly concerned with syntheses and analyses of compounds (he coined the term "analysis"). He postulated the existence of elements that cannot be decomposed by means of chemical reactions as the smallest constituents of compounds. He classified, for instance, the pure metals carbon and sulfur as elements and justified this classification with the rhetorical question: "Who has ever decomposed gold into elements?"

At this point it may be useful for a proper understanding of chemistry in general, and for Boyle's argument in particular, to define the term "chemical reaction." Chemical reactions are all those alterations and modifications of elements and compounds that proceed via reactions of valence electrons. Valence electrons are electrons in the outer orbitals of atoms and molecules in their energetic ground state. Other methods allowing alteration or modification of material properties are:

I. Physical methods, such as dilution, mixing, melting, crystallization, evaporation, or condensation of gases and vapors
II. Reactions involving atomic nuclei, such as splitting of nuclei or combination (fusion) of two nuclei (e.g., $H + H \rightarrow He$)

On the basis of Boyle's definition of elements, the following elements were known in the antique world: carbon, sulfur, iron, tin, lead, zinc, mercury, copper, silver, and gold. In the fifteenth century improved mining methods resulted in the discovery of cobalt, nickel, and tungsten as "impurities" of iron. In 1669, phosphorous was discovered. Before 1751, arsenic, antimony, and bismuth were detected as "impurities" of copper. Between 1751 and 1800, the following elements were obtained in a pure state and characterized: hydrogen, nitrogen, oxygen, chlorine, tellurium, titanium, chromium, manganese, yttrium, zirconium, and uranium. In the period 1800–1830, the following (main group) elements were discovered: lithium, sodium, potassium (alkali metals), magnesium, calcium, strontium, barium (alkaline earth metals), silicon, selenium, and bromine plus iodine (halogens). In the years up to 1869, eleven more elements were isolated: helium, rubidium, cesium, indium, thallium, niobium, ruthenium, lanthanum, cerium, erbium, and terbium. The year 1869 is here selected as the end of this historical excursion, because in that year the periodic table was elaborated on the basis of the elements enumerated above.

In the period between Boyle's work and Mendeleev's publication of the periodic table several chemists, above all John Dalton (1766–1844, see Biography), made major contributions to a better understanding of what the term element means. Dalton was first interested in meteorology and later began to perform experiments with air, water vapor, and mixtures of gases. He found that in a mixture each gas maintains its own characteristic partial pressure. He measured expansion coefficients of gases upon heating and found the inverse relationship of volume and pressure. Furthermore, he determined the weights of various gases relative to hydrogen and published a first short table of atomic weights, with hydrogen conventionally assumed to weigh 1. He concluded that each element is defined by a characteristic type of atom, and that reactions between atoms are responsible for the formation of compounds. His atomic theory was based on the following statements:

(A) Elements are composed of extremely small particles called atoms
(B) Atoms cannot be created or destroyed
(C) Atoms of one element are identical in mass, size, and other properties
(D) Atoms of different elements differ in mass, size, and other properties
(E) Chemical reactions result from combination, separation, or rearrangement of atoms
(F) Atoms of different elements combine in simple whole-numbered ratios when forming compounds

Therefore, Dalton found a consistent explanation for his own law of multiple proportions, Joseph Proust's law of constant proportions (1794), and Jeremias B. Richter's law of equivalent proportions (1791). Dalton's most important publication describing his atomic theory was published in 1808 with the title *A New System of Chemical Philosophy*.

Dalton did not know that gases such as hydrogen, oxygen, and chlorine consist of two atoms. The Italian chemist Stanislao Cannizzaro (1826–1910) solved this problem. Cannizzaro was professor of chemistry in Pisa, Torino, Alessandria (N. Italy), Genoa, Naples and, from 1871, Rome where he died. He studied the chemical structures and properties of gases with a focus on diethylzinc, which was an unusual substance at that time because of its combination of organic groups with a metal. In this connection he discovered that most gases consist of two atoms. At this point it should be mentioned that another Italian chemist, Amedeo Avogadro (1776–1856), coined the term "molecule" in 1811.

Cannizzaro's results fueled a new debate on the weights of atoms and molecules. In 1860, the German professor Friedrich A. Kékulé (1829–1896), famous for his discovery of the structural formula of benzene, arranged in Karlsruhe an international conference with the goal of achieving an international agreement on atomic

Table 9.2 Element symbols and their linguistic origin

Chemical symbol	English name	Linguistic origin
H	Hydrogen	Greek: *Hydrogenium* (water forming)
O	Oxygen	Greek: *Oxygenium* (acid forming)
Cl	Chlorine	Greek: *chloros* (yellowish green)
Br	Bromine	Greek: *bromos* (stink, stench)
I	Iodine	Greek: *ioeides* (violet)
C	Carbon	Latin*: Carbo*
S	Sulfur	Latin*: Sulfur*
Au	Gold	Latin: *Aurum*
Ag	Silver	Latin: *Argentum*
Cu	Copper	Latin: *Cuprum*
Fe	Iron	Latin: *Ferrum*
Pb	Lead	Latin: *Plumbum*
Sn	Tin	Latin: *Stannum*

weights. His conference was successful and, in analogy to Dalton's concept, the mass 1 was attributed to the hydrogen atom. A later revision of this convention is discussed in Sect. 10.1.

After clear definitions of the terms "element" and "molecule" were achieved, a new problem emerged, namely the labeling of elements with symbols suitable for the formulation of complex molecular structures and for the formulation of chemical reactions. Since the early Middle Ages it had been an unofficial law that the discoverer of a new element had the right to give it a name.

The influential Swedish chemist Jöns J. Berzelius (1779–1848) proposed that the symbols be derived from the names of the elements by an abbreviation not exceeding two letters. This proposal was internationally accepted and examples based on the Greek or Latin names of elements are listed in Table 9.2.

Another problem intensively discussed in the nineteenth century was the question of how the increasing number of elements could be arranged in a reasonable order. All experts were conscious of the fact that a simple listing according to the alphabet (the standard method at that time) or according to the mass number was not satisfactory. The German chemist Johann W. Döbereiner (1780–1849), professor at the University of Jena, reported on the observation that triads of elements having quite similar properties existed (e.g., calcium, strontium, barium). In 1865, the English chemist John A. R. Newlands (1837–1896) pointed out that more than two sequences of eight elements should exist showing similarities in chemical reactions. Mendeleev's periodic table integrated all these findings.

The Russian chemist Dmitri Mendeleev (see Biography) is said to have envisioned the periodic table in a dream. Regardless of whether this is a myth or not, in March 1869 he presented a paper to the Russian Chemical Society entitled (translated): "Dependence between the properties of the atomic weights of the elements." In this paper he stated:

1. The elements arranged according to their atomic weights exhibit a periodicity of their properties
2. Elements with similar properties have atomic weights that are either of almost the same value (e.g., Fe, Co, Ni) or that increase regularly (e.g., K, Rb, Cs)
3. The arrangement of the elements in groups according to their atomic weights corresponds to their so-called valences and to some extent to their distinctive chemical properties
4. The elements that are the most widely diffused have small atomic weights
5. The magnitude of the atomic weight determines the character of the element, just as the magnitude of a molecule determines the character of a compound
6. We may expect the discovery of many still unknown elements, for instance, analogs of aluminum and silicon, whose atomic weights would be between 65 and 75
7. Certain characteristic properties of elements can be predicted from their atomic weights

An important aspect of Mendeleev's work was the correct prediction of then unknown elements, such as gallium, germanium, and five more elements. The German chemist Julius L. Meyer (1830–1895) published a similar table of elements only a few months later, but without predicting new elements. Mendeleev was criticized by many chemists for the lack of numerous elements in his periodic table. However, when gallium was detected in 1875 and germanium in 1886, the vast majority of chemists accepted his concept. The periodic table and discovery of the missing elements completed the paradigm shift from Empedokles's understanding of elements to that of modern chemistry.

This paradigm shift is particularly noteworthy for two reasons. First, it was a kind of evolutionary process proceeding via several stages over a period of 200 years; it was not a sudden revolution according to the definition of T. Kuhn (see Chap. 6). Second, it was in many ways the most important paradigm shift in the history of science. The shift from the geocentric to the heliocentric world was, of course, of immense significance for astronomy and for the *Weltanschauung*, the self-confidence and self-esteem of Europeans, but for a long time it had little influence on everyday life and on human civilization. In contrast, the new definition of the term element was the basis of chemistry, pharmacy, and mineralogy, and, as outlined in Chap. 5, almost all the progress accumulated in medicine over the past 150 years is based on chemistry and pharmacy. The new understanding of elements was decisive for the development of biochemistry and molecular biology and, thus, for the proper understanding of genetics and of the theory of evolution (see Sect. 8.4). Furthermore, it was important for elaboration of the modern atomic model and, thus, for large areas of modern physics (see Sect. 10.1).

Robert Boyle

Robert Boyle was born on 25 January 1627 (4 February according to the Gregorian calendar), in Lismore Castle, County Waterford in Ireland, as the fourteenth child and seventh son of Richard Boyle and his wife Catherine Fenton. Richard Boyle, First Earl of Cork, was born in England and arrived in Dublin in 1588, in the position of a deputy escheator, when large parts of Ireland were under the control of the English crown. At the time when Robert was born, his father was a wealthy man who had acquired an enormous mass of landholdings. His mother was the daughter of Sir Geoffrey Fenton, the former Secretary of State for Ireland.

As a boy, Robert and his elder brothers were fostered to a local family where they learned Irish so that they could later help their father to translate Irish into English, and vice versa.

Through private tutoring, the young Robert Boyle also learned French, Latin, and ancient Greek. His mother died when he was 8 years old, and his father sent him to Eton College in England where a friend was provost at that time. Furthermore, his father hired a private tutor, who had knowledge of Irish, to provide his son with an optimum education. After 3 years at Eton, Robert traveled to France, Switzerland, and Italy accompanied by a French tutor. In Florence, he studied law, philosophy, medicine, mathematics, and astronomy in connection with the work of Galilei who lived near Florence and died in 1642.

His father died in 1643 and Robert returned to England, where he arrived in the middle of 1644. His father had left him substantial estates in County Limerick (Ireland) and the manor house in Stalbridge (Dorset). Robert was financially independent and decided to devote his life to scientific research. At first he lived in Stalbridge where he wrote his first book. He became member of the so-called Invisible College, a group of scientists and scholars who devoted themselves to cultivation of the "new philosophy." This group usually met in London, but several members also met in Oxford from time to time.

After having visited his Irish estates several times, he moved to Ireland in 1652, but soon became frustrated because of his inability to make progress in his chemical studies. In 1654, he returned to England and settled down in Oxford to pursue his work more successfully. After reading about Otto Guericke's experiments with vacuum, Boyle became highly interested in air pumps. With the help of the creative and experienced physicist Robert Hooke (1636–1703) he substantially improved Guericke's pump and presented his "pneumatical engine" to the public in 1659. He published his experiments with air and vacuum in 1660 in a book entitled *New Experiments Physico-Mechanical: Touching the Spring of the Air and its Effect*. In this connection he discovered the gas law later named after him, namely the inverse relationship of volume and pressure. The French physicist Edme Mariotte (1620–1684) published the same law in 1676 (perhaps discovered independently) and, thus, this law is called the Boyle–Mariotte law in Europe.

Boyle had many scientific interests including biology and physics, but chemistry remained his favorite study throughout his life. He published his first book on chemistry, *The Sceptical Chymist*, in 1661 criticizing the speculations of contemporary alchemists that their salts sulfur and mercury were the true principles of things. He held that certain materials, such as pure metals or sulfur, were elements that were not decomposable by chemical reactions. He also differentiated between compounds and mixtures. Chemistry was for him the science of synthesizing compounds and analyzing compounds or mixtures. He coined the term "analysis."

Boyle conducted experiments on respiration and in various areas of physiology, but he kept away from anatomical dissections and especially from vivisection.

Boyle emphasized experimental facts and logical explanations and disliked speculations and mystic interpretations. Compared with his contemporary alchemists, he was the first modern chemist. On the basis of the Invisible College, the Royal Society of London was founded in 1663 with the purpose of improving "natural knowledge." Boyle became member of the council and was elected president in 1680, but he declined this honor because of a scruple about oaths. In 1668, he moved from Oxford to London, where he lived in the house of his elder sister Katherine Jones, Lady Ranelagh. It is reported that there was "a lifelong intellectual partnership, where brother and sister shared medical remedies, promoted each other's scientific ideas and cited each other's manuscripts." In the years after 1688 his health began to worsen and Boyle gradually withdrew from all official duties and even from communications with the Royal Society, but he still tried to complete articles and treatises. He died on 31 December 1691, just 1 week after his beloved sister. He was buried in the churchyard of St. Martin-in-the-Fields.

John Dalton

John Dalton was born on 6 September 1766 in the Quaker settlement of Eagersfield near Cockermouth, in the County of Cumberland, UK. His father was a weaver and too poor to support him for many years. He received his early education at a private school of the Quaker John Fletcher at Pardshaw Hall. He had to earn at least part of his living at the age of 10 in the service of a wealthy local Quaker. In 1781, at age 15, he joined his older brother Jonathan in teaching youth at a Quaker school in a village about 40 miles from his home. His attempts to study law or medicine were not successful, because as a Dissenter (a Christian opposed to the state religion) he was barred from entering English universities. However, he improved his scientific knowledge through private instruction by John Gough, a blind philosopher and scientist. At the age of 27, Dalton obtained a position as teacher of mathematics and natural philosophy at the New College in Manchester. Because of the worsening financial situation of this college he had to resign from this position after 7 years. At first, he tried to earn his living as a private tutor of mathematics and natural philosophy. Yet, soon, in the year 1800, he became secretary of the Manchester Literary and Philosophical Society. After becoming member and finally president of this society he maintained this membership until his death.

In the following years, Dalton's scientific reputation steadily increased as a result of numerous publications covering a broad variety of topics. In previous years he had spent many weekends hiking and climbing in the Lake District. He made meteorological studies measuring temperature and humidity, and he measured the altitude of the mountains. Together with his friend Jonathan Otley he became known as an expert on the Lake District. In the years after 1800, he also published essays and treatises on topics such as rain, dew steam, heat, the color of the sky, the reflection and refraction of light, and on the origin of springs. In 1802, he published four essays in the *Memoirs of the Literary and Philosophical Society of Manchester* dealing with the properties of mixed gases, with the thermal expansion of gases, and with the pressure of steam or other vapors at different temperatures. He concluded from the vapor pressure measurements of six different liquids that the variation in vapor pressure for all liquids is equivalent for the same variation in temperature. The following remark in his fourth essay is well known:

> I see no sufficient reason why we may not conclude that all elastic fluids under the same pressure expand equally by heat and that for any given expansion of mercury, the corresponding expansion of air is proportionally something less, the higher the temperature. It seems, therefore, that general laws respecting the absolute quantity and the nature of heat are more likely to be derived from elastic fluids than from other substances.

In 1803, Dalton published several essays on the absorption of gases in water and various liquids whereby he formulated the law of partial pressures, later named Dalton's law. In the years up to 1805 he also developed and published the most important part of his scientific work, the atomic theory described above. From 1817 until his death he wrote 117 contributions to the *Memoirs of the Literary and Philosophical Society of Manchester*.

As a result of his increasing reputation, he was twice invited (1803 and1809/1810) by the Royal Institution to present lectures on natural philosophy. In 1816 he was elected member of the French Academy of Sciences, in 1822 he was elected member of the Royal Academy of Sciences, and in 1834 (at age 68) he became a Foreign Honorary Member of the American Academy of Arts and Sciences.

Dalton was not married and lived a modest life as Quaker. He lived for 26 years until his death in a room in the house of Rev. W. Johns (a biologist) and his wife in Manchester. Dalton suffered from color blindness, which fortunately did not affect his experiments with colorless liquids, vapors, or solid chemicals. He had noticed this deficiency in his youth and learned that his brother Jonathan suffered from the same "disease." Therefore, he concluded that this deficiency was inherited. Because nothing was officially known about color blindness, he began to publish his experience and its interpretation in the years after 1795. He explained his problem by the hypothesis that the liquid medium of his eyeball was discolored. Much later it was found that he suffered from a rare genetic defect that caused a lack of cones in the inner eye. Nonetheless, his systematic reports on color blindness became so widely known that this deficiency was called Daltonism in English-speaking countries.

Although this problem did not affect his scientific career, a major stroke in 1838 following a minor stroke in 1832, resulted in speech impairment. He still continued to conduct experiments, but in May 1844 he had a third stroke, and died on 27 July of the same year, most likely from a fourth stroke. A civic funeral with all honors was arranged in the Town Hall of Manchester and more than 40,000 people filed past his coffin. Dalton was then buried in the Ardwick cemetery in Manchester, a place that serves as playing ground today.

When Dalton was still alive a large statue was crafted and finally placed in the Manchester Town Hall after its construction in 1877. He is probably the only scientist who has been honored by a statue in their lifetime. The Metropolitan University, the University of Manchester, and numerous Quaker schools named buildings after him. Furthermore, several cities in various countries and a lunar crater carry his name. Moreover, chemists use the designation Dalton (Da) to denote one atomic mass unit.

Dmitri Mendeleev

Dmitri Ivanovich Mendeleev was born on 8 February 1934 (Gregorian calendar) in the village of Verkhnie Aremzyani near Tobolsk in Siberia. He was the youngest of numerous children (11–17 depending on the source) of Ivan Pavlovich Mendeleev and his wife Maria Dimitrieva (nee Kornilieva). His grandfather was a priest of the Orthodox Church and therefore Dmitri, like the entire family, was educated as an Orthodox Christian. His father was a teacher of politics, philosophy, and fine arts. When Mendeleev was aged 11–12, his father became blind and lost his position, so that the financial situation of the family deteriorated rapidly. His mother decided to restart the abandoned glass factory of the family. About 1 year later his father died and the factory was destroyed by fire. Nonetheless, Mendeleev continued to attend the high school of Tobolsk.

Despite the bad financial situation, his mother decided in 1849 to take him to Moscow to obtain an optimum education for him. The University of Moscow did not accept him, and so they traveled to St. Petersburg to contact his father's former university. He eventually entered the Main Pedagogical Institute and the entire family moved to St. Petersburg. After graduation he was able to treat his worsening tuberculosis. Because remedies for this disease were not available at that time, he decided in 1855 to move to the Crimean peninsula hoping that the better climate would help him. There he was active as a science master in the gymnasium of Simferopol and, after complete restoration of his heath, he returned to St. Petersburg in 1857.

In 1859, Mendeleev moved to Heidelberg and spent 2 years studying spectroscopic methods. After his return to St. Petersburg he published his first book dealing with spectroscopy. At that time, he had a firm position as teacher at the Nikolaev Engineering Institute, and in April 1862 he married Feozova Nikitichna Leshcheva. In 1876, he fell in love with Anna Ivanova Popova, filed for divorce from Feozova, and married Anna Ivanova in 1882. He had sons and daughters from both wives.

In 1864, Mendeleev was appointed professor at the St. Petersburg Technological Institute and in 1865 full professor at the University of St. Petersburg. In the same year he received the doctor of science degree for his dissertation *On the Combination of Water and Alcohol* and achieved tenure in 1867.

In the years between 1861and 1871, Mendeleev made St. Petersburg an international center of chemical research. His textbook *Principles of Chemistry* published in 1868–1870 in the form of two volumes and publication of his *Periodic Table* raised his reputation to a climax. In addition to his experimental and theoretical research at the University of St. Petersburg, he investigated the composition of petroleum and supported the founding of the first oil refinery in Russia. In 1892, he became head of the Archive of Weights and Measures in St: Petersburg and introduced the metric system in Russia. However, it is a myth popular in Russia that he was engaged in the production of vodka containing 40 vol% of alcohol.

The Bureau of Weight and Measures did not set any production standards and his dissertation was focused on alcohol concentrations above 70 % and did not contain any comment on vodka.

After his periodic table was widely accepted, he was honored by various scientific organizations in Europe. For instance, the Royal Society of London awarded him the Davy Medal in 1882 and the Copley Medal in 1905. In 1905, he was also elected a member of the Royal Swedish Academy of Sciences. In the same year he was proposed for the Nobel Prize in Chemistry for 1906, but as a result of the strong opposing influence of Svante Arrhenius (1859–1927) the Nobel Prize Committee chose Henri Moissan. Although Arrhenius was not a member of the committee, he also prevented Mendeleev's nomination in 1907. It is reported that Arrhenius had a grudge against Mendeleev, because of his critique of Arrhenius's dissociation theory.

Mendeleev retired in August 1890 from the University of St. Petersburg, but continued to write scientific articles. He died from influenza in 1907 in St. Petersburg at the age of 72. Today a number of places and objects commemorate Mendeleev's achievements, for example:

In St. Petersburg, his name was given to a street and the National Meteorological Institute

In Moscow, the University of Chemical Technology is named after him

A large crater on the moon has his name, and the synthetic element No. 101 was named Mendelevium (symbol Mv)

Bibliography

Adams H (2012) A letter to American teachers of history (Classic Reprint Series). Forgotten Books, New York

Bakulis N (2005) Handbook of Greek philosophy: from Thales to the Stoics. Analysis and fragments. Trafford Publishing, ISBN 1-4120-4843-5

Burnet J (1920) Greek philosophy: Thales to Plato, 3rd edn. A&C Black

Dalton J. http://www.chemheritage.org/discover/online-resources/chemistry-in-history/themes/the-path-to-the-periodic-table/dalton.aspx

Dalton J. http://www.quaker.org.uk/john-dalton-1766-1844

Earnshaw A, Greenwood N (1997) Chemistry of the elements, 2nd. edn. Butterworth-Heinemann

Emsley J (2011) Nature's building blocks: an A-Z guide to the elements. Oxford University Press

Friedman RM (2001) The politics of excellence: behind the Nobel Prize in science. Times Books, New York

Gordin MD (2004) A well-ordered thing: Dimitri Mendeleev and the shadow of the periodic table. Basic Books, New York

Greenway F (1966) John Dalton and the atom. Cornell University Press

Hunter M (2000) Robert Boyle, 1627-1691: Scrupulosity and science. The Boydell Press

Hunter M (2009) Between god and science. Yale University Press, New Haven

Mendeleev D, A short CV and a story of life. http://www.mendcomm.org/Mendellev.aspx
Mendeleev DI (1901) Principles of chemistry. Collier, New York. http://www.archive.org/details/principlesofchem00menduoft
Mendeleyev DM, Jensen WB (2005) Mendeleev and the periodic law: Selected writings, 1869–1905. Dover Publications, Mineola, NY
Robert Boyle. In: The Stanford Encyclopedia of Philosophy. http://plato.stanford.edu/entries/boyle
Stewart MA (1991) Selected philosophical papers of Robert Boyle. Hackett, Indianapolis
Thackray A (1972) John Dalton: critical assessments of his life and science. Harvard University Press

9.2 The Phlogiston Theory

An error is the more dangerous in proportion to the degree of truth which it contains.
(Henry F. Amiel)

Emancipation of chemistry as an exact natural science from alchemy, with a history of several thousand years, proceeded by revisions of fallacies and by elaboration of a specific formula language. The development of a formula language was finished in 1874 when the young Dutch chemist Jakobus Henricus van't Hoff published a treatise presenting formulas suitable for description of the three-dimensional structure of organic molecules. For this performance and for other merits van't Hoff was awarded the first Nobel Prize in Chemistry in 1901.

One of the fundamental misunderstandings that had to be overcome was the phlogiston theory. The term "phlogiston" originates from ancient Greek (*phlox* meaning light, flame) and means burnt. In the Middle Ages, the Latin word *caloricum* was also frequently used. Phlogiston was understood as a highly volatile, colorless, tasteless gaseous medium that escaped when a solid product was combusted. When a substance, such as carbon or sulfur, burnt without leaving much residue, it was considered rich in phlogiston. When a substance, such as a metal, left a relatively large residue upon combustion (metal oxide and/or carbonate), it was believed to contain little phlogiston. In the case of some metals it was possible to reverse the combustion (today called oxidation), namely by heating the metal oxide or carbonate with powdered charcoal. This reversion of oxidation was later called reduction.

The first version of the phlogiston theory was described by the German alchemist Georg E. Stahl (1659–1734), who lived in Halle (Saxony). The theoretical foundation of his new concept was the work of the alchemist and mining expert Johann J. Becher (1635–1682). In his book *Physica Subterranea* Becher improved the four element theory of the Greek philosopher Empedokles (490–430 B.C.) by subdividing the element earth into three classes:

Terra fluida (earth of Mercury): Associated with volatility, fluidity, fineness, and metallic properties
Terra pinguis: The principle of greasy and oily earth that made substances oily, sulfurous, and combustible
Terra lapida: The principle of hard, stone-like earth that made substances solid, crystalline, and fusible

This list indicates that the three *Terras* were not primarily understood as materials, but as functional principles. Stahl was concerned with a more detailed elaboration and interpretation of the *Terra pinguis* and presented his considerations in 1697 in his book *Zytomechanica fundamentalis*. The principles sulfurous and combustible became "phlogiston," which not only escaped during combustion of a material, but also during fermentation of organic materials and upon rotting of plants or putrefying of dead animals. Stahl was more concerned than Becher with a clear differentiation between materials, on the one hand, and principles, on the other. He clearly declared that sulfur, metals, and other combustible materials all contain the same volatile principle, the phlogiston.

At this point it should be emphasized that Stahl observed a loss of weight when metals were combusted. In his experiments, the total weight of the remaining oxide (plus carbonate) was lower than the initial weight of the metal. This finding, later revised by Antoine Lavoisier, was for Stahl a particularly strong argument in favor of his phlogiston theory. Furthermore, he discovered that heating of metal oxides with charcoal brought the metals back, a reversion of the combustion that he understood as a reintroduction of phlogiston into the metal. Moreover, he was able to reduce sulfuric acid to sulfur (at least in small amounts). All these findings together confirmed Stahl of the correctness of his theory and its general validity.

An initially confusing observation was the rapid termination of combustion when the burning material was placed under a glass bowl or under a metal hood. The argument found in defense of the phlogiston theory was the hypothesis that the phlogiston concentrates around the burning substance and hinders the escape of more phlogiston. Only dilution with fresh air allows continuous liberation of all phlogiston. Another issue intensively debated among the supporters of the phlogiston theory was the question of whether or not phlogiston is a real, gas-like substance that can be isolated and characterized.

The man who seemed to have a convincing answer to this question was Henry Cavendish (1731–1810). He brought metals such as iron, zinc, or tin in contact with hydrochloric acid or diluted sulfuric acid and observed the evolution of a gas with unusual highly interesting properties (e.g., $Zn + 2HCl \rightarrow ZnCl_2 + H_2$). This new gas, called combustible air or hydrogen (i.e., water-forming gas), was highly volatile, colorless, inodorous, and inflammable. Cavendish observed the appearance of water at the end of its combustion, but believed that this small amount of water was moisture accompanying the hydrogen gas.

Furthermore, Cavendish observed that hydrogen was liberated from various acids, independently of their concentration, and, hence, concluded that this gas originated from the metals and not from the acids. Therefore, Cavendish was convinced that hydrogen was the long-searched-for phlogiston. This finding and conclusion was accepted by almost all supporters of the phlogiston theory, so that the identification of hydrogen as phlogiston became the second version or the second stage of the phlogiston theory. The credibility of this new theory was supported by the following experiments. Cavendish treated metal oxides or carbonates with various acids and found that no phlogiston escaped. In contrast to the metals themselves, their oxides and carbonates were, as expected, free of phlogiston. Another series of experiments, seemingly supporting Cavendish's phlogiston theory, was contributed by Joseph Priestly (1733–1804). He passed hydrogen over hot, finely powdered metal oxides and observed the formation of metals and consumption of hydrogen (e.g., $PbO + H_2 \rightarrow H_2O + Pb$). Obviously, the treatment of metal oxides with hydrogen generated phlogiston-containing metals.

Cavendish was also eager to explain why the water resulting from the combustion of hydrogen in air was contaminated with nitric acid, whereas combustion of hydrogen with pure oxygen yielded pure water. Eventually, he arrived at the conclusion that the nitrogen of air was phlogiston-containing nitric acid, which liberated nitric acid when the phlogiston reacted with oxygen. This conclusion was, in turn, supported by the finding that the (sometimes explosive) reaction of carbon with potassium nitrate (a salt of nitric acid) liberated nitrogen. Obviously, the phlogiston of the carbon powder was transferred to the nitrate ion, yielding nitrogen gas. Furthermore, the gas NO was discovered in 1776 and classified as nitrogen partially deprived of phlogiston (partially "dephlogistinated" nitrogen) or as partially "phlogistinated" nitric acid.

The experiments of four researchers, Daniel Rutherford (1749–1819), Joseph Priestley (1733–1804), Carl W. Scheele (1742–1786), and Antoine Lavoisier (1743–1794) paved the way for a proper understanding of air and water. In 1772, Rutherford discovered that living animals and burning phosphorous consumed only a certain fraction of air. Only few months later, Scheele and Cavendish made similar observations. Lavoisier called it "azote," derived from the Greek word *azoos* meaning inanimate. The later name, nitrogen, means saltpeter yielding (gas). In 1774, Scheele and Priestley obtained almost simultaneously pure oxygen, the life supporting gas, upon heating of mercury oxide ($2HgO \rightarrow 2Hg + O_2$). Lavoisier coined the name "oxygen," meaning acid forming (gas), because the combustion of nonmetals such as nitrogen, sulfur, or phosphorus generates oxides that yield acids upon dissolution in water.

Based on the knowledge of these discoveries Cavendish conducted new experiments that demonstrated that oxygen reacts easily with NO, whereby nitric acid was identified as final reaction product (the intermediates were unknown at that time). He interpreted this result as formation of nitric acid from NO after consumption of the phlogiston by reaction with oxygen. In contrast, he found that oxygen did not react with hydrogen, the apparently pure phlogiston, when both gases were mixed in a poorly illuminated room. He also found that the mixture of both gases was stable on storage in a cold, dark room. This, at first glance, surprising result prompted Cavendish to developed a new (the third) version of the phlogiston theory. He interpreted hydrogen now as phlogiston-containing water (or hydrated phlogiston). The normal water resulting from the reaction of hydrogen with oxygen was, accordingly, phlogiston-free water.

In 1772, Louis B. G. de Moreau (1737–1814) had found that the combustion of metals, when carefully performed, resulted in an increase in the total weight. Lavoisier confirmed these results and extended the experiments to nonmetals. In other words, Stahl's experiments were proven incorrect. Obviously, he had used an open flame, which blew off some of the fine metal oxide particles formed during the combustion. Cavendish used these new results, which were deadly for the first version of the phlogiston theory, as proof of his latest version of the theory. He interpreted the combustion as a reaction of metals with hydrated phlogiston, whereby metal hydrates were formed that had, of course, a higher weight than the metal alone. However, all those experiments were conducted without exact quantification of starting materials and reaction products.

Because of his profession and wealth, Lavoisier had the use of the best scales available at the time, and he attempted to quantify the starting materials and reaction products of all his experiments as exactly as possible. In this way, he detected that the higher mass of the reaction products isolated after combustion of metals or nonmetals exactly corresponded to the consumption of oxygen. Phlogiston was definitely obsolete in explaining these findings. When Lavoisier developed his new "oxidation theory" before 1774, he was confronted with the difficulty of why the reaction of metals with various aqueous acids liberated hydrogen. The detection of pure oxygen and its reaction with hydrogen to yield pure water convinced him that water was the oxide of hydrogen. To understand the problems chemist in the eighteenth century had with identification of simple molecules such as water, one has to keep in mind that exact formulas based on element symbols and stoichiometric considerations were unknown at that time. To prove his new hypothesis, Lavoisier attempted to generate hydrogen by abstraction of oxygen from water. He studied two series of experiments. First, he passed water vapor over red-hot coal and observed formation of hydrogen along with carbon dioxide. (CO was not detectable at that time). Second, he passed water vapor over red-hot iron and again observed formation of hydrogen together with iron oxides. With this information in hand, he revised Cavendish's interpretation of the metal/acid experiments. Lavoisier concluded that the liberated hydrogen stems from the acids and not from the phlogiston-free metals.

Lavoisier continued to improve the quantification of all his experiments and, thus, elaborated the experimental basis on which Joseph L. P. Proust (1754–1816) formulated in 1797 the "law of definite proportions." This law claims that quantitative reactions of pure elements can only be achieved in certain precise proportions. Two years later, Proust also presented the "law of constant composition," which emphasizes that in chemical reactions matter is neither destroyed nor created. However, this law had been previously formulated by other chemists, such as Jean Rey (1583–1645), Mikhail Lomonosov (1711–1765), Joseph Black (1728–1799), Henry Cavendish, and Antoine L. Lavoisier. Both Lavoisier and Proust paved the way for Dalton's law. That law, published in 1805, stated that the total pressure of a mixture of gases is the sum of the partial pressures of the individual gases. At that time, Lavoisier's oxidation theory had definitely proven to be the superior concept and the phlogiston theory, which had lasted for almost 100 years, became a stage in the history of chemistry. However, it should be emphasized that the phlogiston theory made a substantial contribution to the emancipation of chemistry from alchemy. Oxidation and reduction processes were correlated and this correlation formed the basis of what are known today as redox reactions.

The fate of the phlogiston theory is also a characteristic example of a hypothesis that is modified and improved several times to obtain a satisfactory explanation of new experiments until it is necessary to replace it by a new and better theory. With regard to Sect. 9.1, it should be remembered that the identification of nitrogen and oxygen as two different components of air and their identification as elements was an important contribution to the paradigm shift concerning the term element (see Sect. 9.1).

Antoine-Laurent Lavoisier

Antoine-Laurent Lavoisier was born on 26 April 1743 to a wealthy family in Paris. He was the eldest son of Jean Antoine Lavoisier (1715–1775), who served as an attorney at the parliament of Paris. When he was five, his mother died and Lavoisier inherited a large fortune. The family moved to the house of his grandmother (on his mother's side) where he lived until his marriage in 1771. In this house, he installed a small laboratory because he had become interested in science in his earliest youth. In1754, at the age of 11, Lavoisier began his schooling at the College Mazarin where he attended lectures in four disciplines. Chemistry was taught by Guillaume-Francois Rouelle, experimental physics by Jean-Antoine Nollet, botany by Bernard de Jussieu, and mathematics by Nicolas L. de Lacaille. De Lacaille was not only a mathematician, but also an astronomer. He recognized the talents of his disciple and stimulated Lavoisier's interest in astronomy, an enthusiasm that never left him.

Persuaded by his father, Lavoisier entered law school in 1760, received a bachelor's degree in 1763, and finalized his studies with a doctoral degree at the end of 1764. He was registered in the official list of lawyers, but apparently never practiced this profession, partially because it did not meet his interest and partially because he did not need a permanent income from his profession. In the spare time during his studies of law he had continued to study chemistry, attending lectures of his former teacher F. Rouelle. Furthermore, Lavoisier had continued to perform chemical experiments in his private laboratory. His enthusiasm for chemistry was also strongly influenced by Etienne Condillac, a prominent scholar of the eighteenth century. His experiments with calcium sulfate were so successful that he presented his first publication dealing with the properties of gypsum to the French Academy of Sciences in 1774, aged 22 years. However, his scientific interests were broader and in 1766 he was awarded a gold medal by the King for a treatise on urban street lighting. In the years 1763–1767, he also studied geology under Jean-Etienne Guettard and, in collaboration with his supervisor, worked in the summer of 1767 on a geological survey of Alsace-Lorraine. In the following 2 years, he utilized his new experience by working on the first geological map of France. In recognition of his fruitful scientific activities he received a provisional appointment to the Academy of Sciences in 1768.

After 1770, he became interested in the phenomena associated with the combustion of nonmetals, such as sulfur and phosphorus, and with the combustion of metals. In 1772, Lavoisier presented two notes to the Academy of Sciences describing the combustion of phosphorus and sulfur, respectively. In the second note he concluded: "what is observed in the combustion of sulfur and phosphorous may well take place in the case of all substances that gain in weight by combustion and calcination and I am persuaded that the increase in weight of metallic *calces* is due to the same cause."

Through systematic reviews of literature dealing with the nature of air, Lavoisier came across a publication by the Scottish chemist Joseph Black, who had reported the existence of "fixed air." Black had observed that the difference between mild alkali (e.g., chalk, $CaCO_3$) and the caustic form (e.g., quicklime, CaO) lies in the fact that the former contained "fixed air," a special component of normal air. Lavoisier realized that this fixed air (carbon dioxide, CO_2) is also formed when metal oxides or calces were heated with charcoal and that the calcination of metals upon combustion results from the combination with fixed air. These results and Lavoisier's experiments with pure oxygen led him write his famous treatise *On the Nature of the Principle Which Combines with Metals During their Calcination and Increases their Weight*. This work was presented to the Academy in April 1775 and is usually referred to as the "Easter Memoir." A summary of his new theory of combustion was presented to the Academy under the title *Reflexion sur le Phlogistique* in 1783 as a full-scale attack on the phlogiston theory.

In 1771, at the age of 28, Lavoisier married Marie-Anne Pierrette Paulze, the 13-year-old daughter of a senior member of the Ferme Generale (see below). His father–in-law was a rich man whose marriage gift was a new large house, which allowed Lavoisier to install a large, well-equipped laboratory. Even more important were the talents of his wife. Despite her young age, Madame Lavoisier soon began to play an important role in his scientific career. She translated English publications and she improved and edited his memoirs (treatises). She also sketched and carved engravings of his scientific instruments, serving as illustrations in his books and memoirs. Furthermore, she arranged parties at which various scientists discussed ideas and problems related to chemistry. Madame Lavoisier also wrote or improved the notes accompanying all experiments. Lavoisier and his wife developed the system still used today, namely a subdivision into three parts:

1. Description of the experiment
2. Description of the results
3. Discussion and conclusions

In the interpretation of his results, Lavoisier laid great emphasis on the distinction between evidence-based arguments and speculations. Together with colleagues, he also wrote books on nomenclature and the role of stoichiometry.

In 1768, when he was elected to the Academy of Sciences, Lavoisier also started a new professional career. He purchased a share in the Ferme Generale and became an assistant of Francois Boudin one of the leaders of this organization. The Ferme Generale was a kind of "tax farming" company that advanced the estimated taxes to the royal government in return for the right to collect taxes. The collected taxes were considerably higher than the estimated ones and this difference yielded the private income of the *Fermiers*, the members of this organization. These new activities had important consequences for Lavoisier. They improved his income, increased his social reputation, and put him in personal contact with the rich merchant Jean Paulze, his future father-in-law. Lavoisier attempted to improve the French monetary and taxation system to help the peasants. He also helped to introduce the metric system to secure uniform weights and measures throughout France.

As a result of his increasing reputation as chemist and organizer, in 1775 he was appointed one of four commissioners of the royal gunpowder production. This new institution replaced an inefficient and incompetent private organization. Under Lavoisier's influence, both the quality and efficiency of gunpowder production increased rapidly. As a commissioner, he had a house and a laboratory in the Royal Arsenal and here he spent most of his time between 1775 and 1792. In this position, he became acquainted with a young assistant named Eleuthere Irenee du Pont, who emigrated eventually to the USA and founded the E. I. DuPont company, which became the most important supplier of gunpowder and explosives in the USA for many decades.

As the French revolution went on, after 1789 the world around Lavoisier changed completely. Like other liberal intellectuals he believed that the old regime could be reformed step by step. He tried to stimulate reforms, keeping aloof from direct political activities. His last work submitted to the National Convention was a proposal for a reform of French education. Nonetheless, he was forced to leave his house and laboratory in the arsenal before the end of 1792. On November 1793, he and all other deeply unpopular Fermiers were arrested. He was accused of having plundered the people and the treasury of France, of having adulterated the nation's tobacco with water, and of having supported the enemies of France with huge amounts of money from the national treasury. On 8 May 1794, he was guillotined together with 27 other defendants. About 2 years later, the government withdrew the accusation and delivered his private belongings to his widow with the note: "To the widow of Lavoisier who was falsely convicted."

Lavoisier's importance to science was expressed by the lament of his friend, the mathematician Joseph L. Lagrange: "It took them only an instant to cut off his head and one hundred years might not suffice to reproduce it's like." Still, during his lifetime he was honored by being elected member of the Societe Royale de Medicine in 1782, member of the Société Royale d'Agriculture in 1783, head of the Academy de Science in 1784, and member of the Comité d'Ágriculture in 1785. About a century later, a life-size bronze statue was erected in Paris, but the sculptor had not copied Lavoisier's head. This statue was melted down during World War II and not replaced. Smaller statues were later added to the Town Hall and to the façade of the Louvre. A high school and a street in the 8th arrondissement were named after him and, more recently the asteroid C 826. Furthermore, his name is inscribed on the Eiffel Tower as well as on buildings around the Killian Court at MIT in Cambridge, MA.

Bibliography

Bowler PJ (2005) Making modern science: a historical survey. University of Chicago Press

Brock WH (1993) The Norton history of chemistry. W.W. Norton, New York

Conant JB (1950) The overthrow of phlogiston theory: the chemical revolution of 1775–1789. Harvard University Press, Cambridge, MA

Donovan A (1993) Antoine Lavoisier: science, administration and revolution. Cambridge University Press

Grey V (1982) The chemist who lost his head: the story of Antoine Lavoisier. Coward McCann & Geoghegan, New York

Holmes FL (1985) Lavoisier and the chemistry of life. University of Wisconsin Press

Jungnickel C, McComach R (1999) Cavendish: the experimental life. Bucknell University Press

Mason SF (1962) A history of the sciences, revised edn., Chapter 26. Collier Books, New York

McComach (2004) Speculative truth: Henry Cavendish, natural philosophy and the rise of modern theoretical science. Oxford University Press

Partington R, McKie D (1937, 1938, 1939) Historical studies on the phlogiston theory. Ann Sci 2:361–404, 3:1–58, 5:113–149. Reprinted 1981 as ISBN 978-0-405-13895-9

9.3 Vitalism in Chemistry

> Why shouldn't truth be stranger than fiction? After all fiction has to make sense.
> (Mark Twain)

The term "vitalism" is based on the Latin word *vita* (life) and claims that any living organism, regardless of whether plant or animal, contains a vital force or energy that is the driving force behind its evolution and all its biological functions. Hence, it is this vital energy that represents the most profound difference between a living organism and dead matter. This philosophical stance has roots going back to ancient Egypt, but the emergence of this philosophy at the end of the Renaissance was mainly influenced by Aristotle's concept of entelechy. His view of entelechy (from the Greek *telos* meaning target) is directly connected with his theory of matter and form, the two fundamental components of any real thing. In living organisms, the entelechy is an inner concept and target that stimulates and forces matter to realize its final form and function. Aristotle's theory of entelechy has influenced the European understanding of life and living organism for almost 2000 years.

All the different versions of vitalism that were developed in the sixteenth century and later have one fundamental aspect in common, namely the existence of a *vis vitalis* (life energy) in all living organisms. Furthermore, this *vis vitalis* was thought to be inherent in all organic matter that constituted an organism, and it was thought to survive in the matter remaining after the death of an organism. Prominent representatives of the early period of vitalism were Jan Baptist van Helmont (1577–1644), Georg E. Stahl (1659–1734), Francis Glisson (1597–1677), Marcello Malpighi (1628–1694), Albrecht von Haller (1708–1777), Theophil de Bordeu (1722–1776), Caspar F. Wolff (1733–1794), and Johann F. Blumenbach (1752–1844). The physician J. F. Blumenbach cut up freshwater polyps and proved that the removed parts can regenerate, and established the theory of epigenesis in the life sciences. The German physiologist Johannes P. Müller (1801–1858) explained in his book on physiology (*Handbuch der Physiologie*), which became a leading textbook in the nineteenth century, that all the complex functions of the human body cannot be explained by a simple mechanistic concept exclusively based on physical laws, so that a vitalistic concept was unavoidable. Although vitalism was attacked more and more by various scientists in the second half of the nineteenth and throughout the twentieth century, the German biologist Hans Driesch (1867–1941) maintained his vision of entelechy and vitalism until the beginning of World War II. The recent work of the British biologist Rupert Sheldrake on morphogenesis, presented in his book *The Presence of the Past* (1988), may also be understood as a late resonance of vitalism.

Among the early critics, the Scottish physiologist John S. Haldane (1860–1934) should be mentioned. He wrote in 1931: “Biologists have almost unanimously abandoned vitalism as an acknowledged belief.” Ernst Mayr (1904–2005), co-founder of the modern synthetic theory of evolution (see Sects. 8.3 and 8.4), wrote a more balanced comment:

> It would be ahistorical to ridicule vitalists. When one reads the writings of one of the leading vitalists like Driesch one is forced to agree with him that many of the basic problems of biology cannot be solved by a philosophy as that of Descartes, in which the organism is simply considered a machine. . . . The logic of the critique of the vitalists was impeccable. .. . . Vitalism has become so disreputable a belief in the last 50 years that no biologist alive today would want to be classified as a vitalist. Still, the remnants of vitalist thinking can be found in the work of Alistair Hardy, Sewall Wright, and Charles Birch, who seems to believe in some sort of nonmaterial principle in organisms.

Although biology was the main playground for vitalists, vitalism had two important consequences for chemistry. First, all chemical compounds were subdivided into two categories, organic and inorganic. Inorganic chemicals were metals, metal salts including minerals and stones, inorganic acids such as hydrochloric or sulfuric acid, and inorganic bases such as sodium hydroxide. Organic compounds were all the chemicals involved in the metabolism of living organisms, including all the biopolymers such as cellulose, chitin, and proteins. These organic compounds were endowed with the *vis vitalis*, which survived after the death of the organism. This subdivision entailed a second conclusion, namely the existence of two realms of chemical reactions, syntheses, and modifications of organic compounds, on the one hand, and the syntheses and modifications of inorganic compounds, on the other. Hence, synthesis of an organic compound endowed with the *vis vitalis* from inorganic chemicals classified as dead matter was thought to be impossible. This second conclusion was, for instance, advocated by the famous Swedish chemist Jöns J. Berzelius (1777–1844). Although the second conclusion proved incorrect (see below), these consequences of vitalism and the response of chemists in the nineteenth century made a significant contribution to the emancipation of modern chemistry from alchemy.

From the viewpoint of modern chemistry, organic chemicals are compounds of the element carbon and in almost all cases also contain hydrogen atoms. Between 1850 and 1950, numerous chemists demonstrated that carbon differs from all other elements in that it is capable of forming with itself three different types of chemical bond (single, double, and triple bonds) and to form stable bonds with elements such as hydrogen, oxygen, nitrogen, and sulfur, which are present in almost all compounds of a living organism. The unique properties of carbon also enable the formation of large molecules consisting of hundreds or thousands of atoms (see Sects. 9.4 and 9.5). These macromolecules may have a linear, branched, cyclic, or crosslinked (network) architecture. It is this multitude of chemical structures and properties of carbon compounds and not a *vis vitalis* that forms the basis of all living organisms.

The discovery that organic compounds can be synthesized from inorganic starting materials in a laboratory experiment was one of several paradigm changes that promoted the development of modern chemistry. The man who is famous for this discovery is the German chemist Friedrich Wöhler (see Biography). In 1824, he succeeded in preparing oxalic acid (HO-OC-CO-OH) from dicyan (cyanogen; NC-CN). Oxalic acid is an organic acid belonging to the metabolism of many plants. Dicyan is bare of hydrogen, does not belong to the metabolism of living organisms and, hence, may be understood as an inorganic chemical. Even more convincing was the second experiment performed in 1828, the synthesis of urea from ammonia and cyanic acid ($NH_3 + HOCN \rightarrow NH_2CO\text{-}NH_2$; the cyanic acid was used in the form of sodium cyanate). Urea is a metabolite of the human organism and excreted daily via the kidneys, whereas ammonia and cyanic acid or its salts are poisonous inorganic chemicals.

In later decades and up to the twenty-first century, Wöhler's synthesis of urea was hailed as one of the most important chemical reactions of all times. In 1982, for the centenary of the urea synthesis, the German Bundespost printed a stamp displaying the formula of urea. Douglas McKie and more recently Peter J. Ramberg have clearly elaborated and published that Wöhler himself and his contemporary colleagues had a more temperate and prosaic view of these experiments and did not expect or intend to eliminate vitalism from any kind of natural sciences or philosophy.

Wöhler had certainly not devoted his entire life to a fight against vitalism and, as shown below, was successful in various fields of chemistry. In the first years after the urea experiment he was not sure whether the cyanic acid was free of the *vis vitalis*. In a letter to Berzelius, one of the most important and prominent chemist of the early nineteenth century, he wrote: "A philosopher might say that the organic character (meaning the *vis vitalis*) has not completely vanished from the animal charcoal used for the preparation of cyanic, so that an organic compound might well be synthesized from it." Yet about 10 years later he was more determined in his comments. Together with Justus von Liebig (1803–1873), another prominent chemist of that time, he wrote in the introduction of a treatise about the nature of urea (*Über die Natur des Harnstoffs*): "The philosophy may conclude from this work that the production of all organic compounds, if not belonging to a living organism, in a laboratory is not only highly probable, but should be considered a certainty. Sugar, salicene, morphine will artificially be synthesized, although we don't know the synthetic methods that will enable us to achieve these results, because we don't know the starting materials and reaction intermediates needed for their syntheses, but we will certainly acquire the necessary knowledge."

Wöhler's experiments were certainly decisive for the elimination of vitalism from chemistry, but they contributed little to an early eradication of vitalism from biology and philosophy. Biologists and philosophers were either not informed about Wöhler's experiments or they underrated and misunderstood the importance and consequences of his work. It is difficult to estimate to what extent Wöhler's syntheses of oxalic acid and urea stimulated other chemists to attempt syntheses of further organic compounds from various starting materials. However, the number of organic syntheses either from inorganic and/or organic chemicals increased rapidly after 1828, and a few examples should highlight this progress:

In 1831, T. C. Pelouze prepared α-amino acids from hydrocyanic acid
In 1845, L. H. F. Melzer prepared acetic acid from trichloroacetic acid, which was synthesized by A. W.H. Kolbe (1818–1884) from tetrachloroethylene
In 1846, A. W. H. Kolbe and E. Frankland prepared acetic and propionic acid from methyl cyanide and ethyl cyanide, respectively
In 1856, the British chemist W. H. Perkin (1838–1917) synthesized the dyestuff Mauvein from chemicals extracted from coal tar
In 1878, Adolf von Bayer, Nobel laureate in chemistry 1905, synthesized indigo, a dyestuff still used in the twenty-first century for the production of blue jeans
In 1897, Felix Hoffmann and Arthur Eichengrün, members of Farbenfabriken Bayer AG, elaborated the first technically useful synthesis of acetyl salicylic acid (ASA, aspirin)

In the second half of the nineteenth century, numerous chemical companies were founded in various European countries. The main activities of those chemical companies were the invention and production of dyestuffs and drugs. This process was particularly successful in Germany and at the beginning of World War I, Germany owned more than 50 % of all chemical patents worldwide and was nicknamed the "Pharmacy of the world."

Friedrich Wöhler

Friedrich Wöhler was born on 31 July 1800 in Eschersheim, a suburb of Frankfurt am Main, as son of the veterinarian A. A. Wöhler. From 1820, Wöhler studied chemistry in Marburg, but he moved to Heidelberg in 1821 to study medicine and chemistry. As a result of his increasing interest in chemistry, his supervisor Leopold Gmelin arranged a position for him in the laboratory of Berzelius in Stockholm, where he was trained in analytical chemistry. After returning to Germany he taught chemistry at the *Gewerbeschule* (technical high school) in Berlin from 1825 to 1831. During this time he translated Berzelius's textbook *Animal Chemistry* into German. The German version *Tier Chemie* was published in Leipzig in 1831. In 1828, he was honored with the title of Professor and moved in 1831 to Kassel, where he held a professorship at the technical college. In 1836, he was appointed full professor of chemistry, medicine, and pharmacy at the University of Göttingen. He was awarded the title Honorary Citizen, and stayed in Göttingen until his death in 1882. In 1890, the city of Göttigen named a square after him and erected a bronze statue.

Among the numerous merits of Wöhler's work, the following examples are particularly noteworthy. Wöhler and von Liebig analyzed the oil of bitter almonds and published in 1834 a theory claiming that certain groups composed of carbon, oxygen, and hydrogen can behave like an element and be exchanged for elements in chemical compounds. This "theory of radicals" had a profound influence on the formulation and understanding of chemical structures. Wöhler collected and analyzed meteorites and possessed a large private collection of meteorites. In this connection, he discovered that certain meteorites contain organic compounds. After several highly reputed chemists had failed to detect and isolate the element aluminum, in 1827 Wöhler succeeded in preparing it by reduction of aluminum chloride with potassium. In the same way, he isolated for the first time the elements beryllium, titanium, and yttrium. In 1824, Berzelius had produced an impure amorphous form of the element silicon. Wöhler succeeded in preparing it in a purer, crystalline form, similar to that needed (after further purification) for the production of semiconductors and wafers. Hence, Wöhler may be seen as the chemical grandfather of computers and solar cells. Finally, the synthesis of calcium carbide is worth mentioning. In the twentieth century calcium carbide became an important source of acetylene, a versatile chemical enabling syntheses of many useful organic compounds.

Among the honors Wöhler received during his lifetime, the order Pour le Mérite for science and art (1864) is noteworthy. In a suburb of Vienna and on the campus of the Technical University of Darmstadt alleys have been named after him. Furthermore, a high school in the city of Singen and a crater on moon carry his name. Finally, a statement published 1882 in the supplement of the journal *Scientific American* needs citation: "for two or three of his researches he deserves the highest honor scientific man can obtain, but the sum of his work is absolutely overwhelming. Had he never lived, the aspect of chemistry would be very different from that it is now."

Bibliography

Bechtel W, Richardson RC (1998) Vitalism. In: Craig E (Ed) Routledge encyclopedia of philosophy. Routledge, London

Bedau MA, Cleland CE (2010) The nature of life: classical and contemporary perspectives from philosophy and science. Cambridge University Press
Hankinson RJ (1997) Cause and explanation in ancient Greek thought. Oxford University Press
Hoppe B (2007) Review of "The life and work of Friedrich Wöhler" by Robin Keen (Buttner J. ed.). Isis 98(1):195–196
Jidenu P (1996) African philosophy, 2nd edn. Indiana University Press
Keen R (2005) The life and work of Friedrich Wöhler. Bautz
Kie M (1944) Wöhlers synthetic urea and the rejection of vitalism: a chemical legend. Nature 153:606–610
Ramberg PJ (2000) The death of vitalism and the birth of organic chemistry. Wöhler's urea synthesis and the disciplinary identity of organic chemistry. Ambix 47:170–195
Schummer J (2003) The notion of nature in chemistry. Stud Hist Philos Sci 34:705–736
Schwedt G (2000) Der Chemiker Friedrich Wöhler. Hischymia, Seesen
Wöhler F (1828) Über künstliche Bildung des Harnsstoffs. Annalen der Physik und Chemie 88(2):253–256

Links
Dictionary of the history of ideas. http://etext.lib.virginia.edu/cgi-local/DHI/dhicontrib2.egi?id=dv3-69
Mayr E (2002) The Walter Arndt lecture: the autonomy of biology. http://www.biologie.uni-hamburg.de/b-online/e01_2/autonomy.htm

9.4 Do Macromolecules Exist?

Errors tend to be tenacious but slowly truth is eating away at them.
(Rudolf G: Binding)

As already mentioned in Sect. 9.1, the Italian chemist Amedeo Avogadro (1776–1856) coined the term "molecule" for the smallest unit of a substance built up from two or more atoms. Correspondingly, a giant molecule, labeled "makromolecule" by H. Staudinger (1888–1965) in 1922, consists of hundreds or even thousands of small molecules connected by stable covalent bonds. An alternative term, coined by the Swedish chemist Berzelius in 1833 is "polymer," a Greek word meaning a connection of many small molecules, monomers. For a better understanding of the role polymers play in the everyday life of western civilization it is reasonable and advisable to distinguish two large groups, synthetic polymers and biopolymers.

Biopolymers are produced by any kind of living organisms, microbes, plants, animals, or humans. Their mass fraction of the dry weight of an organism usually amounts to approximately 90 %, with bones and teeth making up the remaining

10 %. In the case of humans and vertebrates, about 95 % or more of the dry biopolymers are proteins and peptides, polymers that are built up of α-amino acids. The surface of a human exclusively consists of proteins from the hair down to the toenails. Furthermore, the muscles, cartilage, and tendons also consist of proteins. In contrast, the dry mass of plants mainly consists of poly(saccharide)s, biopolymers built up from monosaccharides (sugars). An alternative name is carbon hydrates, because the structure of polysaccharides can formally be described as a combination of equal numbers of carbon atoms and water molecules ($C + H_2O$). The biopolymer produced in largest amounts is cellulose, which is synthesized by all plants and is responsible for their mechanical stability. Another polysaccharide produced by plants in large quantities is starch, a major component of human nutrition (e.g., corn, grain, rice, potatoes). Both cellulose and starch consist of glucose units (generated by photosynthesis) but the bond between the repeat units is different. Whereas the production and utilization of synthetic polymers only began about 100 years ago (the first commercial polymer was Bakelite), humankind has utilized biopolymers for more than 10,000 years, mainly in the form of clothes.

Among the earliest activities of men in this direction are unhairing and dressing of hides, skins, and pelts for the production of leather clothing, leather straps or belts, and parchment.

More than 1000 years ago, the Chinese developed the spinning of silk fibers and the weaving of silk fabrics, which later became the ultimate fashion in Rome. Shearing of sheep and spinning of wool yielded warm clothes. Production of papyrus and true paper from the cellulose fibers of various plants is another example of the early utilization of biopolymers and it is impossible to underestimate the role of paper in the history of the past 500 years.

However, cellulose fibers have also played an important role in other areas of human civilization. Hemp ropes were decisive for the construction and handling of sailing ships for more than 2000 years, and clothes made from cotton are still attractive and widely used in the twenty-first century. Furthermore, the first stage in the evolution of photographic films and movies was based on chemically modified cellulose. Last but not least, all smokeless gunpowder and dynamite-related explosives since the end of the nineteenth century have been based on cellulose. At this point it should be remembered that explosives play an important role in our modern civilization because mining of coal and metals requires explosives, as well as the construction of roads, railway tracks, tunnels, and canals.

A later development in our civilization was the application of natural rubber, poly(isoprene). The Mayas in Central America used rubber in small quantities more than 1000 years ago. Systematic studies of its usefulness in our modern civilization began in the first half of the nineteenth century. Charles Goodyear's discovery in 1848 of vulcanization, which transforms the soluble natural rubber into a technically useful elastomer, was a decisive step forward. Seals, tires, and electric cables are the most widely used applications today.

These examples suffice to demonstrate that the formation of polymers by stable (covalent) connection of small molecules (monomers) entails specific properties, the utilization of which have had a considerable influence on the progress of our civilization over the past 100 years. Because of this scenario, it is difficult to understand from the viewpoint of modern chemistry and techniques that before 1925 almost all experts, whether chemists or technicians, believed that polymers are nothing but aggregates of small molecules such as monomers, dimers, or trimers. Aggregates means in this context that the small molecules adhere to each other by relatively weak electrical forces, a paradigm called "older micelle theory." The modern view of macromolecules resulting from Staudinger's paradigm change interprets the structure of polymers as one big (macro)molecule where all atoms are connected to each other by strong covalent bonds. Characteristic properties resulting from this structure are, for example, stiffness, high tensile strength, flexibility, and elasticity.

It is, of course, absolutely unlikely that all the experts who were active between 1915 and 1930 were far less intelligent than those who lived later. Hence, the question arises, why did almost all experts favor the older micelle theory and decline Staudinger's new concept. There were many experimental results that looked at first glance like reliable evidence in favor of the micelle theory. Furthermore, there was little information on the nature and strength of weak electronic interactions or partial valences that were assumed to be responsible for the association of small molecules.

A particularly influential enemy of Staudinger's concept was Emil Fischer (1852–1919), professor of organic chemistry at the University of Berlin. Fischer had an extraordinarily high reputation all over Europe and he was the second Nobel laureate in chemistry. He had synthesized peptides (the smaller analogs of proteins) in a stepwise manner up to a length of 18 amino acids, corresponding to a mass of 2100 Da. Furthermore, he had prepared polysaccharides up to a mass of 4000 Da. In 1913, at a conference of scientists in Vienna he declared that the molar mass of his best polysaccharide was higher than the molar mass of any natural protein and stated that living organisms never synthesize biopolymers of higher mass.

Prof. C. Harries, expert in the chemistry and structure of natural rubber, published in 1905 the hypothesis that rubber was an aggregate of dimethyl cyclooctadiene. Thiele's theory of "partial valences" exerted by the double bonds of olefinic compounds supported the hypothesis of Harries. Furthermore, X-ray studies of natural rubber, partially crystallized under mechanical stress, apparently confirmed the low molar mass of the rubber molecule. X-ray studies of crystalline cellulose were also interpreted as evidence in favor of short oligomers as the basic constituents of this polysaccharide. H. Bergmann and E. Knoche reported that chemically modified starch (amylose triacetate) had a mass of 288 Da after dissolution in phenol (rapid degradation of the long amylose chains was not taken into account). Moreover, several polymers studied at that time, such as poly(styrene), poly(formaldehyde), and later poly(methyl methacrylate), could be degraded back to their monomers upon heating in a good vacuum.

Characteristic for the attitude of most experts towards polymer chemistry before 1930 is a statement by Heinrich Wieland (1877–1957). Wieland was professor of organic chemistry at the University of Freiburg im Breisgau, and he was awarded the Nobel Prize in 1937 for his work on steroids and alkaloids. In 1926, Staudinger was his successor as director of the Institute of Organic Chemistry, and when both colleagues met in Freiburg im Breisgau. Wieland gave Staudinger the friendly advice: "Dear colleague, you better abandon the idea of giant molecules; organic molecules having molar masses above 5000 (Da) do not exist. Purify your products, such as natural rubber, intensively and you will see they consist of low molar mass compounds and will crystallize."

However, in the years before 1920 Staudinger had begun to study the polymerizations of isoprene and formaldehyde and the chemical modification of natural rubber, for example by addition of bromine. Based on these results, he published in 1920 a famous article in *Berichte der Deutschen Chemischen Gesellschaft* in which he outlined his revolutionary concept that high molar mass polymerization products (including biopolymers) contain long polymer chains consisting of a large number of covalently connected repeat units. Clear evidence for the existence of long covalent polymer chains was elaborated by Staudinger and Fritschi in 1922 via a first example of a so-called polymer analogous reaction. Natural rubber was thoroughly hydrogenated so that all double bonds and assumed partial valences were saturated, and high molar mass poly(alkane) chains were obtained and not the volatile cycloalkanes (e.g., dimethyl cyclocotane) expected from the theory of Harries. Other examples followed, for example, the perfect hydrogenation of polystyrene, which yielded poly(vinyl cyclohexane).

In 1927, Staudinger and coworkers reported X-ray crystallographic studies of poly(formaldehyde), which definitely proved that these polymers consisted of long chains, although only four repeat units occupied the crystallographic unit cell. In other words, X-ray crystallography proved to be useless for determination of chain lengths.

However, from X-ray fiber patterns of crystalline cellulose and proteins Hermann Mark (1895–1992), an expert in the field of polymer characterization (see Biography), concluded that natural polysaccharides and proteins contain much longer chains than predicted by Fischer and Wieland, Furthermore, it was found that polysaccharides degrade rapidly in warm phenol so that the previously reported low molar masses were incorrect. Molecular weight measurements by means of ultracentrifuge and light scattering also contributed to the increasing acceptance of Staudinger's concept after 1930. However, its final breakthrough was delayed by the "newer micelle theory" of Kurt. H. Meyer (1882–1956) and H. Mark. Meyer was director of polymer research at BASF AG (part of I.G. Farben at that time) and for a short period of time was supervisor of Mark. Both scientists agreed with Staudinger in that covalent macromolecules containing up to 50 repeat units may exist. Yet, they attributed the typical colloidal properties of high molar mass polymers to an aggregation of short polymer chains. It was as late as 1940 before Meyer and Mark eventually conceded that Staudinger was right with his postulate that giant covalent macromolecules exist.

Staudinger's victory over his scientific enemies had tremendous consequences for the chemical industry and for the everyday life of humankind. After 1920, an increasing number of Staudinger's students entered the chemical industry and slowly convinced managers that the production of polymers might become a

Table 9.3 Groups and applications of synthetic polymers

Groups	Applications
Engineering plastics	Boat building, bumpers, cases and boxes for batteries and electric apparatus, household equipment, computers, ball-points, spectacle frames
Fibers and yarns	Textiles, sails, parachutes, awnings, sport bags, ropes, fishing lines
Films and foils	Packaging foils, bags, tilts, magnetic tapes, movies, membranes
Elastomers	Tires, belts for technical applications (e.g., transportation), cables, flexible tubes, pipes, hoses, seals
Solid or flexible foams	Packaging materials, insulating materials for heat storage or refrigerators
Paints and lacquers	Polymers form the protecting layers on wood or metals and fix the dyestuffs and pigments
Adhesives and glues	Fixation of paper on paper or other materials, fixation of wood on wood and other materials, fixation of ceramics on ceramics or other materials, fixation of metals on metals
Solid phases	Production of distilled water and for chromatography

profitable business. Before World War I, Bayer AG had already started to produce the first synthetic rubber (methyl rubber). In 1926, BASF began to explore the feasibility of technical production of polystyrene, which was commercialized in 1929. In the same year, a cooperation of Bayer and BASF succeeded in the synthesis of butadiene–styrene copolymers, elastomers that are still used in the twenty-first century for tires. At almost the same time, the technical production of poly(formaldehyde), poly(vinyl chloride), and poly(vinyl acetate) were explored and the production of nylon-6,6 (see Chap. 10) and nylon-6 followed in 1939. After World War II, the number of commercial polymers increased exponentially and polymers invaded all kinds of industry and all areas of human life. Tables 9.3 and 9.4 provide a short summary of the most important applications.

In summary, Staudinger's work is an excellent example of an insight formulated by the German philosopher Arthur Schopenhauer (1856–1915): "All truth passes through three stages. First, it is ridiculed; second, it is violently opposed; third, it is accepted as if it is self-evident."

Hermann Staudinger

Hermann Staudinger was born on 23 March 1881 in Worms as son of the high school professor Franz Staudinger and his wife Auguste (nee Wenck). Staudinger had one sister and two brothers. His father was active in a labor movement and forced Staudinger to learn joinery for several months for a better understanding of the life and mentality of the working class. After graduating from high school he entered the University of Halle to study chemistry. During the following 3 years, he also attended lectures at the Universities of Darmstadt and Munich, but returned to Halle where he received a Ph.D. in 1903. Immediately afterwards he became assistant to Prof. Johannes Thiele at the University of Straßburg, where he began to develop the chemistry of ketenes. After his habilitation (tenure) in 1907, he was appointed associate professor at the Technical University of Karlsruhe. During this

Table 9.4 Synthetic polymers in medicine and health care

Syringes	
Frames of spectacles	Plasters
Unbreakable lenses	Elastic bandages
Bags for infusion	Insoles and arch supports
Parts of artificial joints	Cardiac valves
Flexible tubes and pipes	Bone cement
Tissue-separating films	Resorbable sutures
Hearing aids	Temporary inlays in dentistry
Stents	Packaging of drugs

working period, he isolated and characterized numerous interesting and useful organic substances, such as a synthetic coffee flavor.

In 1912, he took on a full professor position at the ETH Zürich (Swiss Federal Institute of Technology). Here, he began to work on natural rubber and became more and more interested in polymer science. However, he worked mainly on low molar mass organic compounds and his broad working field encompassed the following topics:

1. Syntheses and reactions of ketenes
2. Diazomethane and reactions of CH_2
3. Reactions of oxalyl chloride
4. Reactions of organic azides with triphenylphosphine
5. Synthetic pepper
6. Insecticides
7. Synthetic flavors of coffee
8. Syntheses of certain drugs
9. Isoprene: synthesis and polymerization

In 1926, he accepted a position at the University of Freiburg im Breisgau as director of the central organic laboratory. Here he spent the rest of his career until his retirement in 1951. During this time he performed numerous important studies of the structure and properties of natural and synthetic polymers, which found more and more acceptance in the international community of chemists, so that he was eventually awarded the Nobel Prize. Furthermore, it is worth mentioning that Staudinger founded in1949 the first journal of polymer chemistry, *Die Makromolekulare Chemie*.

For a better understanding of Staudinger's character and his role as highly reputed professor in Germany, it is necessary to take a look at Staudinger's political convictions. Staudinger was a pacifist and kept away from any political activities. In contrast to many other German professors he systematically refused to sign any nationalistic proclamation before and after World War I. Together with Albert Einstein and Max Born, he also refused to make any contribution to the development of chemical weapons, such as poisonous gases. A fierce dispute with Prof. Fritz Haber about his leading role in the gas war of 1916–1918 is well documented.

When the Nazis came to power in the spring of 1933, Staudinger maintained his pacifistic attitude and refused to become a member of the Nazi party. In contrast, the majority of German professors after 1933 confessed to National Socialism. The University of Freiburg im Breisgau was at that time the focus of public attention, because in May 1933 the famous philosopher Martin Heidegger had been elected Rektor (head) of the university. Heidegger was not a racist, but he glorified Hitler; he admired his rigorous leadership and tried to eliminate all democratic structures and elements from the administration of the university. Furthermore, he favored and stimulated denunciation of nonconformist professors and academic staff and tried, mostly with success, to remove them from their positions. Staudinger as one of the most famous professors of the University of Freiburg im Breisgau was a particular challenge for Heidegger, but in this case Heidegger did not succeed.

When, after the end of World War II, Heidegger did not confess his political aberration and the weakness of his character, he was forced in 1951 to quit his position at the university. In contrast, Staudinger was awarded the "Bundesverdienstkreuz," the highest honor of the new Federal Republic of Germany, and the Nobel Prize followed in 1953. Furthermore, he became Honorary Citizen of Freiburg im Breisgau and several decades later a high school was named after him. Moreover, the new institute of macromolecular chemistry built in Freiburg im Breisgau after 1962 was named "Staudinger Haus." In its entrance hall the Society of German Chemists and the American Chemical Society installed a bronze plaque honoring this place as the *Ursprung der Polymerwissenschaften* (origin of polymer science).

In his long fight against the prejudices of his colleagues, Staudinger was strongly supported by his wife Magda. Dr. Magda Staudinger (1902–1997, nee Woit) was the daughter of the Latvian ambassador and had studied biology. Therefore, she was familiar with the structure and properties of biopolymers and understood the scientific work of her husband. After Staudinger's death on 8 September 1965, she edited seven volumes of the *Collected Works of Hermann Staudinger* and supervised, until her death, editing of the journal *Die Makromolekulare Chemie*. Staudinger and his wife are buried in the Hauptfriedhof (main cemetery) in Freiburg im Breisgau.

Hermann Francis Mark

Hermann Mark was born on 3 May 1895, in Vienna as the oldest of three children of the physician H. C. Mark and his wife Lili (nee Mueller). Mark attended elementary school and high school in Vienna. He became interested in chemistry in the age of 12, because the father of a friend taught science at the university and allowed the boys to tour the chemical and physical laboratories. Furthermore, Franz Holloway, his teacher of mathematics and physics had a strong influence. Because both boys had access to chemicals through their fathers, they soon started to conduct chemical experiments at home. After graduating from high school, World War I broke out and Mark had to serve in the Austro-Hungarian Army. He soon obtained the rank of officer in the elite K.K. Kaiserschützen Regiment Nr. II. He was seriously wounded in the battle of Mount Ortigara (located in the Alps south of the Dolomites), but he was also highly decorated and admired as an Austrian hero.

During his stay in a hospital in Vienna, Mark began to study chemistry. He completed his studies with a Ph.D. at the end of 1921. He then accompanied his supervisor, Prof. Wilhelm Schlenk to the University of Berlin as assistant. One year later, he switched to a position that Prof. Fritz Haber had offered to him at the new Institute of Fiber Chemistry (part of the famous Kaiser-Wilhelm Institute, the forerunner of the Max-Planck Society). In this position he specialized in the new analytical methods, such as X-ray scattering and ultramicroscopy. He focused his studies on biopolymers, such as cellulose, silk, and wool, and became expert in crystallography. The Nobel Prize laureate Linus Pauling learned to apply X-diffraction to proteins from Mark.

In 1926, Mark accepted an offer from Dr. Kurt Meyer, research director of BASF, part of I.G. Farben at that time. He worked as vice director of research on the characterization of polymers such as polystyrene, poly(vinyl chloride), poly(vinyl alcohol), and synthetic rubber to support technical production and commercialization of these new materials. BASF was the first European chemical company which, with the help of former students of Staudinger, attempted to build up the technical production of synthetic plastics. When the Nazis came to power, Mark's plant manager recognized that Mark's existence was endangered because he was a foreigner and son of a Jewish father. Mark followed the advice of his manager and returned to Austria as professor of physical chemistry at the University of Vienna.

In the fall of 1937, in Dresden, Mark met C. B. Thorne, a representative of the Canadian International Pulp and Paper Company, who offered him a position as research manager in Ontario. The company intended to modernize the pulp production for the purpose of making cellulose acetate, cellophane, and rayon. At this time, Mark was not very interested in this position, but he changed his opinion in the following year. In the spring of 1938, Austrian and German Nazis cooperated to arrange the *Anschluss*, the political union of Austria and Germany. Mark was arrested by the Gestapo and stripped of his passport, although he had a Christian education because his father had converted to Christianity after his marriage. A few weeks later he was released with the warning not to contact Jews. After paying a bribe worth a year's salary he retrieved his passport with a visa to enter Canada and to pass through Switzerland, France, and England. In April, Mark and his family drove to England in their own car. In September, Mark reached Montreal and stayed in Canada for nearly 2 years. In 1940, he went to the USA and joined the Polytechnic Institute of Brooklyn, where he was appointed associate professor. He established a polymer program including research and teaching, the first undergraduate polymer education in the USA.

In 1944, Mark obtained a full professor position and in 1946 he established the Polymer Research Institute at Polytechnic Institute of Brooklyn. This was the first research center in the USA dedicated to polymer science, and Mark was later recognized as a pioneer in establishing the curriculum and pedagogy for polymer science. In 2003, the American Chemical Society titled Mark's research institute a Historic Chemical Landmark.

Mark received numerous decorations and awards, for example, the Austrian Decoration for Science and Art (1965), the Elliot Cresson Medal (1966), the William Gibbs Award (1975), the Harvey Prize (1976), the National Medal of Science (USA) (1979), the Wolf Prize in Chemistry (1979), the Perkin Medal (1980), and the Charles Goodyear Medal (1988).

Mark died on 6 April 1992, in Texas, but his ashes were buried in Vienna where an alley was named after him.

Bibliography

Breinbauer R (2004) The Staudinger ligation—a gift to chemical biology. Angew Chem Int Ed 43(24):3106

Chemical Heritage Foundation; Herman Francis Mark. http://www.chemheritage.org/explore/polymark.html

Hermann Mark and the Polymer Research Institute. http://www.qcc.cuny.edu/NYACSReport2003/HistPolymer.htm

John E, Martin B, Mück M, Oh H (1991) Die Freiburger Universität in der Zeit des Nationalsozialismus. Ploetz, Freiburg i.Br

Morawetz H (1985) Polymers—the origins and growth of a science. Wiley, New York

Mülhaupt R (2004) Hermann Staudinger and the origin of macromolecular chemistry. Angew Chem Int Ed 43(9):1054

National Academy of Sciences Biographical Memoir. http://www.nasonline.org/publications/biographical-memoirs/memoir-pdfs/mark-h-f.pdf

Staudinger H (1920) Über polymerisation. Ber Dtsch Chem Ges 53:1073

Staudinger H (1924) Über die Konstitution des Kautschuks. Ber Dtsch Chem Ges 59:3014

Staudinger H (1932) Die hochmolekularen organischen Verbindungen Kautschuk und cellulose. Springer, Heidelberg

Staudinger H (1961) Arbeitserinnerungen. Hüthig & Wepf, Basel

9.5 Invention of Nylon

> A subtle thought that is in error may give rise to fruitful inquiry that can establish truth of great value.
>
> (Isaac Asimov)

All the numerous methods developed for the preparation of polymers from monomers can be subdivided into two categories, chain-growth polymerizations and step-growth polymerizations. Chain-growth polymerizations are characterized by the fact that only one sort of growing steps exists, namely the reaction of monomers with a reactive end group of the growing chain. Monomers cannot react with monomers and polymers cannot react with polymers (termination steps in radical polymerizations are an exception). All the polymerizations performed by Staudinger and coworkers or by members of I.G. Farben before 1930 belong to this category.

In the case of step-growth polymerization, all monomers possess two reactive groups and react with each other to yield dimers. The dimers react with monomers to yield trimers or they react with each other and generate tetramers, which can, in turn, react with monomers, dimers, trimers, and other tetramers. This process continues until long chains are formed, whereby all species in the reaction mixture can react with each other at any time. Hence, the success of a step-growth polymerization is based on the principle that the reactivity of all end groups are nearly identical and independent of the lengths of oligomers and polymers.

However, before 1930, Staudinger, like the vast majority of organic chemist, believed that the reactivity of end groups rapidly decreases with increasing length of the oligomers. If this hypothesis was correct, it would be impossible to synthesize long chains (e.g., more than 20 repeat units) via step-growth polymerizations. In agreement with this fallacy, Staudinger, his coworkers, and admirers never studied step-growth polymerizations. Wallace H. Carothers (1896–1937), who together with his coworker Paul J. Flory (1910–1985), eventually revised this incorrect paradigm, shared Staudinger's view at the beginning of his career at DuPont. In his own words: "there can be no question that the reactivity of functional groups diminishes with the size of the molecules."

The chemical company E. I. DuPont de Nemours was founded in 1801 by a former assistant of Antoine-L. Lavoisier (see Sect. 9.3). Eleuthere Irenee DuPont had learned how to produce gunpowder of prime quality when working in the royal powder mills of France. To escape the French Revolution, he emigrated together with his father from France to the USA in 1799. Because their gunpowder was of better quality than that produced by their competitors, they became the preferred suppliers of gunpowder to the American army. At the end of World War I, DuPont was the most important producer of gunpowder and explosives in the USA.

After the end of World War I, the need for gunpowder and explosives decreased dramatically and DuPont was forced to find new working fields. One of the consequences resulting from this situation was the decision to finance fundamental research and to hire a promising young chemist, Wallace H. Carothers, who held a position as instructor of chemistry at Harvard University. Two renowned professors, R. Adams and C. S. Marvel, had refused to quit their academic positions and, thus, DuPont was left with only a few alternatives. Therefore, DuPont offered a contract that, in addition to a high salary, permitted Carothers to concentrate on fundamental research.

Carothers had a great interest in polycondensations, the most versatile and important part of step-growth polymerizations. His primary ambition was to synthesize a polymer having a molar mass above 4200 g/mol, the maximum achieved by the Nobel laureate Emil Fischer (1852–1919) via stepwise synthesis of polysaccharides. Together with two coworkers (later three), Carothers began in 1927 to study the synthesis of aliphatic polyesters. However, until the end of 1930 the group did not succeed in overcoming the magic limit of 5000 g/mol. He interpreted this failure on the basis of two dogma. First, he still suspected that the reactivity of end groups decreased when the chain length increased. Second, he had doubts about the thermal stability of his polyesters. All the chain-growth polymerizations known at that time were conducted at temperatures below 150 °C, but Carothers needed temperatures around 240 °C for his polycondensations.

At this point, a second fallacy of Staudinger's came into play. From physico-chemical studies of oligoalkanes (diesel, paraffin, etc.) Staudinger had concluded that high molar mass poly(alkane)s should be unstable at temperatures above 200 °C. In a quite different context, Carothers had synthesize poly(alkane)s with molar masses (M_n) in the range of 1000–2000 g/mol. Carothers could demonstrate that these poly(alkane)s are stable for a short time even up to temperatures as high as 400 °C. Carothers and his coworker Dr. Julian Hill improved their condensation apparatus and applied temperatures up to 240 °C. The first experiment proved successful and a polyester having a molar mass above 10,000 g/mol was obtained. Furthermore, this "superpolyester" allowed fibers to be drawn from the melt. Carothers and DuPont now had the vision that production of synthetic textile fibers might be feasible.

This breakthrough had two consequences. First, DuPont hired the young Ph.D. chemist Paul J. Flory as coworker in the research group of Carothers. Flory was a physical chemist and theoretician and had the objective of analyzing the kinetic course of polyester syntheses and characterizing their molar masses in more detail. The syntheses of superpolyesters and Flory's measurements definitely proved that the reactivity of oligo- and polyesters did not vary with chain length. The paradigm change was complete. The realization of this paradigm change, and not the invention of nylons (as frequently reported), was the greatest performance and success of Carothers and Flory because it laid the basis for all successful step-growth polymerizations in the future. After the death of Carothers, Flory had a successful academic career and was awarded the Nobel Prize in 1974.

The physical and mechanical properties of polyesters were disappointing. The melting temperatures were so low that they formed a viscous, sticky melt in contact with boiling water and they were soluble or became swollen in numerous organic solvents, such as acetone, ethyl acetate, or chloroform. This fiasco prompted the second consequence. Carothers now attempted to realize his vision of a synthetic textile fiber by using another chemically more stable and high melting class of polycondensates, namely polyamides, nicknamed "nylons." The group synthesized a broad variety of slightly different nylons to study and compare their properties. Eventually Dr. Bolton and the management of DuPont decided to concentrate their efforts and technical production on nylon-6,6, which allowed relatively inexpensive production from 1,6-diaminohexane and adipic acid. The technical production started in 1938 and the first fabrics with high tensile strength, as well as the first run-proof stockings for women, were commercialized in 1940. The run-proof stockings soon became a big worldwide commercial success. Yet, after the outbreak of World War II against Japan in 1941, the American military forces seized all nylon-6,6 for the production of ropes and parachutes.

After World War II, nylon-6,6 and the German invention nylon-6, began a triumphant march all over the world. However, it soon became evident that nylon fabrics do not "breathe" so that nylon clothes for everyday life favored intensive perspiration. Therefore, in the 1960s nylon articles were withdrawn from the everyday clothes market, but they have maintained a predominant role for sports articles due to their high tensile strength. Windbreakers, sails for boats and surfers, sport bags, sleeping bags, and fishing lines are a few examples of the usefulness of nylon fibers in outdoor activities. The largest quantities of nylon-6,6 were and are still consumed for the production of wear-resistant carpets.

After 1955, DuPont and many other chemical companies in various countries began to develop polycondensates for a broad variety of applications. Useful applications in addition to textile fibers are shockproof helmets, unbreakable lenses, safety glass, insulating lacquers, fireproof clothes for military pilots, and bullet-proof jackets for the police and soldiers. The polyester PET, poly(ethylene terephthalate), which avoids all the disadvantages characteristic of the aliphatic polyesters of Carothers, became a big commercial success. PET was developed in 1939 by the English chemist Rex Winfield (1901–1966) and was first used for the production of magnetic tapes and films. However, its most successful applications in the twenty-first century are as textile fibers and as bottles for soft drinks and mineral water.

Finally, it should be mentioned that neither Carothers, nor Flory were immune to fallacies. In the decades after Carothers ,first synthesized a superpolyester, it became clear that polycondensates having molar masses above 50,000 g/mol, which can easily be produced via chain-growth polymerization, are difficult to obtain via step-growth polymerization. Because polycondesates possess two reactive end groups, Carothers, and Flory discussed the possibility that ring-closing reactions, which prevent further chain growth, might occur. Yet for theoretical and experimental reasons Flory denied the existence of cyclization reactions competing with chain growth. However, in the 1970s the British physicochemists M. Gordon et al. and R. F. T. Stepto et al. published mathematical models of step-growth polymerizations that postulated efficient competition of cyclization and chain-growth reactions. Using the new method of mass spectrometry, H. R. Kricheldorf et al. contributed experimental evidence for the formation of cyclic polymers in step-growth polymerization after 1999. In summary, the history of polymer chemistry, in general, and polycondensation chemistry, in particular, demonstrates that several Nobel laureates and other famous chemists adhered to fundamental mistakes and fallacies, but as usual in science the revision of big mistakes entailed big steps forward.

Wallace Hume Carothers

Wallace H. Carothers was born on 27 April 1896, in Burlington, Iowa, as the first child of Ira Hume Carothers and his wife Mary Evalina (nee McMullin). One brother, John, and two sisters, Isobel and Marie, followed. As a young boy he showed interest in tools and mechanical devices and spent many hours performing experiments. When he was 5 years old, the family moved to Des Moines, Iowa, where he attended North High School. After graduation in 1914, his father, who was teacher and vice-president of the Capital City Commercial College, persuaded him to enroll in this College to complete the accountancy and secretarial curriculum.

In September 1915, Carothers entered Tarkio College in Missouri, where he began to study English. He also attended a course in chemistry and, under the guidance of Arthur Pardee, head of the chemistry department, he became more and more fascinated by chemistry. Eventually he switched from English to chemistry. His performance in chemistry was so excellent that when Pardee left the College to become chairman of the chemistry department at the University of South Dakota, Carothers was made instructor and had to teach the senior course. In 1920, at the age of 24, he graduated from Tarkio College with a bachelor's degree. Afterwards, he joined the University of Illinois, where he studied chemistry under the guidance of Prof. Carl S. Marvel and received the master of art degree in 1921. During the school year 1921/1922 Carothers took a position as chemistry instructor at the University of South Dakota. It was here that he began independent research. Carothers returned to the University of Illinois to acquire a Ph.D. under the supervision of Prof. Roger Adams. His studies were mainly focused on organic chemistry, and to a minor extent on physical chemistry and mathematics. He received the Carr Fellowship for 1923–1924, which was the most prestigious award of the university, and received a Ph.D. in 1924.

Carothers stayed at the University of Illinois for a further 2 years as an instructor in organic chemistry and eventually moved to Harvard University in 1926. There he again held a position as instructor of chemistry until he joined DuPont in 1927. James B. Conant, President of Harvard College in 1933, said in the aftermath: "In his research Dr. Carothers showed even at this time the high degree of originality which marked his later work. He was never content to follow the beaten path or to accept the usual interpretations of organic reactions. His first thinking about polymerization and the structure of substances of high molecular weight began while he was at Harvard."

In 1927, DuPont decided to finance fundamental research, meaning research not directly aimed at the development of a profit-making product, a decision that was quite unusual for a chemical company at that time. After two distinguished professors of organic chemistry had declined to accept a position at DuPont, DuPont contacted Carothers following the advice of Prof. A. Rogers. Carothers was fascinated by fundamental research and suspected that a position at DuPont would sooner or later force him to concentrate on the development of profitable products. Therefore, he refused DuPont's, first offer explaining that "I suffer from neurotic spells of diminished capacity which might constitute a much more serious handicap there than here." Despite this confession, Hamilton Bradshaw traveled to Harvard and convinced Carothers to accept a new offer. His salary was doubled (relative to what he received at Harvard) and DuPont ,definitely permitted him to concentrate his efforts on fundamental research. Later, Carothers confessed in a letter to Wilko Machetanz, a friend from the days at Tarkio College: "I find myself, even now, accepting incalculable benefits proffered out of sheer magnanimity and good will and failing to make even such trivial return as circumstances permit and human feeling and decency demand, out of obtuseness or fear of selfishness and complete lack of feeling."

On 6 February 1928, Carothers began to work at the experimental station of DuPont, in Wilmington, Delaware. His primary aim was the synthesis of a polymer with a molar mass above 4200 g/mol, the maximum mass achieved by Prof. Emil Fischer in Berlin before World War I. His research group consisted of two or three chemists and two consultants, A. Rogers and C. S. Marvel. For the following 6 years, his working field encompassed the following topics:

1. Development of a useful synthetic elastomer
2. Syntheses of poly(alkane)s
3. Syntheses of polyesters
4. Syntheses of poly(anhydride)s
5. Syntheses of cyclic esters and carbonates
6. Syntheses of polyamides

The first 2 years were disappointing because no high molar mass polymer was obtained. The first success came with the synthesis of poly(chloroprene) (trademark Neoprene), a synthetic rubber that is still being produced in the twenty-first century. However, comments in the English *Wikipedia* and other sources, saying that Neoprene was the first synthetic rubber are incorrect, because Dr. Fritz Hofmann (Bayer AG) synthesized poly(isoprene) and poly(2,3-dimethly-butadiene) in 1909 . In the years 1931–1934, Carothers also succeeded with the syntheses of high molar mass polyesters and polyamides (nylons), but despite these successes the frequency and intensity of his depressive moods steadily increased.

Carothers suffered from a depressive nature since his childhood. Any kind of stress worsened his psychical disposition. In 1932 and thereafter he came under pressure from two sides. In January 1930, Dr. Elmer Bolton was appointed vice director of the chemical department and, thus, became the immediate boss of Carothers. Bolton was not much interested in fundamental research, but in profitable products. In 1932, he modified the contract under which Carothers was hired, so that the funding for Carothers's projects was shifted from pure to practical research. In 1935, Bolton made Dr. George Graves leader of the polyamide project instead of Carothers. Stress also accumulated in his private life after 1932. Carothers began an affair with Sylvia Moore, a married woman, who eventually filed for divorce in 1933. Carothers continued this affair, which caused intensive tension in his relationship with his parents. He bought a house in Arden about 16 km from Wilmington and invited his parents to move to Arden and live in his house. However, his parents highly disapproved of his contact with Sylvia Moore, even after her divorce, and decided to return to Des Moines in the spring of 1934.

Immediately thereafter Carothers began to date Helen Sweetman, who had a bachelor's degree in chemistry and worked for DuPont on the formulation of patents. They married in February 1936 and at the end of that year his wife became pregnant.

On 30 April 1936, Carothers was elected to the National Academy of Sciences, an exceptional honor because he was the first industrial organic chemist to receive this honor. Despite this positive event, he fell into a severe depression at the end of May, which prevented him from working. In June, Carothers was admitted to the Philadelphia Institute of the Pennsylvania Hospital, a prestigious mental clinic, where he was treated by the psychiatrist Dr. Kenneth Appel. After 4 weeks he was permitted to leave the clinic with the advice to go hiking together with two friends. However, he started hiking by himself without informing anyone, including his wife. He suddenly returned to Wilmington on 14 September, but was not able to perform any systematic work at the experimental station.

On 8 January 1937, Carothers's sister Isobel died, an event that worsened his depressive mood. Together with his wife, he traveled to Chicago to attend the funeral and then to Des Moines for her burial. He then visited his psychiatrist Dr. Appel in Philadelphia, who immediately afterwards told a friend of Carothers that Carothers was inclined to commit suicide. On 28 April 1937, Carothers came to his laboratory to work. Yet, on the following day he checked into at a hotel in Philadelphia and committed suicide in his hotel room by drinking a solution of potassium cyanide in lemon juice. He did not leave any note or message.

After his death DuPont continued to file patents with Carothers named as inventor.

Paul John Flory

Paul J. Flory was born on 19 June 1910 in Sterling, Illinois, as the sixth generation of European immigrants born in America. His father, Ezra Flory, was a clergyman educator and his mother Martha (nee Brumbaugh) was a school teacher. Both parents were descendants of farmers and were the first of their families to attend college. Flory graduated in 1931 from Manchester College, Indiana, where his interest in chemistry was stimulated by Prof. C. W. Holt. This teacher encouraged him to enter the graduate school of Ohio State University, where his interest focused on physical chemistry. He accomplished his Ph.D. under the guidance of H. L. Johnston.

After completion of his Ph.D. Flory joined the Central Research Department of DuPont in Wilmington, Delaware, where he was assigned to the research group of Carothers. His supervisor inspired him to explore the fundamentals of polycondensation processes and the structural characteristics of polymers. In 1938 (i.e., after the death of Carothers), he joined the Basic Science Research Laboratory of the University of Connecticut, the beginning of a long and successful academic career.

Two years later, the outbreak of World War II forced him to return to industry. At that time, the USA badly needed the rapid development of technical production of synthetic rubber, because the supply of natural rubber was endangered by the occupation of south Asia by Japanese troops. After 3 years in the Esso laboratories of the Standard Oil Company, he moved to the laboratories of the Goodyear Tire and Rubber Company. In the spring of 1948, he was hired by the chemistry department of Cornell University for a lectureship, which had two important consequences. First, his lectures formed the basis of his famous book *Principles of Polymer Chemistry*, which was published by Cornell University Press in 1953. Second, he was offered a professorship, which he accepted in the autumn of 1948.

In 1957, Flory moved with his family to Pittsburgh to establish a broad program of fundamental research at the Mellon Institute, but some of the promised funds were later withdrawn. Therefore, he accepted a professorship in the chemistry department of Stanford University in 1961. This move led him to reshape the concept of his future research activities. In his own words: "Two areas have dominated the interests of my coworkers and myself since 1961. The one concerns the spatial configuration of chain molecules and the treatment of the configuration-dependent properties by rigorous mathematical methods; the other constitutes a new approach to an old subject, namely the thermodynamics of solutions. Our investigation in the former area have proceeded from foundations laid by Prof. M. V. Volkenstein and his collaborators in the Soviet Union and were supplemented by major contributions of the late Prof. K. Nagai in Japan." The success of his work is summarized in another important book entitled *Statistical Mechanics of Chain Molecules*, which was published in 1969.

Flory married the former Emily Catherine Tabor in 1938. They had three children, two daughters, and a son, who pursued a career in medicine. Flory died on 8 September 1985, and his son died in the same year of a heart attack.

In addition to the working fields mentioned above, the following areas of interest and achievement should be mentioned:

1. Fundamentals of step-growth polymerizations (mainly before 1952) and formulation of the classical theory of polycondensation.
2. The role of the excluded volume effect, above all, its influence on the configuration of polymer chains.
3. Formulation of the hydrodynamic constant "theta" and definition of the "theta point," where excluded volume effects are neutralized. These results were particularly important.
4. Flory's published work comprises more than 400 papers in addition to the aforementioned books. In 1985, Stanford University Press published the book *Selected Works of Paul J. Flory*. Of the numerous awards and honors he received (e.g., Priestley Medal in 1974, Perkin Medal in 1977, Elliot H. Cresson Medal in 1971), the Nobel Prize awarded in 1974 was the most prestigious. Flory used his prestige and his award to campaign for international human rights, mainly with regard to the treatment of scientists in the former Soviet Union.

Bibiography

Adams R (1939) Wallace Hume Carothers 1896-1937. Biographical memories of the National Academy of Sciences of the USA. St. Barbara, USA

Carothers WH (1929) An introduction to the general theory of condensation polymers. J Am Chem Soc 51:2560

Flory JP (1953) Principles of polymer chemistry. Cornell University Press, Ithaca

Johnson WS, Stockmayer WH, Taube H (2002) Paul John Flory—a biographical memoir, vol 82. The National Academy of Sciences Press http://www.nobelprize.org/nobel_prizes/chemistry/laureates/1974/flory

Kricheldorf HR (2014) Polycondensation—history and new results. Springer, Heidelberg

Matthew H (1996) Enough for a lifetime. Wallace Carothers, the inventor of nylons. Chemical Heritage Foundation

Morawetz H (1985) Polymers—the origins and growth of a science. Wiley, New York

Roberts RM (1989) Serendipity. Accidental discoveries in science. Wiley, Hoboken, NJ

Smith JK, Hounshell DA (1985) Wallace Carothers and the fundamental research at DuPont. Science 229:436

Chapter 10
Physics and Geology

10.1 The Heliocentric World

> The most important scientific revolutions all include as their only common feature, the dethronement of human arrogance from one pedestal after another of previous convictions about our centrality in the cosmos.
> (Stephen J. Gould)

The term "heliocentric" is derived from the Greek term $\eta\lambda\iota o\sigma$ (*helios* meaning sun) and the Latin *centrum* (center). The concept put forward by Copernicus (see Biography) of a heliocentric world is now considered to be one of the greatest revolutions in the history of science. However, this vision of a universe having the sun at its center was not absolutely new, because the geocentric model had prevailed for almost 2000 years beginning with Thales of Miletus (c.624–546 B. C.). The geocentric model stood on three foundations. First, almost all Greek philosophers from Thales to Aristotle (385–322 B.C.) adhered to the geocentric system. Second, the prominent Roman astronomer Ptolemy (c. 90–168 A.D.), descendant of a Greek family in Alexandria, also adhered to geocentrism and his model of the universe prevailed in Europe until the end of the sixteenth century. Third, all Christian Churches, including the Protestants, favored a geocentric world. Although the Bible does not describe the structure of the universe, a few sentences suggest a geocentric world. For instance, in the battle of Gibeon, Joshua stops the sun, because a longer day was favorable for a complete victory. Furthermore, the daily revolving of the heavens is mentioned in Psalms. Moreover, man is described as being in God's image in the first chapter of the Bible (Genesis) and it was a simple logic to place the "crown of the creation" at the center of the world.

Ptolemy was the most influential astronomer, mathematician, and geographer of the ancient world. His most famous multivolume work, entitled *Almagest* (a term of Arabian origin meaning mathematical compilation), summarized all his mathematical calculations and observations concerning astronomy. His view of the universe was based on the geometric system described by Aristotle. The Earth is fixed at the

H.R. Kricheldorf, *Getting It Right in Science and Medicine*,
DOI 10.1007/978-3-319-30388-8_10

center surrounded by numerous spheres that coordinated the motions of Moon, Sun, planets, and stars. Ptolemy calculated the orbits of Moon and Sun and he also calculated the dates of future eclipses. This part of his work was mainly based on the calculations and observations of the Greek astronomer Hipparchus (c. 190–120 B.C.). Five volumes of the *Almagest* deal with the motions of planets (five were known at that time), his greatest achievement in terms of an original contribution. He described the path of a planet as a circular motion on an epicycle, the center of which itself moved on a cycle, the center of which was offset from the center of the Earth. This complex model allowed the most precise prediction of planetary positions until the Danish astronomer Tycho Brahe (1546–1611) presented his new system based on more accurate observations.

In the 2000 years before Copernicus, a handful of Europeans and a few Indian and Islamic astronomers speculated about a universe that was more or less different from the Ptolemaic model. The Pythagorean Philolaos (c. 480–385 B.C.) favored a system having a central fire that was circled by Earth, Moon, Sun, planets, and stars (in that order outwards). His hypothesis was mentioned by Copernicus. For Heraklides Ponticus (387–312 B.C.) it is well documented that he postulated an Earth rotating around its axis, whereas other frequently cited comments on a heliocentric model are unproven. Aristarchos of Samos (c. 310–230 B.C.) was the first to describe a heliocentric world, and he concluded that the fixed stars are at a great distance from the Earth and Sun. Only a small fragment of his work has survived and most information about his work was handed down by Archimedes of Syracuse (287–212 A.D.), whose complete work, the *Opera Archimedes*, was published in 1544, a year after the death of Copernicus. Nonetheless, Copernicus mentioned Aristarchos in an early unpublished version of his manuscript *De revolutionibus orbium coelestium*, but removed this citation from the published version. Hence, it is not clear what Copernicus knew about this work and to what extent he was influenced by the concept of Aristarchos. The theologian Nikolaus von Kues (Nicolas of Cusa, 1401–1404 A.D.) formulated the hypothesis that neither Earth nor Sun were fixed in the center of the universe and that all heavenly bodies were in permanent motion.

The Indian mathematician and astronomer Aryabhata (476-c. 550 A.D.) postulated an Earth rotating about its axis in his comprehensive work *Aryabhatiya* (c. 499), and his model of planetary motions included epicycles quite analogous to the Ptolemaic system.

The Maghara Observatory (today Maragheh in eastern Iran) was constructed around 1259 and was the leading observatory of the Islamic world until its decline in the fourteenth century. At least two astronomers from the team working in Maghara discussed a heliocentric model of the universe, without formulating and advocating a consistent system. Another member of this observatory, Ibn-al Shatir (born 1304) elaborated a mathematical model of planetary motions using mathematical methods such as the Tusi couple and *Urdi lemna*, closely resembling methods used by Copernicus. Once again it is unknown to what extent Copernicus was informed about the knowledge of his predecessors.

In their biography *Kopernik Mikolaj*, the authors Jerzy Dobrzycki and Leszek Hajdukiewicz suggest that Copernicus got the idea of a heliocentric world during his studies in Padua. Yet, it remains obscure why he was convinced from those early years that the heliocentric system was superior to the Ptolemaic model, in as much as the calculations of planetary motions were not more accurate than those realized by Ptolemy. Between 1509 and 1512 Copernicus wrote a 40-page manuscript commonly cited with the shortened title *Commentariolus* (Little Commentary). It was a first sketch of the planned book *De Revolutionibus Orbium Coelestium* and did not contain his mathematical approach. The basic assumptions of his new theory are summarized as follows:

1. There is no one center of all celestial circles or spheres.
2. The center of the Earth is not the center of the universe, but only of gravity and of the lunar sphere.
3. All spheres revolve around the Sun as their midpoint, and therefore the Sun is the center of the universe.
4. The ratio of the Earth's distance from the Sun to the height of the firmament is so much smaller than the ratio of the Earth's radius to its distance from the Sun that the distance from the Earth to the Sun is imperceptible in comparison with the height of the firmament.
5. Whatever motion appears in the firmament arises not from any motion of the firmament, but from the Earth's motion. The Earth together with its circumjacent elements performs a complete rotation on its fixed poles in a daily motion, while the firmament and highest heaven abide unchanged.
6. What appears to us as motions of the Sun arise not from its motion, but from the motion of the Earth and our sphere, with which we revolve about the Sun like any other planet.
7. The apparent retrograde and direct motion of the planets arises not from their motion but from the Earth's. The motion of the Earth alone, therefore, suffices to explain many apparent inequalities in the heavens.

This manuscript was not intended to be printed and only a dozen handwritten copies were distributed among friends and Polish astronomers. In the following years, Copernicus collected more astronomical data and improved his mathematical methods. In principle, his work was complete in 1532, but he hesitated to publish it, being afraid of the scorn "to which he would expose himself on account of the novelty and incomprehensibility of his theses."

Influential friends such as Bishop Johannes Danticus von Hofen, Bishop Tiedemann Giese, and Cardinal Nikolaus von Schomburg urged him to publish his work (and even offered funding), but Copernicus resisted. In 1539, George J. Rheticus (1514–1576), a mathematician in Wittenberg (Germany), arrived in Frauenburg on a round trip visiting several astronomers. He became fascinated by the work of Copernicus, continued his stay for 2 years, and finally convinced Copernicus to concede printing of his manuscript. It was eventually printed by Johannes Petrus in Nürnberg (Nuremberg) under the supervision of Rheticus, but printing of the final pages was supervised by the Lutheran theologian Osiander who added an unauthorized preface defending the new theory. The complete work, subdivided into six books, appeared in the year of Copernicus's death.

Intensive research on the acceptance of the work of Copernicus by the Catholic Church revealed that his mathematical approach was frequently studied and even accepted. The Catholic Church had problems with the reformation of the Gregorian calendar, and any new mathematical approach that simplified calculations of lunar, solar, and planetary motions was welcome. Furthermore, the work of Copernicus was not read by laics and the majority of scientists rejected his heliocentric model. Hence, the first harsh criticism by theologians came from Protestants. Philip Melanchthon (1497–1560) wrote: "Some people believe that it is excellent and correct to work out things as absurd as did that Samartian [i.e., Polish] astronomer who moves the Earth and stops the Sun. Indeed, wise rulers should have curbed such light-mindedness." Martin Luther is reported to have called him a fool. In contrast, the theory of Copernicus was not discussed by Catholic theologians at the Council of Trent (1545–1563).

However, the situation changed after 1600 for several reasons. First, the reform of the calendar was completed in 1583. Second, the Counter-Reformation headed by the Jesuits made rapid progress everywhere in central Europe and the Jesuits favored the theory of Tycho Brahe. Brahe's theory assumed circular planetary motions around the Sun, but that the Moon and Sun orbited the Earth, the fixed center of the universe. It was a modification of the Ptolemaic model, with the advantage that the calculations of planetary motions were more accurate than those of Ptolemy or Copernicus.

Finally, Galilei's conspicuous astronomic observations, achieved by means of an improved telescope, made astronomy more popular, even among laics, and Galilei (1564–1642) was a militant advocate of the heliocentric theory of Copernicus. The Catholic Church began to understand the theory as a challenge to its authority. Hence, in March 1616 the Catholic Church's Congregation of the Index issued a decree suspending the book by Copernicus until it could be "corrected." In the same decree, any other work defending the mobility of the Earth and a central position of the Sun was also prohibited.

Substantial support for the theory of Copernicus came, at least in the eyes of scientists, from the German mathematician Johannes Kepler (see Biography), a Protestant who worked in Graz, Linz, Prague, and Ulm and was permanently fleeing the Counter-Reformation. During his studies in Tübingen (southern Germany), Kepler had become a dedicated Copernican, mainly for theological reasons, because he considered the universe as an image of the Trinity with the Sun as equivalent to God-the-Father in central position.

During his stay in Graz (1594–1600) Kepler began to elaborate a first sketch of his new theory of astronomy, which was published in 1597 under the title *Mysterium Cosmographicum*.

In this treatise, Kepler described an attempt to substitute the cycle–epicycle concept that Copernicus had borrowed from Ptolemy by a new geometrical and mathematical approach. He improved and completed his work during his stay in Prague (1699–1612), where he had been appointed imperial mathematician by

Emperor Rudolph II since 1601. As successor to Tycho Brahe, Kepler had access to his numerous accurate data, which allowed him a comparison with the calculations of Copernicus. He was eventually convinced that the orbits of all planets were elliptical and not combinations of cycles and epicycles. A major problem for the success of his calculations was the unknown distance between Earth and Sun or Mars and Sun. However, he was able to determine the ratio of both distances and this result sufficed to calculate the elliptical orbit of Mars with the Sun in one of the two foci. He published his new theory in a book entitled *Astronomia Nova* (New Astronomy), which appeared in 1609 after a legal dispute with the heirs of Tycho Brahe. In this work, Kepler presented two famous laws:

1. The path of a planet about the Sun is elliptical, with the Sun at one focus of the ellipse (law of ellipses).
2. If an imaginary line is drawn between the center of the Sun and center of a planet, that line would sweep out the same area in equal periods of time. This law of equal areas reflects the fact that the speed of a planet is continually changing, being highest when the planet is closest to the Sun.

   Kepler also found out that the time a planet needs to orbit the Sun increases with the distance from the Sun. He related this (true) result to the wrong assumption that the force binding a planet to Sun or Moon to Earth decreases proportionally with the distance. Based on the book *De Magnete* of William Gilbert (1544–1603), Kepler also adhered to the wrong assumption that the driving force behind the motion of the planets is the magnetic field of the Earth.

   In the years 1615–1621, Kepler wrote and edited a three-volume textbook of astronomy with the title *Epitome Astronomiae Copernicanae* (Epitome of Copernican Astronomy). More important was his next book entitled *Harmoniae Mundi* (Harmony of the World), which appeared in 1619. In that work he attempted to explain the proportions of nature, particularly the astronomical aspects, in terms of music and he presented his third law, the law of harmonies:
3. The ratio of the square of the time (T) needed for orbiting the Sun and the cube of the distance (R) from the Sun is almost identical for all planets: T^2/R^3 of planet A is equal to T^2/R^3 of planet B and so on.

   This law is also valid for the Moon and artificial objects orbiting the Earth, such as satellites.

Kepler's laws and calculations helped Isaac Newton (1643–1727, see Biography Sect. 10.5) to develop his laws of mechanics, notably the law of gravity. That law says that the attractive force (F) between two masses (M_1 and M_2) is proportional to their product (via a constant G) and decreases with the square of their distance (R): $F = G \times M_1 \times M_2 / R^2$. Initially, Newton believed, like Kepler and other scientists of previous centuries, that any force between two bodies is inversely proportional to the distance between them. The dependence on the square of the distance was a new insight contributed by the genial British physicist Robert Hooke (1635–1703). Newton's law of gravity was not only great progress in the proper understanding of various physical phenomena on earth, it was a mental and even religious revolution. In the ancient world and in the Middle Ages the laws responsible for motions and changes of heavenly bodies had a divine character that was quite different from the laws regulating human life and natural phenomena on earth. The law of gravity demonstrated for the first time that the same laws operated everywhere in the universe.

In summary, the works of Copernicus, Kepler, and Newton caused three paradigm changes. First, the structure of the universe changed from a geocentric to a heliocentric system. Second, the motions of the heavenly bodies were explained by elliptic orbits and not by a combination of cycles and epicycles. Third, the laws of nature were shown to be the same everywhere in the universe.

Nicolaus Copernicus

Nicolaus Copernicus (German: Nikolaus Kopernikus) was born on 19 February 1437 in Thorn (today Torun, Poland) in a province of Prussia under the supremacy of the King of Poland. He was the youngest of four children of the wealthy merchant Nicolaus Copernicus and his wife Barbara (nee Watzenrode). The ancestors of both parents originated from Silesia and belonged to the patrician families of Thorn. Copernicus spoke German and Latin fluently and he also spoke Polish and Italian. Most of his writings that have survived were written in Latin (the language of academia at that time) and a few documents and letters in German. After the early death of his father (1443) a maternal uncle, Lucas Watzenrode (1447–1512), took care of his education and career. This uncle was a wealthy and influential man, because he was advisor to the Polish King.

Unfortunately, no reliable information exists about the early stages of Copernicus's education.

At the end of 1491, Nicolaus and his brother Andreas entered the University of Krakow (today Jagiellonian University) and began their studies in the Department of Arts. These studies included mathematics, and Copernicus became a disciple of Albert Brudzewski (1445–1497), who was professor of Aristotelian philosophy and private teacher of astronomy. He also attended lectures by other scientists and acquired a thorough grounding in arithmetic, geometry, optics, cosmography, and astronomy. Due to the financial support of his family he began to collect a large library on astronomy, which was later stolen by Swedish soldiers as war booty and is now stored at the library of the University of Uppsala.

He left the University of Krakow without a degree, because his uncle planned to install him as canon at Frauenburg Cathedral in the province of Ermland (Warmia) where Lucas Watzenrode had become Prince-Bishop in 1484. A good knowledge of law was required for this ecclesiastical career and, thus, Copernicus was sent to Bologna (the oldest university in Europe) to study canon law. However, he also pursued his interest in astronomy and became disciple and eventually assistant of the famous astronomer Domenico Maria Novara da Ferrara. After 3 years, he earned the degree of *Magister Artium*. Together with his brother, Copernicus spent the jubilee year 1500 in Rome and returned via Bologna to Ermland in 1501. In October 1497, he was officially appointed to the position of canon. A few months later he received from the chapter a 2-year extension of leave to finish his studies in law and to study medicine. He studied at the University of Padua for

2 years and moved in 1503 to Ferrara, where he obtained his doctorate in canon law. In the fall of 1503, at the age of 30, he returned to Ermland where he lived for the rest of his life, apart from brief journeys to cities in Prussia and Poland.

Copernicus served as secretary and physician in the administration of his uncle, who enjoyed substantial autonomy having his own parliament, treasury, and monetary unit. Copernicus resided in the Castle of Heilsberg (Litzelbark) and took part in most of his uncle's ecclesiastical, administrative, economic, and political duties. The political situation was difficult, because the province of Ermland together with the King of Poland were fighting the influence of the Teutonic Order, but it also tried to defend its autonomy against the Polish Crown.

After the death of his uncle in 1512, Copernicus moved to Frauenburg (Frombork) a small town on the coast of the Baltic Sea. The chapter gave him a house outside the walls of the cathedral mount, but in 1514 he purchased a tower inside the walls. Nonetheless, his home and his instruments were damaged during an attack by the Teutonic Order in October 1520. However, Copernicus continued to live and to work in Frauenburg until the end of his life in 1543. In 1511 he was elected chancellor of the chapter and from 1512 was responsible for administration of the economic enterprises. In his role as administrator of the Prussian province he was also concerned with monetary reform, a major issue in regional Prussian politics. In 1517, he wrote down a quantity theory of money including a theory in economics, a text that is still useful today. In 1526, he published a treatise on the value of money entitled *Monetae Cudendae Ratio* in which he formulated an early version of what was later called Gresham's law. This law says that bad (debased) money drives good coinage out of circulation. The analyses and recommendations of Copernicus were widely accepted by the politicians of both Prussia and Poland. Furthermore, it is worth mentioning that Copernicus was active as a cartographer. He drew an improved map of the Duchy of Prussia, which was edited in 1529. Together with Bernard Wapowski he also edited a new map of the unified Kingdom of Lithuania and Poland.

After his return from Italy, Copernicus worked on his new concept of the heliocentric world in parallel to his numerous duties as administrator and physician. His work also included the elaboration of tables listing the future of positions of planets. These positions were calculated on the basis of their past positions. Such tables were of great interest for astrologers and for political leaders (poor people could not afford such expensive work), because most people of that time believed that the planets had a significant influence on their life. Therefore, Duke Albert of Prussia financed the completion of these tables, although he had turned Protestant, unlike Copernicus. The work was finally edited by the astronomer Erasmus Reinhold under the title *Prussian Tables* 8 years after the death of Copernicus.

At the end of 1542, the health of Copernicus was affected by apoplexy and paralysis and he died on 24 May 1543. Legend says that he awoke from stroke-induced coma, looked at the final printed pages of his work *De Revolutionibus*, and then died peacefully. He was buried in the Cathedral of Frauenburg. An epitaph financed by Prince-Bishop Martin Cromer in 1580 was later destroyed and a new epitaph was installed by the chapter of Frauenburg in 1735. A bust followed in

1973. For over two centuries, archeologists tried in vain to find his burial place and to identify his skeleton. A new campaign by archeologists and anthropologists led by Jerzy Gassowski was finally successful in 2005. The computer-aided reconstruction of the face of Copernicus from the skull was almost identical with that of a self-portrait, and DNA recovered from the bones was identical with that of a hair found in a book from his library in Uppsala. On 22 May 2010, the remains of Copernicus were buried a second time in exactly the place where they were found in 2005. Now, a black granite tombstone adorned with a golden model of his solar system indicates his grave.

In the nineteenth and twentieth century, Copernicus was honored in various ways. In 1807 a marble bust was placed in the Walhalla (Bavaria). His portrait appeared on German and Polish stamps. Streets, schools, and institutes in Germany and Poland are named after him. Craters on the Moon and Mars bear his name and the new element (no. 112) discovered in 2009 was named copernicium (Cn).

Johannes Kepler

Johannes Kepler was born on 27 December 1571 in Weil der Stadt, a small town about 30 km west of Stuttgart (capital of the German state of Baden-Württemberg). His father was a mercenary soldier and left the family when Johannes was 5 years old. His mother Katharina (nee Guldenmann) was a healer and herbalist and earned a precarious living mainly by helping her father, an inn keeper. Kepler was born prematurely and, thus, was a weak and sickly child. Furthermore, a bout of smallpox left him with weakened vision and crippled hands, limiting his ability in experimental astronomy. However, he had an extraordinary mathematical talent and tried to impress travelers in his grandfather's inn.

He came into contact with astronomy in an early age and was fascinated by it throughout his life. In 1577, at the age of six, he observed the Great Comet and wrote later that he "was taken by his mother to a high place to look at it." Three years later, he had the opportunity to observe another astronomical event, the moon eclipse of 1580, and he remembered that the moon "appeared quite red."

In the years 1579–1584, he and his mother lived in the village of Ellmendingen where his grandfather had rented an inn named "The Sun." There he entered a grammar school to learn Latin. After 1584, he attended the monastic school in Adelberg and after having successfully passed the examination he joined the monastic school of Maulbronn (a kind of high school). High marks in the final examination allowed him to enter the University of Tübingen, a stronghold of the Protestant religion for centuries. Initially, he studied theology under Jacob Herbrand (a former student of Philipp Melanchthon) and philosophy under Vitus Müller. According to his interest, he also attended courses in astronomy and mathematics under the instruction of Michael Maestlin (1550–1631), who was professor of mathematics in Tübingen from 1583 to 1631. From him Copernicus learned both the Ptolemaic and the Copernican system of planetary motion and he became a "Copernican" for the rest of his life.

In discussions with fellow students, Kepler learned to defend the heliocentric system from both a theological and a scientific perspective. He also earned a reputation as astrologer, casting horoscopes for other students. Originally, Kepler intended to make an ecclesiastical career, but when he was recommended for a position as teacher of astronomy and mathematics at the Protestant school in Graz (Austria), he accepted this offer in April 1594, aged 23. Kepler taught in Graz until 1600 and during this period he elaborated and edited his first work *Mysterium Cosmographicum* (The Cosmographic Mystery), the first published defense of the Copernican theory. However, the majority of the residents of Graz were Catholics and most schools and colleges were in the hands of Jesuits. Therefore, Kepler's first manuscript contained a long chapter reconciling the heliocentric system with biblical passages that seemingly supported a geocentric world. Nonetheless, Kepler attempted to publish his work in Tübingen and, with support of Michael Maestlin, he received permission from the senate of the University. However, the Protestant senate ordered him to remove the Bible exegesis and to simplify the explanation of Copernican theory and of his own new ideas. In this work, Kepler attempted to interpret the planetary orbits with the five Platonic polyhedral forms and, even in later years when he developed his revolutionary mathematical description of planetary motions, he never abandoned the Platonic system described in his first work.

Kepler's book was not widely read, but it established his reputation as an excellent astronomer.

During his stay in Graz Kepler became acquainted with Barbara Müller, a 23-year-old widow with a young daughter, and in 1596 he began to court her. She had inherited the estates of her former husband, and her father was a successful mill owner. The Müller family initially opposed the marriage because of Kepler's poverty. However, when Kepler obtained permission to publish his work *Mysterium Cosmographicum*, the Müller family agreed and Barbara and Johannes married on 27 April 1597. The first two children died in infancy, but four others survived childhood.

Kepler sent copies of his book *Mysterium Cosmographicum* to numerous astronomers and maintained correspondence with them. Among the responding colleagues was the famous Danish astronomer Tycho Brahe (1546–1601), who was the official astronomer and mathematician of Emperor Rudolph II (1546–1612) in Prague from 1599. He had followed Reimarus Ursus, who had been court astronomer since 1591, but who became sick in 1599 and died in the summer of 1600. Initially, Brahe criticized part of Kepler's work, but they continued their correspondence with a broad discussion of various astronomic issues. Brahe invited Kepler for a short visit and they met near Prague at the site where Brahe's new observatory was under construction. Brahe eventually offered him a post as assistant and Kepler returned to Graz to clarify his position.

At that time, the influence of the Jesuits was increasing rapidly and when Kepler refused to convert to Catholicism he was banished from Graz. At the end of 1600, Kepler accepted Brahe's offer and moved with his family to Prague. His first duty was to elaborate the *Rudolphine Tables*, a listing of future planetary positions to replace the less accurate *Prussian Tables* of Copernicus and Erasmus Reinhold.

Immediately after the unexpected and mysterious death of Tycho Brahe in October 1601, Kepler was hired by Rudolf II as imperial mathematician with the duty of completing the tables and to serve as astrological advisor to the Emperor. In this strong position, Kepler was able to continue practicing his Lutheran faith, although the only official doctrines in Prague were Catholic and Utraquist. The court life brought Kepler into contact with various other scientists and proved fruitful for the progress of his astronomical studies.

However, the political-religious tensions in Prague increased as the Counter-Reformation made more and more progress everywhere in central Europe. In 1611, Emperor Rudolph II became sick and his brother Matthias forced him to retire. The new King Matthias was less inclined to support Kepler and after Rudolph's death in August 1612 Kepler considered moving to another city. Around the same time, his son Friedrich died from smallpox and his wife Barbara died a few months later. When Matthias was elected as the new Emperor of the Holy Roman Empire, he reaffirmed Kepler's position as imperial mathematician but allowed him to move to Linz (Austria), where he obtained a position as teacher and district mathematician.

For the first few years in Linz, Kepler had financial security and religious freedom and he published two more books. After the death of his wife Barbara, Kepler needed a woman to care for his household and the children who had survived smallpox. After 11 possible matches, he married the 24-year-old Susanna Reuttinger on 30 October 1603. The first three children of this marriage died early, but three more children reached adulthood.

In 1615, a woman accused Kepler's mother Katharina of making her sick with a poisonous brew. The dispute continued for 2 years and escalated in 1617. Katharina was accused of witchcraft and imprisoned in 1620. She was released after 14 months, partly due to the extensive legal defense initiated and financed by her son. The only evidence from the accusers were rumors, but witchcraft trials were common in Europe at that time.

In 1623, Kepler completed the *Rudolphine Tables*, his major work in the eyes of his contemporaries. Yet, this work was mainly based on the measurements of Tycho Brahe and because of negotiations with Brahe's heirs, printing was not feasible before 1627. Meanwhile, the political and religious tensions had culminated in the outbreak of the Thirty Years War (1618–1648) and the progress of the Counter-Reformation in Linz endangered Kepler's life. In 1626 he moved to Ulm, where he printed the Tables at his own expense. In 1628, he became official astrological advisor to General Wallenstein, leader of the Catholic troops. During the war Kepler had to spend a lot of time traveling and during a stay in Regensburg (Bavaria) he fell ill and died on 15 November 1630. Unfortunately, his burial site was lost when Swedish soldiers destroyed the churchyard, but his self-authorized epitaph survived:

Mensus eram coelos, nunc terrae metior umbras
Mens coelestis erat, corporis umbra jacet

I measured the skies, now the shadows I measure
Sky-bound was the mind, earth-bound the body rests

In addition to the numerous article and books dealing with astronomy, Kepler wrote two treatises on different topics. In 1611, he published *De Nive Sexangula* (On the Six-Cornered Snowflake), which described the hexagonal symmetry of ice crystals. He assumed that this symmetry is based on a regular packing of atoms, a visionary and correct hypothesis known as Kepler's conjecture. In the same year, he also described the construction of an improved telescope in a work titled *Dioptrice*.

In the nineteenth and twentieth centuries, Kepler was honored in numerous countries in various ways. Statues were erected in Linz and Prague. A stamp of the GDR and an Austrian 10 euro coin display his portrait. The supernova of 1604, the asteroid 1134, craters on the Moon and Mars, and a NASA mission (2009) were named after him. Numerous streets, squares, schools, and institutes in Central Europe and in North America have his name and even a ridge-way in the mountains of New Zealand is named after him. An opera by Philip Glass and an opera by Paul Hindemith were dedicated to him, and the Episcopal Church of the USA has honored Copernicus and Kepler with a feast day on 23 May.

Bibliography

Anderson H, Barker P, Xiang C (2006) The cognitive structure of scientific revolution, Chapter 6: The Copernican revolution. Cambridge University Press

Armitage A (1990) Copernicus, the founder of modern astronomy. Dorset Press

Caspar M (1993) Kepler. Translated and edited by Hellman CD with bibliographic citations by Gingerich O and Segonds A. Dover, New York

Gassendi P, Olivir T (2002) The life of Copernicus 1473-1573. Xylon Press

Gilder J, Gildr A-L (2004) Heavenly intrigue: Johannes Kepler, Tycho Brahe and the murder behind one of history's greatest discoveries. Doubleday

Gingerich O (1973) Kepler, Johannes. In: Gillispie CC (ed) Dictionary of scientific biography, vol VII. Charles Scribner's Sons, New York

Goodman DC, Russel CA (1991) The rise of scientific Europe, 1500-1800. Hodder Arnold H&S

Moore P (1994) The great astronomical revolution 1543-1687 and the space age epilogue. Albion

Pannekoek A (1989) A history of astronomy. Dover, New York

10.2 The Expanding Universe

> Not only is the universe stranger than we imagine, it is stranger than we can imagine.
> (Carl Sagan)

For many centuries the cosmology and world view of the Jewish and Christian religions held that the creation of the world required 6 days, and afterwards all properties and features of the world remained almost unchanged. The modern natural sciences that emerged step by step from the Renaissance destroyed the dogma of a constant world in three stages. The first stage was elaborated by the geologists of the eighteenth century, who had the insight that the earth's surface was subject to frequent and significant changes over millions of years (see Sect. 10.6). The biologists and naturalists followed in the nineteenth century, postulating that all living organisms were the products of a long-lasting (and still proceeding) evolution. This hypothesis was supported by an increasing number of experimental results, partly resulting from cooperation with geologists (a discipline called paleontology). The third stage was contributed by astronomers and physicists in the twentieth century. They found more and more experimental evidence for dynamic evolution of the entire universe.

Interestingly, not only deeply religious people with little knowledge of natural sciences, but also astronomers and other scientists were convinced until the 1930s that the structure and size of the universe were almost constant. Before 1930, the standard paradigm or dogma was based on the assumption that the size of the universe was almost identical with the size of our galaxy, the Milky Way. Other galaxies having a size similar to the Milky Way were, in principle visible, but due to the low quality of the telescopes they looked like nebulous spheres or disks. These nebulae were considered to be aggregates of gases and dust belonging to the Milky Way system. The first doubts and skeptical comments came from two directions, theoretical physics and experimental astronomy.

At this point, Albert Einstein's general theory of relativity, published in 1915, needs to be mentioned. This theory dealing with the correlation of gravity, space, and time predicted that the universe was either in a process of expansion or contraction, but was certainly not static. At first, Einstein (see Biography Sect. 10.4) did not believe his own mathematical equations and modified them by introduction of a "cosmic constant," with the consequence that his equations indicated a static universe. He published this new theory in 1917. After having learned from Edwin Hubble (see Biography) in 1929 that the latest astronomic measurements clearly indicated an expanding universe, he called the revision of his initially correct general theory of relativity the greatest folly of his life. It was in fact not the only fallacy in his life and Einstein later confessed in a letter to Paul Ehrenfest: "It is convenient with this fellow Einstein, every year he retracts what he wrote the year before."

Starting from Einstein's original theory of general relativity, the Dutch astronomer Wilhelm de Sitter (1872–1934) developed a mathematical model of a mass-free universe that allowed him to predict both a dynamic and a static universe depending on certain constants in his equations. Before 1930, this "de Sitter model" and Einstein's wrong version of his general theory of relativity were considered as decisive confirmation of a static universe.

Again starting out from Einstein's general theory of relativity, the Russian physicist and mathematician Alexander A. Friedman (1888–1925) published in 1922 an article dealing with the curvature of space. His mathematical approach predicted either a steadily expanding universe or cosmic cycles of expansion and contraction. An improved version of his theory was published in 1924, but both articles were ignored by the international community of scientists.

In 1927, the Belgian physicist and theologian George E. Lemaitre (1894–1966) published a treatise on the expansion of the universe. His theory was stimulated by a first report from the Mount Wilson Observatory describing a red shift in the spectra of distant galaxies. Lemaitre's work included an expansion coefficient, a forerunner of the Hubble constant. However, his treatise was written in French and it appeared in the little-known *Annales de la Société Scientifique de Bruxelles*, so it was largely ignored in the international community of astronomers. In the English version, published in 1931, the expansion coefficient was omitted and, finally, it was Hubble who was honored with the unofficial title "father of the Big Bang theory." Yet, Hubble's publication in 1929 describing a permanent expansion of the universe had the decisive merit that it included new experimental facts supporting his theory.

Hubble's experimental data resulted from measurements of cosmic distances with the best telescope available at that time. This part of Hubble's work also had a previous history that deserves comment. Measurements of cosmic distances before World War II were mainly based on two quite different methods, namely parallax measurements and luminosity measurements of a certain type of star, the so-called standard candles. Parallax measurements are only suitable for relatively short cosmic distances, such as those of stars neighboring the sun. The standard candle method enables measurements over far greater distances, even beyond the local group, and therefore merits a short description. Stars suitable as standard candles have to meet the following requirements:

1. Their absolute luminosity must be much higher than that of the Sun to allow their observation over long distances
2. They should exist in various segments of the Milky Way and in other components of the local group
3. They should be easily and unambiguously identifiable

Variable stars such as the R,R-Lyrae stars and the Cepheids meet all these requirements, and thus played a decisive role in Hubble's work and in elaboration of the Big Bang theory.

The first example of a R,R-Lyrae star was detected in 1895, but the name originated from the later discovery of a standard type R,R-Lyrae star in the constellation Lyra. R,R-Lyrae variable stars are relatively old (Population II) stars having 40–60 % of the Sun's mass. They are red giants with a diameter five times larger than that of the Sun and their luminosity is about 40–50 times higher. They display short pulsation periods, which fall into the range of 0.2–1.2 days. The strict relationship between pulsation period and absolute luminosity makes them good standard candles for distance measurements within the Milky Way. An additional advantage is their existence in all areas of the Milky Way, notably in globular clusters.

Distance measurements with standard candles require the comparison of their absolute luminosity with their relative (apparent) luminosity as observed on earth. The number of photons per square unit decreases with the inverse square of the distance, a general law valid for any kind of radiation emitted from a punctual source. A major problem for application of the standard candle method is calibration of the absolute luminosity. Based on their spectral properties, stars resembling the Sun were identified in globular clusters and their luminosity served, in turn, for calibration of all neighboring stars, including R,R-Lyrae variables.

Cepheids were named after the star δ-Cephei, the first star of this type discovered in 1784. The properties of Cepheids are different from, but complementary to, those of R,R-Lyrae stars.

Classical Cepheids (the most important subclass) are relatively young (Population I) stars. They are yellow giants, 4–20 times more massive than the Sun, and their luminosity can be up to 100,000 times higher. Hence, Cepheids can be used to determine the large cosmic distances between components of the local group and beyond. The pulsation period can vary between 1 and 130 days, but typically falls into the range of 2–10 days. In 1912, the American astronomer Henrietta Swan Leavitt (1868–1921) elaborated an equation for Cepheids of the Small Magellan Cloud, correlating the pulsation period (*P*/*D*), their absolute luminosity (M), and their apparent luminosity (m):

$$M = 2.8\ \log(P/D) - 1.43$$

For distance measurements (D) the following simple equation was derived:

$$D = 10^{(m-M-5)/5}$$

Further methods enabling distance measurements between remote galaxies and quasars were developed after World War II.

In 1914, the astronomer Harlow Shapley (1885–1972) at the Mount Wilson Observatory (near Pasadena, California) began using Cepheids to determine the dimensions of the Milky Way. His measurements included several nebular objects and he concluded that all observed cosmic objects belonged to the Milky Way. He obtained a value of 300,000 light years for the diameter of our home galaxy, overestimating the true value by a factor of three. Shapley also concluded that the Milky Way filled the entire universe. The American astronomer Herbert D. Curtis (1872–1942, since 1920 director of the Allegheny Observatory) strongly criticized Shapley's "big-galaxy hypothesis." He postulated a smaller Milky Way and considered the nebulae to be remote galaxies. Both scientists met on 26 April 1920, during a conference at the Smithsonian Museum of Natural History and exchanged their arguments in a heated debate, later called the "Great Debate." This debate was also fueled by another controversy: Shapley correctly located the Sun in an outer sphere of the Milky Way, whereas Curtis adhered to the model of a heliocentric galaxy. In 1923, Hubble reported new distance measurements based on Cepheids that clearly supported Curtis's hypothesis of a giant universe containing numerous galaxies.

Following the wish of his father, Hubble originally studied law, but after his father's death he turned to astronomy. After World War I, George E. Hale (1868–1938), director of the Mount Wilson Observatory, offered him a permanent position. In 1917, the new Hooker Telescope was installed, which, with a mirror of

2.5 m diameter, was the best instrument of that time. In 1919, Hubble began to work with this telescope and soon detected that most nebulae were galaxies outside the local group.

From 1917, Milton Humason (1891–1972) was a member of the staff of the Mount Wilson Observatory as an unskilled temporary worker. G. E. Wilson recognized his talent and offered him a permanent position. Furthermore, he was allowed to work as assistant to Hubble, although he did not have any scientific education or training. He spent most of the nights with the Hooker Telescope and he was the first to detect the red shift in the light of distant galaxies (in a few cases, a blue shift was also observed). In this connection, a red shift has the following meaning. In the spectra of sunlight or starlight, sharp black or bright lines are detectable. The black lines result from the absorption of wave bundles when white light passes through gas clouds in interstellar space or surrounding the star under investigation. The bright lines result from emission of certain wave bundles by exited atoms or molecules located between the light source and Earth. The exact wavelengths of absorption and emission lines can be determined for fixed distances of light source and observer by laboratory experiments or by analysis of sunlight. Yet, when the distance steadily increases, the wavelengths of these lines also increase, approaching those of red light (see Sect. 10.5). A steady shrinking of the distance has, of course, the consequence of a blue shift. In other words, this so-called Doppler effect reliably indicates the growing or shrinking of distances between light source and observer.

The red shift observed by Humason was published in 1925 without any interpretation, but it stimulated, as mentioned above, Lemaitre to postulate an expansion of the universe. In his famous 1929 publication, Hubble presented two more aspects. First, he presented experimental data proving that the rate of expansion increases with the distance of the observed object in a linear mode. The expansion coefficient became famous as the Hubble constant. Second, when the expansion rates of all galaxies were extrapolated back to the past, the same time, the same moment was found, the birth of the universe. The basis of the Big Bang theory was born. The international reaction to Hubble's publication exactly obeyed the scheme described by the theoretician and historian Thomas Kuhn (see Chap. 5):

A revolutionary concept is at first criticized and/or ridiculed by most representatives of the established paradigm.

The highly esteemed British astronomer Fred Hoyle (1915–2001) fought the Big Bang theory until the end of his life; yet, it was Fred Hoyle who coined this term in a BBC broadcast on 28 March 1949. He accepted the expansion of the universe, but postulated that the expansion is a steady process combined with permanent creation of new matter in interstellar space. However, his hypothesis and similar concepts cannot explain the existence of "background radiation" (see below). At this point, the skeptical question of a prominent laic, Woody Allen, should be mentioned: "If the universe really expands, why is it increasingly difficult to find a parking spot?"

An important piece of evidence in favor of the Big Bang theory resulted from microwave research after 1948. The Russian professor of mathematics George Gamow (1904–1968, immigrated to the USA in 1933) calculated that the first moment after the Big Bang, prior to the formation of matter, was a state of immensely intensive radiation, some of which should have survived. He and his Ph.D. students Ralph Alpher (1921–2007) and Robert Hermann (1914–1997) published a mathematical treatise in 1948 predicting that radiation from the Big Bang should have cooled down to low frequency microwave radiation as a result of the expansion and cooling of the universe.

In 1964, the background radiation was unexpectedly found by the physicists Arno A. Penzias (born 1933) and Robert W. Wilson (born 1936). These members of the Bell Laboratories used a big horn-shaped antenna to analyze microwave radiation emitted by the Milky Way or other members of the local group. They detected "microwave noise" with a wavelength of 7.5 cm that came uniformly from all directions. All attempts to identify a terrestrial or extraterrestrial source failed and the detection of this microwave noise was reproducible in numerous measurements. Because Penzias and Wilson were not able to explain their discovery, they contacted the professors of physics Robert Dicke (1916–1997) and James Peebles (born 1935) at Princeton University. Independently of Gamow's work, these physicists had elaborated new calculations for the background radiation, and Dicke's research group had constructed a special antenna for pertinent measurements. Dicke immediately understood that Penzias and Wilson had serendipitously discovered this radiation and that his measurements were obsolete. It was hard for him to see that the luck of Penzias and Wilson was awarded the Nobel Prize (in 1978). In 1990, the horn-antenna of the Bell Laboratories was designated a National Historic Landmark.

The further research activities of numerous astronomers and physicists have contributed additional evidence for the Big Bang theory. However, one important problem remained unsolved until 2015. The gravitation of the entire universe requires much more matter than detected, regardless of what kind of cosmic radiation was being studied. Obviously, the properties of this so-called dark matter differ largely from those of normal matter. As long as the nature of this dark matter remains unknown and unexplained, the Big Bang theory is not definitely proven.

Edwin Powell Hubble

Edwin Hubble was born on 20 August 1889 in Marshfield, Missouri. Some 11 years later, his father John P. Hubble, an insurance executive, and his mother Virginia Lee James, moved to Wheaton, Illinois, where he attended the high school. Hubble earned good grades in every subject except spelling, but he was better known for his athletic abilities. He played baseball, basketball, football, and ran track in both high school and college. In 1907, in his first year at the University of Chicago he led the basketball team to its first conference title. In Chicago, he studied astronomy and mathematics and earned the bachelor of science degree in 1910.

In 1909, his father moved with the entire family (including two sisters and two brothers) to Shelbyville, Kentucky. In this year, still before completing his bachelor's degree in science, Hubble began to study law obeying the wish of his sick father. Supported by a Rhodes Fellowship, he went to Oxford, UK, for 3 years where he continued to study law. He received a bachelor's degree in law there, added studies in literature and Spanish, and earned the masters degree, but he also attended lectures in astronomy (his main interest) and in mathematics.

His father died in the winter of 1913 and Hubble returned in the summer of 1914 to care for his mother and his younger sisters. He did not want to practice law, so in order to earn money he took a position at the New Albany High School in New Albany, Indiana, as teacher of Spanish, mathematics, and physics. However, about a year later he decided to become a professional astronomer. With the help of his former professor he obtained a position at the Yerkes Observatory of the University of Chicago, which allowed him to study astronomy and to achieve a Ph.D. in 1917. His dissertation was entitled *Photographic Investigations of Faint Nebulae*.

When the USA declared war on Germany, he volunteered for the army and was assigned to the newly established 86th division. He rose to the rank of a major, but the war was over before the 86th division reached the front. After the end of the war, Hubble spent a year in Cambridge to renew his studies of astronomy. In 1919, George E. Hale, founder and director of the Mount Wilson Observatory near Pasadena, offered Hubble a staff position. He remained a member of Mount Wilson's staff until his death, but when the new giant Hale Telescope (5.1 m refractor) became operational in 1947, he also worked at Mount Palomar and was the first to use this extraordinary instrument.

In the years before and during World War II, Hubble also served in the US Army at the Aberdeen Proving Ground and for this work he received the Legion of Merit award. During vacation in Colorado, he had a heart attack in 1949 but, supported by his wife Grace, he continued to work on the basis of a modified diet and work schedule. He died on 28 September 1953 of cerebral thrombosis (a spontaneous blood clot in his brain) in San Marino, California. He was buried without a funeral and his wife never revealed his burial site.

Bibliography

Alpher RA, Bethe H, Gamov G (1948) The origin of chemical elements. Phys Rev 73(7):803

Bechler Z (2013) Contemporary Netonian research (studies in the history of modern science), vol 9. Springer, New York, NY

Bernstein J (1984) Three degrees above zero: Bell Labs in the information age. Charles Scribner's Sons, New York

Christian GF (1996) Isaac Newton and the scientific revolution. Oxford University Press

Christianson GE (1996) Edwin Hubble: mariner of the nebulae. University of Chicago Press

deGrijs R (2011) An introduction to distance measurement in astronomy. Wiley, Chichester

Hoyle F (1950) The nature of the universe—a series of broadcast lectures. Basil Blackwell, Oxford

Hoyle F (1955) Frontiers of the universe. Heinemann Education Books, London

Hubble EP (1925) Cepheids in spiral nebulae. The Observatory 48:139

Hubble E (1929) A relation between distance and radial velocity among extragalactic nebulae. PNAS 15(3):168
Majaess DJ (2010) RR-Lyrae and type II Cepheid variables adhere to a common distance relation. J Am Assoc Variab Star Observ 38:100
Nussbaumer H, Bieri L (2009) Discovering the expanding universe. Cambridge University Press
Penzias A, Wilson R. http://www.aps.org/programs/outreach/history/historicsites/penzias.

10.3 Atoms: What Does Indivisible Mean?

The glance of the researcher frequently finds more than he wished to find.
(Gotthold E. Lessing)

The term "atom" is derived from the Greek verb τεμνειν (temnein, signifying cut or divide) and means indivisible. Speculations about the existence of tiny, indivisible particles as fundamental constituents of any matter were noted down in India as early as the sixth century B.C. These speculations also included the formation of pairs of atoms or of more complex compounds (today called molecules) consisting of three pairs of atoms. In Europe, several pre-Socratic philosophers designed hypotheses postulating tiny, indivisible particles as the basic components of elements or of all matter.

Empedokles of Akragas (Agrigenti, Sicily, c. 495–c. 430 B.C.) assumed that the four elements earth, water, fire, and air (the material basis of his cosmology) consisted of minute particles, different mixtures of which were responsible for all variations of dead objects and living organisms. Anaxagoras, who was born around 462 B.C. in Klazomenai, Asia Minor (west cost of Turkey) declared that all natural phenomena resulted from mixtures of tiny, indivisible particles. However, in contrast to Empedokles, he postulated a large number of different atoms, but made no comment on their shape or properties. Nonetheless, for more than 2000 years Demokritos and his mentor Leukippos were considered to be the fathers of atomism and the fathers of modern science.

Demokritos was born in Abdera, Thrace, around 460 B.C. and died perhaps around 370 B.C. Nothing is known about the life of Leukippos and it has proved impossible to disentangle the contributions of Leukippos and Demokritos to ancient atomism, because even at the time of Aristotle (384–322 B.C.) no original text by Leukippos was available. In the case of Demokritos, numerous small fragments of his work survived until the twenty-first century, but most of what is known about his life and work was handed down by Aristotle and his successors. Demokritos inherited the wealth of his father, which allowed him to travel to foreign countries including Egypt and parts of Asia. After returning to Thrace, he also traveled throughout Greece to acquire a better knowledge of its culture. His wealth allowed him to buy the writings of other philosophers and to build up his own comprehensive library.

He was highly esteemed by his fellow citizens because he had predicted several natural phenomena that became true. He was also known as the "Mocker" and as the "Laughing Philosopher" because he liked to laugh frequently at the follies of his contemporaries.

Demokritos was the first to use the term *ατομοσ* (atom) for the tiny particles that he believed were the basis of all matter and the origin of all natural phenomena. According to his understanding of atoms, these particles were indivisible and indestructible and their number was infinite, but they varied in shape and size. Demokritos is reported to have said about their mass: "The more any indivisible exceeds the heavier it is." He also postulated that their shape and surface structure was decisive for their connectivity and solidness. For instance, iron atoms had hooks that lock them into a solid, water atoms were smooth and slippery, salt atoms were sharp and pointed, and air atoms were light and whirly. All atoms were endowed with mobility and, when not temporarily fixed, they were in permanent motion. Aristotle later criticized Demokritos for not saying anything about the origin of their motion and about the driving force. However, this critique misunderstood that Demokritos considered mobility and motion as fundamental and eternal properties of atoms. The world view of Demokritos was so consistent that he not only considered the human body as a combination of certain atoms, but also the human soul. Hence, from the viewpoint of modern philosophy, the world view of Demokritos might be classified as atomistic materialism. Therefore, it is necessary to mention that it would be incorrect to classify Demokritos as a materialist, because his entire philosophy also included thoughts and discussions on ethics.

Roman philosophers, above all Titus Lucretius Carus (c. 99–c.55 B.C.), picked up the concept of Demokritos and added further details and arguments The only known work of Lucretius is a philosophical poem (written in 7400 hexameters) with the title *De Rerum Natura* (On the Nature of Things), which survived in the library of a German monastery. In this poem, he presents the principles of atomism, explains numerous natural phenomena, and describes the nature of mind, soul, and thoughts. His universe operated according to physical principles without any intervention from gods. He presented an experimental argument in favor of the existence of atoms. He observed that all organic materials (including hard wood) are subject to irreversible decay, and even rocks are eroded under the impact of drops or flowing water. Yet, the re-creation of materials (e.g., metals) can also be observed and new trees grow from their seeds.

He concluded that all materials contain a basic component that preserves and maintains characteristic properties for eternity, and these basic components are the atoms. The Roman poet Virgil wrote about Lucretius: "Blessed is he who has been able to know the cause of things."

An evidence-based definition and characterization of atoms began in the seventeenth century with the work of Robert Boyle (1627–1691). He redefined elements, such as gold or sulfur, as pure materials and postulated that their fundamental constituents were indivisible and indestructible by chemical methods (see Sect.

9.1). A more detailed description of atoms and concrete experimental evidence was published by John Dalton (1766–1844, see Biography Sect. 9.1).

For Dalton and all chemists of the nineteenth and twentieth centuries, atoms had a sphere-like shape, in contrast to the concept of Demokritos. Another characteristic difference concerned the formation of materials. For Demokritos, all materials were physical mixtures of different atoms, whereas Dalton and later scientists interpreted all modifications of elements and materials in terms of chemical reactions (for definition, see Sect. 10.1). Yet, for Demokritos and other Greek philosophers chemical reactions were unknown and unimaginable. Dalton found that pure elements react with each other in exact, well-defined mass ratios and considered simple number ratios of atoms (e.g., 1:1 or 1:2) as the basis of this law.

In 1827, the botanist Robert Brown (1773–1858) observed the erratic motions of dust particles suspended in water. This "Brownian motion" was thought to result from motions of water molecules knocking against the dust grains. In 1905, Albert Einstein published a mathematical analysis of this phenomenon, which inspired the French physicist Jean Perrin (1870–1958) to determine the masses and dimensions of atoms, thereby proving Dalton's atom theory. Perrin was awarded the Nobel Prize in Physics in 1926.

A quite different contribution on the structure and stability of atoms came from the German-Austrian professor Julius Plücker (1801–1861), who discovered cathode rays in 1858. Part of a wire was introduced into a glass bulb (analogous to a normal light bulb) filled with a gas. The wire was heated and an electric voltage was applied between the wire and the opposite end of the bulb (which served as positive pole). A radiation was observed between both poles, which was later identified as an electron beam. Joachim Hittorf (1824–1914), a Ph.D. student of Plücker, followed a university career up to full professor at the University of Münster and continued the experiments of Plücker after 1864. He observed that a small object placed somewhere between the wire and opposite glass wall generated a shadow on the fluorescent glass wall. Furthermore, he observed bending of the cathode ray when a magnetic field was installed in its neighborhood, and he observed emission of light when certain gases were present in the bulb. This so-called Hittorf-tube was the forerunner of the neon tube, which was invented by the French engineer George Claude in 1910.

Joseph J. Thomson (1856–1940; Nobel Prize, 1906) professor of physics at the University of Cambridge, studied the influence of cathode rays on the conductivity of gases. Apparently without knowing the German publications of Hittorf, he discovered the influence of magnetic and electric fields on cathode rays. He concluded that these rays consisted of negatively charged corpuscles (Thomson's terminology). He succeeded in determining their mass and found the incredibly low, but correct, value of 10^{-27} g, the lowest mass of an electric charge (later called electron). Several decades later, the components of an atomic nucleus (protons and neutrons) were shown to be 2000 times heavier. With his cathode ray experiments Thomson not only elaborated the basis of mass spectrometry, he also demonstrated for the first time that atoms can be split into two components, electrons and a positively charged ion (from *ιονειν* meaning moving in an electric field).

On the basis of these results, Thomson designed an atomic model consisting of electrons embedded in an amorphous positive mass, analogous to raisins in a plum pudding. This model was significantly improved by his assistant Sir Ernest Rutherford (see Biography). In 1898, Rutherford joined McGill University in Montreal and began to work with the radioactive metals radium and uranium. He discovered two quite different kinds of radiation, which he called α-rays (radium) and β-rays (uranium). A third type of radiation emitted by radium was discovered (but not identified) in 1900 by the French chemist Paul Villard. In 1903, Rutherford identified this radiation as a kind of electromagnetic wave and called it γ-rays. From experiments performed together with his German assistant Hans Geiger (1882–1945) and the British coworker Ernest Marsden (1889–1970), Rutherford concluded that α-particles must have two positive charges. His coworker Thomas Royds eventually proved that α-particles were nothing other than completely ionized atomic nuclei of the element helium, a noble gas. This finding proved again that atoms were divisible into electrons and positive ions.

Rutherford then decided to use α-particles as projectiles to investigate the density and homogeneity of atoms. A thin foil of hammered gold served as target. In contradiction to the plum pudding model, he found that approximately 99 % of the α-particles passed undisturbed through the gold, whereas 1 % were deflected. In 1911, he designed a new atomic model consisting of a minute, but heavy, positively charged nucleus orbited by electrons. Later measurements showed that the radius of a complete atom was about 10^5–10^7 times larger than that of the nucleus alone.

This new atomic model prompted coworkers of Rutherford and other physicists to study the structure and stability of atomic nuclei. The first important contributions came from Rutherford's group. In 1917, they succeeded in identifying the positively charged protons as one component of the nuclei. This insight was achieved by bombardment of nitrogen atoms with α-radiation ($^{14}N + \alpha \rightarrow {}^{17}O + \text{proton}$). In 1930, a new penetrating radiation was discovered by the German physicist Walther Bothe (1891–1957; Nobel Prize 1954) upon bombardment of beryllium with α-particles. This radiation was identified in 1932 by Rutherford's coworker James Chadwick (1891–1974; Nobel Prize 1932) as neutrons, the second component of atomic nuclei.

Further research by other scientists revealed that, with exception of the hydrogen atom, the atomic nuclei of all elements consisted of protons and an equal or higher number of neutrons. The number of protons was found to be identical with the order number of an element in the periodic table (see Sect. 9.1). For one and the same element, the number of neutrons can vary, and these variants are called isotopes because they occupy the "same position" in the periodic table. The existence and definition of isotopes was discovered and elaborated by another coworker of Rutherford, Frederic Soddy (1877–1956; Nobel Prize in Chemistry 1921), but the term was coined by Margaret Todd.

The question of whether or not atomic nuclei are indivisible was first articulated by Marie Curie (1867–1934), who discovered the radioactivity of radium and polonium (awarded the Nobel Prize in 1903 and in 1911). Rutherford and Soddy discussed this problem in 1902 in an article entitled "Theory of Atomic

Disintegration." They proposed that radioactivity should be considered as that transformation of one sort of atomic nuclei into another. About 30 years later, the German chemist Ida Noddack (1896–1978) published a prophetic statement: "It is imaginable that upon bombardment of heavy nuclei with neutrons the nuclei decay into several fragments that are isotopes of known elements but not neighbors [in the periodic table] of the irradiated element." Ida Noddack was one of the first women to study chemistry in Germany and discovered the elements rhenium and possibly technetium. She was nominated several times for the Nobel Prize, but without success. She was the first female scientist awarded the Liebig Medal and the Great Order of Merit of the Federal Republic of Germany.

Before 1939, nobody believed in the speculation of Ida Noddack and several scientists, above all Enrico Fermi (1901–1954), who began to study the irradiation of heavy elements with neutrons hoping to synthesize isotopes of new elements with an order number higher than that of uranium (so-called transuranium elements). These experiments failed, but in 1938 Otto Hahn (see Biography) resumed the irradiation of uranium with neutrons. He and his assistant Fritz Strassmann were quite surprised to find an isotope of barium (^{145}Ba) as reaction product and not a transuranium element. The second reaction product, an isotope of krypton (^{88}Kr), a noble gas, was not detectable at that time. This experiment proved reproducible and Hahn concluded that nuclear fission had indeed occurred. This result, published in January 1939 in the journal *Naturwissenschaften*, represented the second paradigm shift in the understanding of atoms, namely the cleavage of a nucleus.

Prior to this publication, Hahn informed Lise Meitner, his colleague and coworker of many years, who because of her Jewish parents had escaped to Sweden in the summer of 1938. Together with her nephew Robert O. Frisch, she published a theoretical analysis in February 1939 in the journal *Nature*. They concluded that this nuclear fission (a term coined by Frisch) released an enormous amount of energy and three neutrons (from one uranium atom). These neutrons were expected to initiate a chain reaction provided that a sufficient amount of uranium was present. In other words, Meitner and Frisch described the theoretical basis for peaceful utilization of atomic energy, but also the basis for an atomic bomb. The construction of atomic bombs was indeed realized in the USA between 1943 and 1945 under the code name "Manhattan Project" and ended with the bombing of Hiroshima and Nagasaki in August 1945.

Further studies of the bombardment of heavy atomic nuclei with neutrons or smaller nuclei yielded results in two directions. First, further nuclear fissions were discovered and, second, transuranium elements were synthesized . Up to 2012, short-lived isotopes of sixteen new elements have been detected. A third type of collision experiment was conducted with nucleons, for instance with extremely accelerated protons. Such experiments, currently performed at CERN in Geneva, have, for instance, the purpose of studying the structure and stability of nucleons. These experiments have proved that even nucleons are divisible and consist of three quarks. In the beginning of the twenty-first century, it is unpredictable how far such collision experiments will go and what kind of mass or energy packet will be the smallest constituent of matter. Anyway, the experiments of the past 150 years have clearly demonstrated that Demokritos's vision of atoms as indivisible tiny articles that are the fundamental constituents of all matter was nothing but a fiction.

Sir Ernest Rutherford

Ernest Rutherford (First Baron Rutherford of Nelson) was born on 30 August 1871, in Brightwater near Nelson in New Zealand. His father was a farmer and emigrated from Perth, Scotland, while his mother Martha (nee Thompson) originated from Hornchurch, Essex, UK. Rutherford attended Havelock School in Nelson and afterwards Nelson College. He won a scholarship, which allowed him to study at Canterbury College, University of New Zealand. He gained bachelor's and master's degrees followed by the bachelor of science. In 1895, after 2 years of research, Rutherford invented a new type of radio receiver. For this performance, he was awarded a research fellowship from the Royal Commission of the Exhibition of 1851 for postgraduate studies at the Cavendish Laboratories, University of Cambridge. He worked under the leadership of Professor Joseph J. Thomson and, with his help, managed to detect radio waves at half a mile. For a short time he held the world record for distances over which electromagnetic waves were detectable, but in 1896 he was outdone by the Italian engineer Guglielmo Marconi (1874–1934).

As consequence of a recommendation by Thomson, Rutherford was appointed professor at McGill University in Montreal, where he replaced L. Callender. Now on a firm financial basis, he dared to marry his fiancé Mary G. Newton whom he had left behind in New Zealand. They only had one daughter who died at the age of 29. In 1900, he received the doctor of science degree from the University of New Zealand and returned in 1907 to England as the Chair of Physics at the University of Manchester. In 1914, he was knighted by King George V, and in 1916 he was awarded the Hector Memorial Medal. During the war he worked on technical problems related to the detection of submarines by sonar. In 1919, Rutherford returned to Cambridge as successor of J. J. Thomsen and as director of the Cavendish Laboratory.

In addition to Michael Faraday (1781–1867), Rutherford was one of the most successful and productive experimental physicists of the past centuries. Certainly, no other scientist advised and supervised as many successful students and coworkers. The following short list illustrates this point:

Frederik Soddy (1877–1956); Nobel Prize in Chemistry 1921 for his contribution to the definition and identification of isotopes.

James Chadwick (1897–1967); Nobel Prize in Physics 1932 for the discovery of neutrons.

Otto Hahn (1879–1968); Nobel Prize in Chemistry 1944 for the cleavage of uranium atoms by means of neutrons.

Edward V. Appleton (1892–1965); Nobel Prize in Physics 1947 for the discovery of the ionosphere and for his contributions to radar technology.

John Cockroft (1897–1967) and Ernest Walton (1903–1995); Nobel Prize in Physics 1951 for the cleavage of atom nuclei by means of accelerated protons.

Together with the German Hans Geiger (1882–1945) Rutherford developed an apparatus allowing the detection of α-radiation. This mobile apparatus was the origin of what later (after 1928) became known as the Geiger–Müller counter (Walther Müller was Ph.D. student of Professor Geiger). This versatile and transportable instrument enables the detection of α-, β- and γ-radiation.

In the years 1925–1930, Rutherford was president of the Royal Society and thereafter became president of the Academic Assistance Council, which helped almost 1000 academic refugees from Germany. Rutherford had a small hernia for several years, which he neglected. After strangulation, this hernia caused a violent illness and despite an emergency operation in London he died 4 days later on 19 October 1937. After cremation he was buried in Westminster Abbey in the neighborhood of other famous scientists.

Among the numerous honors received by Rutherford during his lifetime, the Nobel Prize awarded in 1907, was one of the first and certainly the most important award. To Rutherford's surprise he received the Nobel Prize in Chemistry, not in physics. Furthermore, he received the Rumford Medal in 1905, the Elliot Cresson Medal in 1910, the Matteucci Medal in 1913, the Copley Medal in 1922, and the Franklin Medal in 1934. He also received the Order of Merit in 1922 and was elevated to the peerage as Baron Rutherford of Nelson, a title that became extinct upon his death.

Finally, it should be mentioned that the synthetic element rutherfordium (Rf) with the atomic number 104 was named after him.

Otto Hahn

Otto Hahn was born on 8 March 1897, in Frankfurt am Main as the youngest son of Heinrich Hahn and his wife Charlotte (nee Giese). Otto and his three brothers Karl, Heinrich, and Julius were raised and educated in sheltered surroundings. At the age of 15 he became interested in chemistry and performed simple experiments in the laundry of his home. In 1897, Hahn graduated from the Klinger Oberrealschule in Frankfurt (a high school). Although his father wanted him to become architect, Hahn entered the University of Marburg to study chemistry and mineralogy, supplemented by physics and philosophy. He studied the second year organic chemistry course in Munich under Adolf von Bayer (1835–1917; Nobel Prize 1905), but returned to Marburg. In 1901, Hahn obtained the doctor's degree for the dissertation *On Bromine Derivatives of Isoeugenol.*

After a year of military training, Hahn returned again to Marburg and worked for 2 years as assistant to his doctoral supervisor Theodor Zincke (1843–1928). Originally Hahn intended to work in the chemical industry. To broaden his knowledge in chemistry and in English he moved to London for a position at University College. He worked there under the guidance of Sir William Ramsey, who was awarded the Nobel Prize in 1904 for discovery of the noble gases. In 1905, Hahn began to work with radium salts and radioactive minerals and discovered a new substance, radiothorium, which was believed to be a new radioactive element. Later it was identified as an isotope of thorium (thorium-228), but the term isotope and its definition were coined later (in 1913) by the chemist Frederik Soddy.

The publication "New Radioactive Element Which Evolves Thorium Emanation" appeared in March 1905 in the *Proceedings of the Royal Society* and was the first of 250 papers published by Hahn in the field of radiochemistry. Following a recommendation by Sir Ramsey, Sir Ernest Rutherford offered Hahn a position in his research group at McGill University in Montreal. Between September 1905 and mid-1906, Hahn discovered three new radioactive isotopes: thorium-C (later identified as polonium-212), radium-D (lead-210), and radioactinium (thorium-227). In those days Rutherford used to say: "Hahn has a special nose for the discovery of new elements."

After his return to Germany, Hahn obtained a position at the University of Berlin in the research group of the organic chemist Emil Fischer (1852–1919; Nobel Prize in 1902). Despite primitive working conditions, he discovered the new isotopes mesothorium-I, mesothorium-II, and ionium (later identified as thorium-230). In the following years, mesothorium-I (radium-228) became important for medical radiation treatment, in as much as the costs of manufacture were much lower than those of the previously used radium-226 (discovered by Marie and Pierre Curie). The year 1907 saw two important events: first, completion of the habilitation that allowed Hahn to teach at the university and, second, meeting the Austrian physicist Lise Meitner. With this acquaintance began a 30-year collaboration and a life-long friendship.

In 1910, Hahn was appointed professor and in 1912 he became head of the Radioactivity Department of the Kaiser-Wilhelm Institut für Chemie in Berlin-Dahlem (today Hahn-Meitner-Institut of the Freie Universität Berlin). In the years 1908 and 1909, Hahn discovered the so-called radioactive recoil as consequence of the α-emission. This effect was described in 1904 by the physicist Harriet Brooks, but it was wrongly interpreted. Hahn's interpretation was correct and allowed him to develop the "emanation (recoil) method" as a new technique for the detection of new radioactive isotopes. In this way, he found polonium-214, thallium-207, thallium-208, and thallium-210. His work on radiochemistry was interrupted during the first two years of World War I, because Hahn like other chemistry professors was ordered to help Professor Fritz Haber develop new poisonous chemicals for the gas war.

In December 1916, Hahn was allowed to return to his institute in Berlin. Together with Lise Meitner he isolated a new long-lived radioactive isotope called proto-actinium (Pa-231). In 1913, K. Fajans (1887–1975) and O. Göhring had isolated a short-lived radioactive isotope, which they called "brevium," later identified as Pa-234. The International Union of Pure and Applied Chemistry confirmed in 1949 that both isotopes represented the new element no. 91, gave it the name protactinium (Pa), and confirmed Hahn and Meitner as discoverers. Hahn also isolated Pa-234 from uranium and, thus, was able to isolate for the first time two different isotopes of a new element. In 1936, Hahn published a book in English entitled *Applied Radiochemistry*, in which he described the emanation (recoil) method and other techniques useful for the detection, isolation, and identification of radioactive isotopes. This book had a great influence on all nuclear chemists and physicists worldwide.

By bombardment of uranium with neutrons, Hahn, Meitner, and Strassmann found in the years 1934–1938 a large number of radioactive reaction products.

Because the existence of the actinide elements had not yet been established and because uranium was believed to be an element of group 6 (of the periodic table), they incorrectly interpreted some of these new isotopes as elements 93–96. After Meitner's escape to the Netherlands in July 1938, Hahn and Strassmann eventually discovered the fission of uranium nuclei with neutrons by the detection of barium isotopes in the reaction products. First Hahn informed Meitner and asked her for an explanation. The results were published in January 1939. As a chemist, Hahn originally hesitated about claiming a revolutionary discovery in physics, but Meitner and her nephew Otto R. Frisch (both active in Sweden at that time) elaborated a theoretical interpretation of the nuclear fission (a term coined by Frisch) supporting Hahn's experiments and hypothesis. In 1963, Meitner wrote:

> The discovery of nuclear fission by O. Hahn and F. Strassmann opened up a new area in human history. It seems to me that what made the science behind this discovery so remarkable is that it was achieved by purely chemical means.

During World War II, Hahn, Strassmann, and coworkers continued to work on uranium fission, but they never had the development of an atomic bomb in mind. In August 1945, when Hahn and nine other German physicists were imprisoned in Farm Hall (UK), they were informed about the American atom bombs and their devastating effect on Hiroshima and Nagasaki. Hahn was on the brink of despair for these disastrous events and Max von Laue (1879–1960; Nobel Prize 1914) took care of Hahn to prevent his suicide.

In 1928, Hahn became director of the Kaiser-Wilhelm Institut für Chemie, a position he held until 1946. After his return to Germany in January 1946, Hahn (like Heisenberg and von Laue) had to live in Göttingen under the control of British authorities. In 1948 he became founding president of the new Max Planck Society, which succeeded the once-famous Kaiser Wilhelm Society. Hanh became the national and even international spokesperson for social responsibility, supported various movements against the use of nuclear energy for military purposes, and signed several manifestos.

On 22 March 1913, Hahn married in Stettin (today Szczecin, Poland) an art student at the Königliche Kunstschule (Royal Academy of Arts) in Berlin. They had only one son, Hanno, who became a distinguished art historian. Hanno and his wife Ilse died on August 1960 in a car accident and only Hahn's grandson, Dietrich, survived. Hahn lived in Göttingen until his death on 28 July 1968. Otto R. Frisch remembered:

> Hahn remained modest and informal all his life. His disarming frankness, unfailing kindness, good common sense, and impish humor will be remembered by his many friends all over the world.

The Royal Society, London, wrote in an obituary:

> Otto Hahn's achievements are known universally and will hold a special place in the history of science. He is remembered too for his whole character, his generosity of spirit, his belief in the proper use of scientific discovery, and for his humanity.

Hahn has received so many honors, for instance 37 of the highest national and international orders and medals, that a full account is beyond the scope of this biography. He was awarded the Nobel Prize in 1944 after he was twice nominated. He was made an Honorary Citizen of Frankfurt am Main, Göttingen, and Berlin (he refused an offer from Marburg). In German-speaking countries, numerous streets and buildings were named after him. The first European nuclear-powered civilian ship, two high-speed trains in Germany, a crater on the Moon, a crater on Mars, an asteroid, and even a cocktail have his name.

The original table and equipment Hahn and Strassmann used for their first nuclear fission experiments is on display in the Deutsche Museum in Munich.

Lise Meitner

Lise Meitner was born on 7 November 1878 in Vienna as the third of eight children of Phillip Meitner and his wife Hedwig Meitner-Schwan. The Meitners were a Jewish family, but her mother possibly had a Protestant background, because L. Meitner converted as an adult to Lutheranism in 1908. Because women were not admitted to a high-ranking gymnasium (high school) at that time, she was educated in a Bürgerschule (a second-class high school). After graduating from this school she passed an examination as teacher of French. Taking private lessons financed by her parents, she also prepared for an "Externa Matura" examination at a Gymnasium. After receiving the leaving certificate of the reputed Akademisches Gymnasium in 1901, she was accepted as student of physics at the University of Vienna. She also studied mathematics and philosophy, but became soon interested in radioactivity. However, her thesis, completed in 1905, was *Heat Conduction in a Homogeneous Body*. Women studying physics were extremely rare in Austria and she was the second women to obtain a doctoral degree in physics.

Afterwards, Meitner attempted to get a position in the research group of Marie Curie in Paris, but did not succeed. Therefore, she decided in the following year to move to Berlin to attend the lectures by Max Planck. In parallel, she worked in the group of Otto Hahn at the chemical institute of the university without any payment. During the first 3 years she had to use the side entrance, because female students were not accepted at the universities of Prussia. However, this ban was lifted in 1909. Together with Hahn she discovered several new isotopes, became gradually recognized by experts in foreign countries, and met Marie Curie and Albert Einstein. In the years 1912 through 1915 she worked as a fully paid assistant to Max Planck and was elected member of the Kaiser-Wilhelm Institut.

When war broke out, Meitner completed training as nurse and X-ray expert and served in the Austrian army until 1916. She was then allowed to return to Berlin and to continue her work in Hahn's group. In 1918, she was appointed head of the new physical-radioactive section of the Kaiser-Wilhelm Institut. In 1922, Meitner completed her habilitation in physics and was appointed associate professor for experimental nuclear physics in 1926. She was the first female professor of physics in Germany.

In 1933, after Hitler had come to power, the situation of Jews in Germany changed dramatically. All Jews and descendants of Jews who held positions in state-controlled institutions were dismissed or forced to resign from their posts. Therefore, many scientists, including prominent persons such as the Nobel laureate Fritz Haber, the physician Leo Szilard, and her nephew Otto Frisch, decided to emigrate. Meitner felt relatively safe, because she was a registered member of the Lutheran church and an Austrian citizen. Yet when the Nazis occupied Austria (*Anschluss Österreichs*) in 1938, all Austrians became citizens of Hitler's empire. Now Meitner was seriously endangered and with the help of Hahn she and the Dutch chemist Dirk Coster escaped to the Netherlands, but without all her possessions. Hahn had given her a diamond ring inherited from his mother to bribe a frontier guard if necessary, but it was not necessary.

Unfortunately, an attempt to obtain an appointment at the University of Groningen failed and Meitner went to Stockholm. There, she found a position in the Nobel Institute of Physics, although its director, Manne Siegbahn, had a prejudice against women in science. In the following years she obtained a laboratory at the Royal Institute of Technology and participated in research on Sweden's first nuclear reactor. In 1947, a position with the salary of a full professor was created for her at the University College of Stockholm and she received Swedish citizenship in 1949.

After the end of World War II, Meitner travelled several times to the USA, gave lectures at famous universities (e.g., Princeton and Harvard), and received in 1946 the honor "Woman of the Year" by the National Press Club. The American Press also called her "Mother of the Atom Bomb," a title she considered a shame. In 1942, when the Manhattan Project was launched, she was invited to support construction of the first atom bomb, but she declined declaring, "I will have nothing to do with a bomb." Because of her consequent pacifism and her deep humanity she was extremely critical of German scientists who had collaborated with the Nazis without any protest against Hitler's regime. She said, "Heisenberg and many millions with him should be forced to see these [concentration] camps and the martyred people." To Hahn she wrote: "You all worked for Nazi Germany. And you tried to offer only a passive resistance. Certainly, to buy off your conscience you helped here and there a persecuted person, but millions of innocent human beings were allowed to be murdered without any kind of protest being uttered. . . . [It is said that] first you betrayed your friends, then your children in that you let them stake their lives on a criminal war—and finally that you betrayed Germany itself, because when the war was already quite hopeless, you did not once arm yourselves against senseless destruction of Germany."

After her retirement in 1960, Meitner moved to Cambridge because most of her relatives lived in the UK. In 1964, during a strenuous trip to the USA she had a first heart attack. Several small strokes followed, because she suffered from atherosclerosis. In 1967, she broke a hip in a fall and recovered only partially. She died on 27 October 1968, without knowing that Hahn had died just 3 months earlier.

Bibliography

Burnet J (1914) Greek philosophy: Thales to Plato. Macmillan, London, Reprint (2012) Forgotten Books, http://books.google.com/books?id=9vc_AAAAYAAJ

Campbell J (1999) Rutherford: scientist supreme. AAS Publications, Christchurch
Clark RW (1980) The greatest pioneer on earth: the story of nuclear fission. Sedgwick & Jackson, London
Demröder W (2002) Atoms, molecules and photons: an introduction to atomic-molecular- and quantum physics. Springer, Berlin, Heidelberg
Ernest Rutherford-Biography: http://www.nobelprize.org/nobel_prizees/chemistry/laureates/1908/rutherford.html
Hahn O (1936) Applied radiochemistry. Cornell University Press, Ithaca, NY and Humphrey Milford, London
Hahn O (1958) The discovery of fission. Scientific American 198(2):76
Hahn O (1966) A scientific autobiography. Translated and edited by Ley W., Charles Scribner's Sons, NY; British edition (1967), McGibbon and Kee, London
Moran BT (2005) Distilling knowledge: alchemy, chemistry and the scientific revolution. Harvard University Press
Mrinalkanti G (1981) Indian atomism: history and sources. Humanities Press, Atlantic Highlands, NJ
Reeves R (2008) A force of nature: the frontier genius of Ernest Rutherford. W.W. Norton, New York
Rhodes R (1986) The making of the atomic bomb. Simon & Schuster, New York
Rife P (1999) Lise Meitner and the dawn of the nuclear age. Birkhäuser
Shea WR (1983) Otto Hahn and the rise of nuclear power. Reidel, Dordrecht
Siegfried R (2002) From elements to atoms: a history of chemical composition. DIANE. https://www.worldcat.org/oclc/186607849
Sime RL (1996) Lise Meitner: a life in physics. University of California Press, Berkeley
Thomson JJ (1906) Nobel Foundation. http://www.nobelprize.org/nobel_prizes/physics/laureates/1906/thomson-bio.html

10.4 What Is Light?

The incomprehensible is the kingdom of errors
(Luc de Clapier, Marquis de Vauvenargues)

Although everybody who is not blind knows light from their first day of life, the question "What is light?" has been discussed and experimentally investigated for almost 2000 years. Errors, fallacies, and their corrections occurred in two dimensions that will be discussed here separately. The first dimension discussed here concerns the question of whether light is a corpuscular radiation (particle theory) or a kind of wave (wave theory). The second discussion is focused on the question of whether the propagation of light needs a space-filling medium and how this medium should be defined and characterized.

The discussion on the nature of light can be traced back to the Greek philosopher Empedokles of Akragas (Agrigenti, Sicily, ca. 495–435 B.C.) who believed that fire in the eyes emits light, which scans an object like radar radiation. Euclid, who worked in Alexandria around 323–283 B.C. (the dates of birth and death are unknown), postulated that light travels in straight lines and wrote a book entitled *Optica* that included a mathematical treatment of the reflection of light by mirrors. Ptolemy (c. 90–188 A.D., full name Klaudios Ptolemaios) had Greek ancestors but lived as a Roman citizen in Alexandria. He worked as astronomer, physicist, and mathematician. Like Empedokles, he believed that seeing is based on the emission of light from the eyes.

In the first centuries A.D., the nature of light was also discussed in India, in the Hindu schools of Samkhya and Vaisheshika. The Samkhya school taught that light is one of the five fundamental "subtle" elements that underlie all gross elements, but a more detailed explanation in terms of particle rays or waves was not given. In contrast, the Vaisheshika school advocated an atomistic theory of the world and light rays were considered a high velocity stream of fire atoms.

When modern science emerged from the Renaissance in Europe, a new discussion on the nature of light was initiated by the philosopher René Descartes (1596–1650). He postulated that light propagates from the object to the eye in a space-filling luminous medium like a pressure wave. Consequently, he erroneously assumed that light travels faster in a denser medium than in a less dense medium, but he correctly concluded that refraction results from the different speeds in different media. This theory of Descartes may be understood as a wave theory favoring a kind of longitudinal wave instead of transversal waves.

The corpuscular theory of Lucretius was first resumed by the atomist Pierre Gassendi (1592–1655), but his work was only published in 1660 after his death. Isaac Newton (see Biography in Sect. 10.5) had read Gassendi's work in his youth and adhered to the corpuscular theory throughout his life. He published his optical studies first in a treatise entitled *Hypothesis of Light* (1675) and in an improved and expanded version in the book *Opticks*, which appeared in 1704. He believed that a light ray is a group of particles flying along a straight line, a theory that allowed him to explain and predict the phenomenon of reflection. His theory had the additional advantage that it could for the first time qualitatively explain why light could be polarized. In the years 1810–1812, the mathematicians Etienne-Louis Malus (1775–1812) and (independently) Jean-Baptiste Biot (1774–1862) published a mathematical treatment of the particle theory of polarization, which was considered as proof of the particle theory. However, the particle theory did not explain the refraction of light and Newton incorrectly assumed that refraction results from an acceleration of the particles caused by gravitation when a light ray entered a denser medium. Consequently, he explained the different colors of the spectrum not by refraction but by different sizes of the particles (see Sect. 10.5).

At the same time, the foundations of modern wave theory were elaborated by the English physicist Robert Hooke (1635–1703) and by the Dutch scientist Christian Huygens (1625–1695). In his book *Micrographia* (1665) and in a later treatise (1772) Hooke postulated that light propagates in the form of pulses or vibrations similar to waves in water, but in contrast to surface waves he speculated that the vibrations of light might be perpendicular to the direction of propagation. Huygens developed a mathematical theory of light focused on the phenomenon of refraction and published it in 1690 under the title *Treatise on Light*. His experiments showed that light penetrating tiny holes in a screen propagated behind the hole as a cone-shaped bundle, as if each hole was itself an original source of light. In contrast, the corpuscle theory predicted that the particles would continue their flight behind the screen exclusively along a straight line. Again in contrast to Newton, Huygens believed that light is not affected by gravity and, thus, should slow down when entering a denser medium. Leonard Euler (1707–1783) was another supporter of the wave theory and agued in his book *Nova Theoria Lucis et Colorum* (1746) that diffraction is best explained by the wave theory.

Decisive experiments in favor of the wave theory were contributed by the oculist and physicist Thomas Young (1773–1829). He studied the phenomenon of interference, which occurs when two waves meet or cross each other. He divided light from a single source into two beams that passed two closely neighboring narrow slits in a screen. The overlapping of the fan-shaped light bundles formed behind the screen generated interference patterns such as those known from sound waves and surface waves. Against the opposition of the physicist and mathematician Siméon D. Poisson (1781–1840) and other members of the Académie Royale des Sciences, Augustin-Jean Fresnel (1788–1827) experimentally demonstrated in 1818 that the particle theory was wrong. Furthermore, he presented a convincing mathematical argumentation explaining the polarization of light by the wave theory, provided that light was a transversal type of wave. After 1820, the wave theory began to predominate over the particle theory.

The last nail in the coffin of the particle theory was contributed by Armand Hippolyte Fizeau (1819–1896) and Léon Foucault (1819–1868), who in 1849 were the first to measure the speed of light. They measured a value of 298,000 m/s and their experiments suggested that light traveled slower in a denser medium, in contradiction to Newton's theory.

In the second half of the nineteenth century, the wave theory was further improved and confirmed by the physicists Michael Faraday (1791–1867), James Clerk Maxwell (1831–1879), and Heinrich Hertz (1857–1894). Faraday discovered in 1845 that the plane of polarized light rotated when it was emitted parallel to a magnetic field, an effect later called Faraday rotation. In 1847, Faraday published the hypothesis that light was a kind of high-frequency electromagnetic wave. This hypothesis prompted Maxwell to compare the properties of light with those of other electromagnetic waves. He concluded that electromagnetic waves traveled through space at a constant speed, namely at the same speed as light, regardless of their wavelength. Maxwell published in 1871 a comprehensive mathematical description of electric and magnetic fields under the title *A Treatise of Electricity and Magnetism*. The summary of his calculations became famous as Maxwell's equations. The experimental evidence for Maxwell's theory was then elaborated by the German physicist Heinrich Hertz around 1885–1886. Hertz generated radio waves in the laboratory and proved that they exhibited the typical properties of light, such as reflection, refraction, and interference. Maxwell's theory and Hertz's experiments yielded the foundation for the development of all kinds of wireless communication, including television.

Despite the enormous progress of Maxwell's theory, it had weak points. It did not explain the photoelectric effect or the existence of spectral lines, regardless of whether absorption or emission spectra were concerned. The missing explanations were provided by the German physicists Phillip Lenard (1862–1942), Max Planck (1858–1947), and Albert Einstein (see Biography) in the years 1800–1805. Max Planck (Nobel Prize in Physics 1918) calculated the theoretical radiation of a black body upon heating (or cooling) and concluded that the radiation can change its energy only in finite amounts that he called "quanta." The energy differences were multiples of what was later called Planck's constant h ($\Delta E = h \times \nu$). This result was a revolution, because it was previously believed that energy was infinitely divisible and that the flow of energy from one object to another was a steady, continuous process.

Phillip Lenard (Nobel Prize in Physics 1905) studied the generation and properties of cathode rays (later defined as beams of electrons). He found that the energy of electrons emitted from a suitable metal surface upon irradiation with UV light increased with the frequency of the radiation, whereby the slope equaled Planck's constant. Albert Einstein (Nobel Prize in Physics 1921) formulated the equivalence of mass (m) and energy ($E = m \times c^2$, where c is the speed of light in vacuo). Arthur Holly Compton (1896–1962) demonstrated in 1923 that the wavelength shift of low intensity X-rays scattered from electrons can best be explained by a particle theory. He was awarded the Nobel Prize in Physics in 1927. All these findings and calculations formed the basis of the modern dualistic understanding of electromagnetic radiation. Depending on the experiment, light is sometimes best described as a transversal three-dimensional electromagnetic wave and sometimes as a kind of particle having the mass E/c^2 when "flying." In 1926, this "wave packet" was named a photon by Gilbert N. Lewis. These results can be understood as a combination of wave theory and particle theory, but in contrast to Newton's particles, photons only exist when flying but not in a fixed position.

These results also required a revision of Rutherford's atomic model (see Sect. 10.1). This model postulated that the tiny nucleus (later identified as a combination of protons and neutrons) was embedded in a homogeneous, isotropic cloud of electrons. Taking into account Planck's quantum theory, the Danish physicist Niels Bohr (1885–1962; Nobel Prize 1922) proposed in 1913 a new model, which postulated that electrons orbited the nucleus in a finite set of orbits. The energy of an electron in a certain orbit increased with the distance from the positively charged nucleus.

The electrons could jump between these orbits only in discrete changes of energy, whereby a loss of energy corresponded to the emission of a photon. Absorption of a photon enhanced the energy of an electron by jumping to a higher orbit. This model explained why different elements emitted or absorbed monochromatic light corresponding to the spectral lines observed in their emission or absorption spectra.

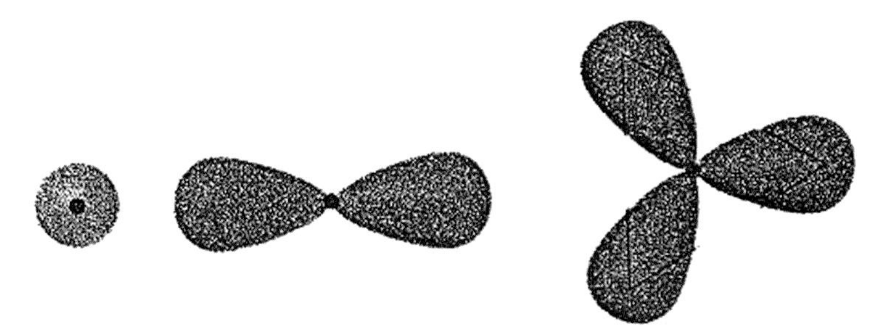

Fig. 10.1 Examples of simple atomic orbitals (the *black dots* represent atomic nuclei)

A shortcoming of Bohr's first atomic model was the assumption of orbiting electric charges, because it was known that an electric charge moving on a circular path emitted the energy invested in its permanent acceleration in the form of electromagnetic radiation, the so-called synchrotron radiation. Hence, the stability of the electrons in their orbits around an atomic nucleus was not compatible with the properties of an electric particle moving on a circular path and, thus, a modification of Bohr's model was required. The following physicists contributed to the solution of this problem: Louis de Broglie (1892–1987), Erwin Schrödinger (1887–1960), Pascual Jordan (1902–1980), Enrico Fermi (1904–1955), Werner Heisenberg (1901–1976), and Max Born (1892–1970). According to the new quantum theory, the electrons occupied so-called orbitals that might be understood as negatively charged electronic clouds or as segments of the space around the nucleus where the probability of encountering an electron was extraordinarily high. Two-dimensional examples of orbitals are presented in Fig. 10.1.

As mentioned at the beginning of this chapter, the nature of light should be discussed in two directions. The second direction concerns the following questions: Does light need a medium for its propagation? If so, is this medium a kind of classical matter or is it a special medium with unusual properties that do not obey the laws of classical physics?

In physics, this hypothetical medium was named "aether" (in English as well as in German), a word derived from the Greek $\alpha\iota\theta\eta\rho$ (aether) meaning upper air or fresh air. Chemists used the same word to denote a class of substances characterized by a C–O–C group. Most widely known from its medical application is diethyl ether (modern spelling in chemistry), which is a colorless, highly fluid, volatile liquid (boiling p. 36 °C), properties that are representative and symbolic for the meaning of aetheric character.

Aristotle (384–322 B.C.) was the first to propose an aether as a fundamental component of the universe. In his cosmology and world view, aether had two aspects. On the one hand, it was a kind of element (the fifth element, see Sect. 9.1) endowed with permanent motion in a sphere beyond the moon. On the other hand, it was a driving force stimulating the motion of all other elements and their mixtures.

In modern physics, Christian Huygens was the first to postulate a luminiferous (light-bearing) aether as medium for the propagation of light. The wave theory required such a hypothesis, because propagation of a wave is based on an excitation and energy transfer between neighboring "particles" of a medium. Hence, all physicists who favored the wave theory necessarily postulated the existence of an aether. Furthermore, Robert Boyle (see Sect. 9.1) and later physicists believed that the "subtle particles" of the aether were responsible for the mechanical interactions of liquid and solid objects. In contrast, Newton and other supporters of the corpuscle theory did not need an aether. Yet, surprisingly, Newton did not refute the existence of an aether and wrote: "I don't know what this aether is, but if it consists of particles then they must be exceedingly smaller than those of air, or even than those of light."

In the early nineteenth century, the structure and mechanical properties of the aether were discussed in more detail. Fresnel calculated that the aether should possess the properties of an elastic solid to enable the rapid propagation of a transversal wave. The British mathematician George Green (1793–1841) elaborated an aether model predicting a rigid structure at high frequency, but a fluid nature at low frequency and low speed of electromagnetic radiation. French physicists modified Fresnel's and Green's hypotheses in various directions.

In 1861, a new aether model was published by Maxwell in his treatise *On Physical Lines*. He described his aether as a "sea of molecular vortices," partly consisting of ordinary matter and partly of an ill-defined medium. He developed mathematical formulas for the dielectric constant and for the permeability of the aether. Later, Maxwell wrote in the *Encyclopedia Britannica*: "Aethers were invented for the planets to swim in, to constitute electric atmospheres and magnetic effluvia, to convey sensations from one part of our body to another and so on, until all space had been filled three or four times with aethers. The only aether which has survived is that which was invented by Huygens to explain the propagation of light."

In the decades before World War I, the number of aether theories almost equaled the number of physicists studying electromagnetic phenomena. Furthermore, numerous experiments were performed to prove or disprove the existence of a luminiferous aether. Most famous became the Michelson–Morley experiment, which led those scientists to conclude that an aether having properties similar to those of ordinary matter did not exist. This negative result was later confirmed by other physicists and more sophisticated experiments. However, after 1895 the Dutch theoretician Hendrik A. Lorentz published a theory strictly differentiating between matter (e.g., electrons) and aether. His aether, later called Lorentzian aether, was absolutely motionless and acted just as a mediator between electrons. His theory, a forerunner of Einstein's special relativity, explained the negative results of the Michelson–Morley experiment by the physical length contraction of all objects (including observer and apparatus) in the direction of the motion. His theory was improved by the mathematician Henri Poincaré (1854–1912).

In 1905, Albert Einstein (see Biography) published the theory of special relativity and concluded that an aether is not necessary to explain the extant experimental results and to confirm the wave theory. In 1909, Einstein summarized his view in a (German) lecture entitled: "The development of our view on the composition and essence of radiation." However, a decade later when Einstein published the general theory of relativity, he postulated an aether enabling propagation of gravity waves. Lorentz sent him a letter complaining about the inconsistency of both relativity theories and Einstein answered:

> We may say that according to the general theory of relativity, space is endowed with physical qualities: in this sense, therefore, there exists an aether. According to the general theory of relativity, space without aether is unthinkable: for in such space there not only would be no propagation of light, but also no possibility of existence for standards of space and time (measuring rods and clocks), nor therefore any space-time intervals in the physical sense. But this aether may not be thought of as endowed with the quality characteristic of ponderable media, as consisting of parts which may be tracked through time. The idea of motion may not be applied to it.

In 1951, the founder of the quantum field theory, Paul Dirac (1902–1984; Nobel Prize 1933) wrote an article in the journal *Nature* supporting the concept of a Lorentz- or Einstein-type of aether: "We have now the velocity of all points of space-time, playing a fundamental part in electrodynamics. It is natural to regard it as velocity of some real physical thing. Thus with the new theory of electrodynamics we are rather forced to have an aether."

Finally, the recent comment (2005) of the American physicist Robert B. Laughlin (born 1950, Nobel Prize 1998) should be cited:

> It is ironic that Einstein's most creative work, the general theory of relativity, should boil down the conceptualizing space as a medium when his original premise [in special relativity] was that no such medium existed. ... The word aether has extremely negative connotations in theoretical physics, because of its past association with opposition to relativity. This is unfortunate because, stripped of these connotations, it rather nicely captures the way most physicist actually think about the vacuum. ... Relativity actually says nothing about the existence or nonexistence of matter pervading the universe, only that any such matter must have relativistic symmetry. ... It turns out that such matter exists. About the time relativity was becoming accepted, studies of radioactivity began showing that the empty vacuum of space had spectroscopic structure similar to that of ordinary quantum solids and fluids. Subsequent studies with large particle accelerators have now led us to understand that space is more like a piece of window glass than ideal Newtonian emptiness. It is filled with "stuff" that is normally transparent but can be made visible by hitting it sufficiently hard to knock out a part. The modern concept of the vacuum of space, confirmed every day by experiment, is a relativistic aether, but we do not call it this, because it is a taboo.

This short review illustrates the enormous difficulties that physicists experienced over the past 400 years in achieving a proper understanding of light and other electromagnetic waves. Numerous correct and incorrect measurements, numerous half-truths, fallacies, and their revisions were published until a satisfactory agreement on the nature of electromagnetic phenomena was reached.

Albert Einstein

Albert Einstein was born on 14 March 1879 in Ulm, in the Kingdom of Württemberg, part of the German empire. His father Herman Einstein, a salesman and engineer, and his mother Pauline (nee Koch) were both descents of Jewish families that had lived for several generations in Germany. In 1880, his parents and an uncle moved to Munich to found a company that manufactured electrical machines and equipment under the name Elektrotechnische Fabrik J. Einstein & Cie. In 1884, Einstein joined a Catholic elementary school, but 3 years later he was transferred to the Luitpold Gymnasium, a high school named Albert Einstein Gymnasium today.

In 1894, his family's company went bankrupt, because their machines based on direct current were not competitive with the rapidly increasing number of machines using alternating current. His family moved to Milan and a few months later to Pavia (Italy). Einstein stayed in Munich to complete his studies with the final examination (Abitur, Matura), but due to a clash with the authorities of the Luitpold Gymnasium he left Munich in December 1894 without the Abitur. At the age of 16 (in 1985), he sat the entrance examination for the Swiss Federal Polytechnic (later ETH Zürich), but failed to achieve the required standard in the general part of the examination. Following the advice of the principal of the ETH he entered the cantonal school in Aarau (Switzerland), where in 1896 he completed his secondary education with the Abitur. The myth that Einstein had low grades in mathematics and physics is a fallacy of his first biographer who misunderstood the Swiss system of grading. In reality, he had top grades. Early in 1896 Einstein abandoned his German citizenship to avoid military service.

Although only aged 17, he was now allowed to study mathematics and physics at the Zürich polytechnic. That same year Mileva Marić, his future wife, also enrolled in the same teaching program. She was the only female student in the mathematics and physics section of the diploma course. In 1900, Einstein received the diploma (master's degree), whereas M. Marić failed because of a poor grade in mathematics. There were rumors that Marić made substantial contributions to Einstein's famous 1905 papers, but historians who studied this issue in detail did not find any evidence for this speculation. By 1900, Einstein's and Maric's friendship had turned into a romance and their marriage followed in January 1903. However, correspondence between Einstein and Marić discovered in 1987 revealed that they had a daughter called Lieserl before the marriage. This daughter was born in the spring of 1902 in Novi Sad, where Marić spent some time in the home of her parents.

Marić returned to Switzerland and left her daughter behind. Einstein apparently never saw his daughter and her further fate remains unknown. The couple later had a son, Hans Albert, born in Bern in 1904, and a second son, Eduard, born in Zürich in 1910. In 1914, Einstein moved to Berlin without his family and in 1919 Albert and Mileva were divorced. The second son was later diagnosed with schizophrenia and after the death of his mother spent all his life in an asylum.

After receiving the master's degree, Einstein acquired Swiss citizenship in 1901, but had difficulties in finding a teaching position. With the help of a friend he eventually found a job as assistant examiner at the Federal Patent Office in Bern. This position became permanent in 1903, but it left enough time to write a thesis entitled *A New Determination of Molecular Dimensions*. On the basis of this thesis, Einstein was awarded a Ph.D. of the University of Zürich in April 1905. In the same year (later called the miracle year), he published four groundbreaking papers (three of them in *Annalen der Physik*) that brought him recognition by the international scientific community. These papers dealt with the theory of special relativity, with the equivalence of mass and energy (including the famous formula $E = mc^2$), with the photoelectric effect, and with Brownian motion.

In 1907, a first attempt to achieve habilitation (tenure) failed, but a second attempt in 1908 was successful and Einstein was appointed lecturer at the University of Bern. As a result of his increasing international recognition, Einstein became associate professor in the same year. In April 1911, Einstein accepted a position as full professor at the Karl-Ferdinand University of Prague and acquired Austro-Hungarian citizenship for this purpose. However, in July 1912 he returned to Switzerland as full professor at the ETH Zürich, teaching analytical mechanics and thermodynamics. He continued his research on the molecular theory of heat and on a new theory of gravitation in cooperation with his friend, the mathematician Marcel Grossman.

In 1914, Max Planck and Walter Nernst (1864–1941; Nobel Prize in Chemistry 1920) visited Einstein and persuaded him to accept a position as director of the Kaiser-Wilhelm Institut für Physik in Berlin. This position was combined with a professorship at the A. v. Humboldt University (but was free of teaching duties). Furthermore, he was elected member of the Prussian Academy of Sciences and, in 1916, was appointed President of the German Physical Society.

In the years 1911–1916, Einstein elaborated and published his theory of general relativity. This theory predicted that light was attracted by huge masses such as the mass of the sun. This prediction was experimentally verified during the solar eclipse of 29 May 1919. The astronomer Sir Arthur Eddington could prove that the light ray of a star behind the sun was slightly bent when passing by the sun. Publication of this result in numerous media made Einstein famous worldwide. The British newspaper *The Times* presented on 7 November 1919 the following headline: "Revolution in Science—New Theory of the Universe—Newtonian Ideas Overthrown." Yet, when Einstein was awarded the Nobel Prize in 1921, he received it for explanation of the photoelectric effect and not for the theory of general relativity, because controversial debates about this theory were not yet settled.

In the years 1921–1932, Einstein traveled to many countries including the USA and Japan. Whenever he was back in Germany, he gradually learned that his position and perhaps even his live might be endangered by the rising power of the Nazis, because of his Jewish background. Furthermore, Einstein had become an active pacifist and an acquaintance, Karl Blumenfeld, had stimulated his interest in Zionism. Therefore, in the years 1921–1925 Einstein helped to establish the Hebrew University in Jerusalem. Early in 1930, Einstein decided to spend one half of every year at the University of Princeton in the USA. After returning to Princeton in December 1932 he decided to cancel his return to Germany because of Hitler's *Machtergreifung* in January1933. During a stay in Antwerp in March 1933 he went to the German consulate and renounced his German citizenship. Some months later his publications were included in Nazi book burnings; according to the proclamation of Minister Joseph Goebbels: "Jewish intellectualism is dead."

Einstein and his second wife were at first without a permanent home and continued their stay in Belgium. They then followed an invitation from the naval officer Oliver Locke-Campson to live in his cottage in England. There Einstein met Winston Churchill and the former prime minister, Lloyd George. Einstein asked them to bring Jewish scientists from Germany to England and to give them suitable positions in British institutions. Einstein also contacted other leaders of European countries, with considerable success in the case of Turkey's prime minister, Ismet Inönü. More than 1000 Jewish scientists eventually emigrated from Germany and Austria to Turkey. In October 1933, Einstein returned to Princeton and took a position at the Institute for Advanced Study, an institute that became a refuge for scientists fleeing Nazi Germany. Einstein maintained his position at Princeton for the rest of his life, applied for US citizenship in 1935, and received it in 1940.

During his time at Princeton, Einstein tried to disprove the accepted interpretation of quantum physics and to elaborate a unified field theory, but he failed in both cases. Together with his coworker Satyendranath Bose he developed the Bose–Einstein theory concerning the properties of condensed matter at extremely low temperatures. Einstein was also politically active in various directions. Famous is his letter to President Roosevelt, co-signed by the emigree Hungarian physicist Leo Szilárd, in which the authors warned the President that Hitler might win the race to build an atomic bomb. Einstein and Szilárd recommended that the USA should launch its own nuclear weapon research as soon as possible. The consequence of this letter was the Manhattan Project and the atom bombs on Hiroshima and Nagasaki. However, in the following years Einstein was engaged in various pacifist activities and in 1954 he confessed to his friend Linus Pauling (1901–1994; Nobel Prize in Chemistry 1954 and Nobel Peace Prize 1962): "I made a one great mistake in my life—when I signed the letter to President Roosevelt recommending that atom bombs be made, but there was some justification—the danger the Germans would make them." Furthermore, Einstein was a passionate antiracist, who joined the National Association for the Advancement of Colored People in Princeton and fought for the civil rights of African Americans. He considered racism to be America's worst disease. Moreover, he was permanently engaged in Zionist activities. When Israel's first president, Chaim Weizmann, died, the prime minister David Ben Gurion offered Einstein the position of president, but Einstein declined.

During his first years in Berlin, Einstein had frequently felt sick and was nursed by his cousin Elsa Löwenthal. This relationship turned into a romance and, shortly after his divorce from Mileva, he married Elsa on 2 June 1919. Einstein also adopted two stepdaughters. Unfortunately, Elsa died relatively young (in December 1936) due to heart and kidney problems. Einstein himself died on 17 April 1955, from internal bleeding caused by an aortic aneurism. Einstein refused a second surgery (the first surgery had been some years before) with the words: "I want to go when I want. It is tasteless to prolong life artificially. I have done my share. It is time to go. I will do it elegantly."

Einstein received numerous honors, medals, and awards and only a few of them can be mentioned here. The DDR edited a coin to commemorate 100 years since his birth. The *Deutsche Post* edited a stamp with his portrait in 2005 on the 50th anniversary of his death. In 1999, *Time Magazine* called him "Man of the Century" and in the same year 100 leading scientists elected him the most important physicist of all time. The synthetic (transuranium) element no. 99 was named Einsteinium, and a crater on the Moon as well as an asteroid have his name. It is trivial to say that numerous streets, schools, institutes, and other buildings in various countries were named after him.

Bibliography

Andriessen CD (2006) Huygens. The man behind the principle. Cambridge University Press.
Bell AE (1947) Christian Huygens and the development of science in the 17 century. Edward Arnold & Cie, London
Brian D (1996) Einstein: a life. Wiley, New York
Dirac P (1951) Is there an aether? Nature 168:906
Einstein A (1920) Ether and the theory of relativity. Republished 1922 in Sidelights of relativity. Methuen, London
Isaacson W. (2007) Einstein his life and universe. Simon & Schuster, New York Paperback
Laughlin RB (2005) A different universe: reinventing physics from the bottom down. Basic Books, New York
Loyd GR (1968) Aristotle, the growth and structure of his thought. Cambridge University Press
Mahon B (2003) The man who changed everything. The life of James C. Maxwell, Wiley, Hoboken, NJ
Renn J (2005) Albert Einstein: Chief engineer of the universe. One hundred authors for Einstein. Wiley VCH, Weinheim
Whittacker ET (1910) A history of the theories of ether and electricity. 1st edn. Longman & Green & Co, Dublin
Schaffner KF (1972) Nineteenth-century aether theories. Pergamon Press, Oxford

10.5 Why Do We See Colors?

> Man will err while yet he strives
> (Johann W. v. Goethe)

Almost all important natural phenomena that were observable for the humans of the ancient world were interpreted and discussed by Greek scientists and philosophers. Perhaps the most important phenomena humans are confronted with, even immediately after birth, are light and color. Therefore, it is somewhat surprising that only two pre-Socratic philosophers, namely Empedocles and Demokritos, were reported as commenting on seeing, in general, and colors, in particular. However, it should be kept in mind that only part of the information and knowledge that was available in the ancient world has been handed down to the twenty-first century.

Furthermore, the fragments of knowledge that have survived were handed down by secondary and tertiary sources, such as Aristotle and his interpreters.

Empedokles of Akragas commented on seeing and vision power, but not on colors. Amazingly, he speculated on two contradictory concepts of seeing. On the one hand, he postulated that fire in the eyes of humans and animals emits light that scans the observed object like radar radiation. On the other hand, he was thought to believe that an emanation or evaporation of an object penetrates the pores of an eye and generates there the image of the object. Apparently, both principles of seeing coexisted until the sixteenth century, but it is impossible to find out now to what extent scientist and laics adhered to either of these principles.

Demokritos of Abdera, Thrace (ca. 460–370 B.C.), explained the origin of colors but he did not comment on the origin and nature of seeing. Aristotle reports that Demokritos denied the real existence of colors and attributed the vision of colors to different arrangements of atoms on the surface of an object. Aristotle also said that Demokritos attributed the color white to a smooth surface and black to a rough surface. These comments agree with the general concept of Demokritos that any kind of perception is based on the different forms and arrangements of atoms.

Prior to a description of how the understanding of color vision has changed in the course of the past four centuries, the modern state of knowledge is summarized to allow a proper understanding of the history. This summary and all further comments are focused on the physical and chemical aspects of color vision.

Light, as explained in Sect. 10.4, is an electromagnetic radiation that proceeds in the form of an electromagnetic wave that, depending on its source, can adopt various wavelengths. The correlation between wavelength (λ), frequency (v), and speed (c) is given by the simple equation $c = v \times \lambda$. Visible light is defined by wavelengths in the range of 380–780 nm (1 nm $= 10^{-9}$ m). The different wavelengths within this range are interpreted by the human brain in terms of different colors. Without need of any physical instruments, the complete spectrum of colors is visible in nature when a rainbow appears. In this spectrum, violet represents the shortest wavelength and dark red the longest

From the physical point of view, a pure color means monochromatic light, which in the ideal case means one single wavelength. However, in experimental practice, monochromatic light is a bundle of closely neighboring wavelengths that cannot be split into a single wavelength by standard physical methods such as prisms or optical grids. Mixed colors result from the simultaneous stimulation of the retina by two or more beams of monochromatic light. Modern studies have revealed that the human eye can distinguish up to 200 colors in the range of 380–780 nm, whereas a prism spectrograph can differentiate between 2000 tones.

In the seventeenth century, most scientists agreed that seeing results from light emitted or reflected by an object and perceived by the human eye. However, at the end of that century a fierce discussion arose concerning the question, "What is white light?" The exponents of this controversy were the British mathematician and physicist Isaac Newton (see Biography) and the German poet, writer, and naturalist Johann W. von Goethe (1749–1832). Most scientist and laics of the seventeenth and eighteenth century believed that black and white defined the ends of the color scale.

All other colors were considered intermediate states resulting from partial shadowing of white light or partial weakening and illumination, respectively, of darkness.

Goethe, who adhered to this theory even at the end of the eighteenth century, wrote in his book *Zur Farbenlehre* (Theory of Colors):

> The highest degree of light, such as that of the sun ... is for the most part colorless. This light, however, seen through a medium but very slightly thickened, appears to us yellow. If the density of such a medium is increased, or if its volume becomes greater, we shall see the light gradually assume a yellow-red hue, which at last deepens to a ruby color. If, on the other hand, darkness is seen through a semi-transparent medium, which is itself illuminated by a light striking on it, a blue color appears: this becomes lighter and paler as the density of the medium is increased, but on the contrary appears darker and deeper the more transparent the medium becomes: in the least degree of dimness, short of absolute transparence, always supposing a perfectly colorless medium, this deep blue approaches the most beautiful violet.

Goethe also formulated this short version of his view: "Yellow is light which has been damped by darkness. Blue is darkness weakened by light." The philosopher and naturalist Rudolf Steiner, science editor of Kurschner's edition of Goethe's works, explained in 1897:

> Modern natural science sees darkness as a complete nothingness. According to this view, the light that streams into a dark space has no resistance from the darkness to overcome. Goethe pictures to himself that light and darkness relate to each other like the north and south pole of a magnet. The darkness can weaken light and its working power. Conversely, the light can limit the energy of the darkness, in both cases color arises.

The physicist John Tyndall emphasized in 1880 another aspect: "The action of turbid media was to Goethe the ultimate fact—the *Urphänomen*—of the world of colors."

The view that white light is the most original kind of light (sunlight) and the purest and most homogeneous kind of light goes back to Rene Descartes (1595–1640) and finally to Aristotle. White light was for many centuries associated with emotional attributes such as purity and innocence. Hence, Isaac Newton provoked an outcry from scientists and laics when he reported experiments proving that the established understanding of light was a fallacy.

Newton, a multitalented and avid researcher, began to perform optical experiments on white light in 1666. When he let a beam of white light penetrate a dispersive prism he observed, to his own surprise, the color spectrum of the rainbow on a screen behind the prism. Furthermore, he could recompose the white light from the multicolored spectrum by means of a lens and a second prism. He concluded that different refraction angles are responsible for the appearance of the color spectrum behind a prism. Moreover, he observed that the color does not change when a colored beam is reflected from different objects. Hence, he concluded that color is an intrinsic property of light and that the observed color of an object results from the interaction of its surface with already colored light. He also concluded that white light is composed of all colors and that the white color is not a physical property of light, but an interpretation of the human eye and brain. This conclusion is known as Newton's theory of colors. Newton published his results for the first time in 1672 in a journal of the Royal Society and he published a full account of all his optical experiments and theories in 1704 in a book entitled *Opticks: or, a Treatise of the Reflections, Refractions, Inflections and Colours of Light.*

As might be expected for a revolutionary theory, Newton's publication provoked numerous critical comments. However, most of the critique concerned Newton's understanding of light as corpuscular radiation (see Sect. 10.1). His theory also stated that different colors were a consequence of corpuscles of different size. Newton rejected the wave theory of light favored by Christian Huygens (1629–1695), but ironically it is the wave character of light that is responsible for refraction, yielding the color spectrum observed by Newton. Nonetheless, it remains true for all time that white light is a combination of numerous monochromatic, colored bundles of light. Towards the end of the eighteenth century more and more scientists and laics accepted this aspect of Newton's theory of light.

An early criticism of Newton's refraction experiment appeared in a publication by the French mathematician Louis B. Castel (1688–1753) in 1740, in which he notes that the sequence of colors split by a prism depends on the distance of the screen from the prism. He complained that Newton over-interpreted a special case. Goethe repeated these experiments and joined the criticism of Castel. Goethe's own comment on his first experiments with a prism was as follows:

> Along with the rest of the world I was convinced that all the colors were contained in the light; no one had ever told me anything different, and I had never doubt it, because I had no further interest in the subject. ... But how I was astonished as I looked at a white wall through the prism, that it stayed white! That only where it came upon some darkened area, it showed some color, the at last, around the window sill all the colors shone. ... It did not take long before I knew here was something about color to be brought forth, and I spoke as through an instinct out loud, that Newtonian teaching was false.

Finally, Goethe adhered to the old-fashioned theory of light and colors cited at the beginning of this chapter. He was even convinced that his theory of light and colors was the most important achievement of his life. His secretary J. Eckermann recalled the following statement:

> As to what I have done as a poet... I take no pride in it... but that in my century I am the only person who knows the truth in the difficult science of colors—of that, I say I am not a little proud, and here I have a consciousness of a superiority of many.

At this point it is necessary to emphasize that Goethe's work on colors was not confined to the controversy with Newton. His theory of colors, including his famous "wheel of colors" concerned various aspects, such as allegorical, symbolic, mystic, emotional, and psychological aspects. Therefore, Goethe's work on colors impressed numerous painters, poets, philosophers, and even naturalists. Numerous more or less positive comments were published over the following two centuries. Criticism mainly came from the side of scientists. Best known are the *Remarks on Colors* published by the science theoretician Ludwig Wittgenstein in 1977. One example should be cited:

> Goethe's theory of the origin of the spectrum is not a theory of its origin that has proven unsatisfactory, it is really not a theory at all. Nothing can be predicted by means of it. It is rather a vague schematic outline of the sort we find in James's psychology. There is no *experimentum crucis* for Goethe's theory of colors.

A well balanced comment was contributed by Werner Heisenberg in1952:

> Goethe's color theory has in many ways born fruit in art, psychology and aesthetics. But victory, and hence influence on the research of the following century, has been Newton's.

Finally, the question asked in the title of this chapter should be answered. First, the difference between a mixing of colored lights and a mixing of dyestuffs or pigments should be explained. When monochromatic light of complementary colors, such as red and green, is mixed, the human eye sees white light although real white light is a combination of all wavelength between 380 and 780 nm. When red and green dyestuffs (or pigments) are mixed, the result is usually a dark, dirty "color" reflecting almost no light. A mixture of blue and yellow dyestuffs usually appears green. The yellow dyestuff absorbs blue and violet light, while the blue dyestuff absorbs the yellow and red light, but both dyestuffs enable at least a partial refection of green wavelengths. Goethe did not know this difference and this lack of knowledge is one reason why Goethe misunderstood the physical nature of colors.

The question of how the human eye responds to incident colored light was first answered by the British oculist and physicist Thomas Young (1773–1829). He assumed that all observable color tones are based on three fundamental colors, which three different types of receptor cells are responsible for. The naturalist and physicist Hermann von Helmholtz (1821–1894) elaborated a somewhat modified and more detailed version of this "three-color theory." In the second half of the twentieth century, this speculative theory was confirmed and most details of the complex process of color vision were elucidated.

The photosensitive layer of the retina contains two types of photosensitive cells, the rods and the cones. The rods are located outside the *fovea centralis*, a pit in the center of the retina responsible for sharp central vision. The rods operate in dim light (and are saturated in full daylight) and their function is the distinction between light and dark or white and black. The cones are concentrated in and around the fovea and their function is the perception of colors.

The cones exist in three versions, which differ by their absorption maximum (AM):

L-cones: AM = 560 nm (yellow)
M-cones: AM = 530 nm (yellowish green)
S-cones: AM = 420 nm (blue)

Correct color vision requires excitation of all three versions of cones. Depending on the color of the incident light, the extent of the excitation of the three different versions is different. This variation in excitation is interpreted by the brain in terms of different colors. The photosensitive molecule responsible for the interaction with incident light is the same in all cones, namely *cis*-11-retinal (see Fig. 10.2). The aldehyde group (–CH = O) is attached to a polypeptide (called opsin) in the form of

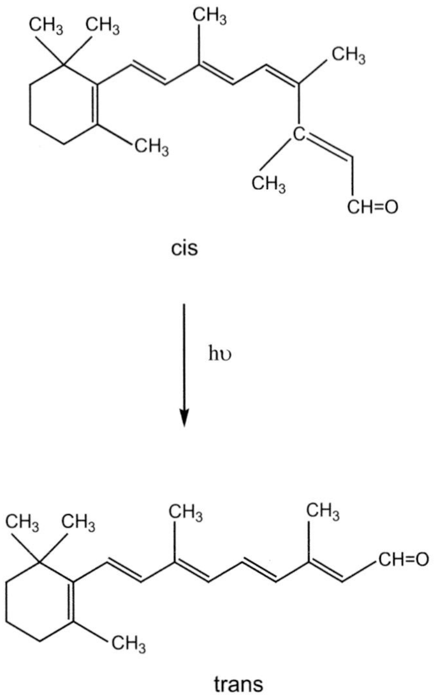

Fig. 10.2 Light-induced *cis–trans* isomerization of retinal

a Schiff base (–CH = N–). The absorption maxima of the three different versions of cones depend on the structure of the opsin peptides surrounding the retina. When the retinal absorbs a photon, it undergoes a *cis–trans* rearrangement, which does not change the chemical composition (see Fig. 10.2). Yet, this *cis–trans* isomerization changes the steric (spatial) structure of the retinal and this change, this motion, is transmitted to an attached opsin molecule. There follows a cascade of reactions resulting in excitation of the optical nerve. The *cis*-form of retinal is restored in the absence of light, and the production of retinal in the human body requires a supply of A-type vitamins.

Isaac Newton

Isaac Newton was born on 25 December 1642, according to the Julian calendar used in England at that time. His birth place was Woolsthorpe Manor in Woolsthorpe-by-Colsterworth, a village in Lincolnshire. He was born 3 months after the death of his father, a wealthy farmer, and because he was prematurely born, he was a small child. When Newton was 3 years old, his mother Hannah Ayscough married the Reverend Barnabas Smith and left her son in the care of his grandmother Margery Ayscough. The young Newton maintained some enmity against his mother and his stepfather for many years, and, whatever the reason, he never married. At the age of 12, Newton entered The King's School, Grantham, where he learned Latin but no mathematics. His mother (again a widow) removed him from the school to make a farmer of him. However, Newton disliked farming and his former teacher persuaded his mother to send him back to school, where he became a top-ranking student.

In June 1661, he joined Trinity College, Cambridge, paying his way by performing valet's duties until he received a scholarship in 1684, which funded his studies for another 4 years. His classes were originally focused on Aristotle's work, but Newton also studied modern philosophers (e.g., Descartes) and became interested in the work of Johann Kepler (1571–1630) and Galileo Galilei (1564–1642). Early in 1665, he developed the generalized binomial theorem and began to develop a mathematical theory that formed the basis of what was later called "fluxions" or "calculus." Shortly after earning his bachelor's degree in August 1665, Trinity College temporarily closed because of the Great Plague (Black Death; infection with *Yersinia pestis*). Newton continued his private studies of mathematical theories, gravitation, and optics at home until he could return in April 1667. He received the master's degree and in October 1667 he was elected Fellow of Trinity College.

Fellows of Trinity College had to become ordained priests, but in the Restoration years an assertion of conformation to the Anglican Church sufficed. He signed the commitment: "I will either set theology as the object of my studies and will take holy orders when the time prescribed by these statutes [7 years] arrives, or I shall resign from the college." Because Newton's religious views did not fully agree with those of the Anglican Church, his position was endangered. However, the Lucasian professor Isaac Barrow, who was impressed by Newton's work, helped him to become his successor when Barrow himself was elected Master of the College. The terms of the Lucasian professorship required that he not be active in the church and so Newton asked for exemption from ordination as priest. King Charles II gave permission and so conflict with the Anglican orthodoxy was avoided.

At almost the same time, Newton developed the first functional reflecting telescope, which was completed in 1668 and presented to the Royal Society in 1671. However, the mirror of his telescope did not possess a parabolic curvature, thus, the performance of his telescope did not exceed that of the best refractometers of his time. In the following years, he published his work on mathematics, gravitation, and optical phenomena in several books:

1671, *Method of Fluxions*
1675, *Hypotheses of Light*
1684, *De Motu Corporum in Gyrum*
1687, *Philosophiae Naturalis Principia Mathematica*
1704, *Opticks*, an expansion of a previous article entitled "Of Colors"
1707, *Arithmetica Universalis*
1733, *De Mundi Systemate* (posthumous)

Newton also published after 1690 several religious tracts dealing with interpretations of the Bible. However, it was his work on mathematics, *Principia*, that was responsible for his international recognition.

Newton's official career and reputation was not only based on his work as mathematician and physicist, it also included a political role as Member of the

Parliament of England, representing the University of Cambridge in the years 1689–1690 and 1701–1702. More important for the rest of his life was an appointment as Warden of the Royal Mint, which he received in 1696 by the patronage of the First Earl of Halifax. He retired from his duties in Cambridge and held this position for 30 years until his death in 1727. Newton received a high income from this position. He moved to London and purchased a house where he could install a small laboratory, allowing him to continue his alchemistic experiments. However, Newton spent most of his time and energy on his new duties, exercising his power to reform the currency. Furthermore, he consequently attempted to identify and punish clippers and counterfeiters and, thus, Newton became the best-known Master of the Mint.

During the last years of his life Newton lived together with his niece and her husband in Cranbury Park, near Winchester, while a half-niece cared for his house in London. Newton died on 20 March 1727 while sleeping. He was buried in Westminster Abbey. A later analysis of his hair revealed poisoning by mercury, probably resulting from his chemical experiments.

Newton's fame is mainly based on his contributions to mathematics and mechanics, including his work on gravitation. In the case of mathematics, he was and is praised in the UK as the inventor of calculus. However, the German philosopher and scientist Wilhelm G. Leibnitz (1646–1716) developed a similar and, for mathematicians, more convenient method at the same time. Leibnitz published a full account of his work in 1684, whereas Newton made his approach public in 1693 and 1704. However, his book *Principia* (1687) contains elements of calculus in geometric form. After 1698, members of the Royal Society including Newton himself accused Leibnitz of plagiarism, because Leibnitz had visited Newton before 1684, but this accusation lacks any evidence. This bitter controversy lasted until the death of Leibnitz.

Almost all modern historians agree that both Newton and Leibnitz developed their different mathematical approaches independently. A modern summary of Newton's mathematics and its impact was written by Tom Whiteside and published in 2013 as a chapter in a book edited by Z. Bechler.

Newton's interest in (celestial) mechanics was stimulated by two events. First, in the winter of 1680/1681 a comet appeared, about which he corresponded with the astronomer John Flamsteed (1646–1719). Second, Newton exchanged letters with the physicist Robert Hooke (1635–1703), who was in charge of managing the Royal Society's correspondence at that time. He learned from Hooke that the attractive force between two masses decreases with the square of their distance. Yet, when Newton published his famous law of gravitation (a great advance and paradigm shift in physics and cosmology, see Sect. 10.1) he did not mention Hooke's contribution and Hooke accused him of plagiarism. Previously, he had a fierce dispute with Christian Huygens (1629–1695) and Robert Hooke about the nature of light. There followed a severe controversy with Flamsteed, resulting in a law suit. Newton, President of the Royal Society at that time, intended to publish a book containing maps of stars that were based on the work of Flamsteed without having the agreement of this author in hand. Flamsteed won the judgment, proving that

Newton's egoistic behavior was incorrect. Another controversy concerned the Jesuits in Liege (Belgium), and when his mother died in the same year Newton had a nervous breakdown. All these controversies and his aggressive and sometimes unfair attitude towards colleagues shed a negative light on Newton's character.

Nonetheless, Newton always was and is an outstanding scientific hero of the British nation. He was knighted by Queen Anne in 1705, mainly for his political activities and not so much for his scientific achievements. After his death, the English poet Alexander Pope (1688–1744) wrote on his marble epitaph the spectacular words:

> Nature and nature's laws lay hid in night
> God said "Let Newton be" and all was light

Newton himself was more moderate about his own work and wrote:

> I do not know what I may appear to the world, but to myself I seem to have been only like a boy playing on the sea-shore, and diverting myself in now and then finding a smoother pebble or a prettier shell than ordinary, whilst the great ocean of truth lay all undiscovered before me.

A statue of Newton was erected at the Oxford University Museum of Natural History and another statue adorns the piazza of the British Library in London. An image of Newton was printed from 1978 to 1988 on £1 banknotes of the Bank of England.

Bibliography

Critical review of Goethe's theory of colours. http://www.handprint.com/HP/WCL/book3.html#goethe

De Villamil R (1931) Newton the man. G. D. Knox, Preface by Albert Einstein, Reprint by Johnson Reprint Corporation, New York (1972)

Duck M (1988) Newton and Goethe on colour. Ann Sci 45(5): 507

Goethe JW (1982) Theory of colours. Translated by Eastlake CL. MIT Press, Cambridge, MA

Manuel FE (1968) A portrait of Isaac Newton. Belknap Press of Harvard University, Cambridge, MA

Sepper DL (1986) Goethe contra Newton. Cambridge University Press, Cambridge.

Wandell BA(1995) Foundation of vision. Sinauer Ass., Sunderland, MA

Wässle H, Boycott BB (1991) Functional architecture of the mammalian retina. Physiol Rev 71(2):447

Westfall RS (2007) Isaac Newton. Cambridge University Press

10.6 How Stable Is the Earth's Crust?

> The ultimate court of appeal is observation and experiment and not authority.
> (Thomas Huxley)

Until the second half of the nineteenth century, the Catholic and the Orthodox Churches advised their faithful that God had created the world a few thousand years before. Based on the dynasties and kings mentioned in the Bible, the Irish Archbishop James Ussher (1581–1656) had calculated the exact date of the creation as 22 October 4004 B.C. according to the Julian calendar. As described in the first chapter of the Bible (Genesis), all the innumerable living and dead components of the world were created within one week and remained essentially unchanged over the past thousands of years. This religious cosmology not only ignored the evolution of plants, animals, and humans, it also denied any significant change or modification in the earth's surface. The existence of volcanoes was undeniable. However, the few active volcanoes known before the nineteenth century were not assumed to have a noteworthy influence on the earth's crust.

Nonetheless, numerous facts and observations concerning the structure and dynamics of the earth's surface had accumulated before the middle of the eighteenth century and formed the basis of a new branch of the natural sciences called geology (*gaia* means earth in ancient Greek). This term was coined by two Swiss naturalists, Jean-Andre Deluc (1727–1817) and Horace-Benedict de Saussure (1740–1799). The roots of this new science can be traced back to the beginning of copper mining and copper production in Anatolia around 7000 B.C. when humankind began to explore the usefulness of inorganic materials, such as minerals and metal ores, hidden under the earth's surface. The ancient Greeks acquired information and knowledge about the earth's surface for two reasons. On the one hand, they were interested in mineral resources like other nation and, on the other hand, they were interested in all kinds of geographical information because of their extensive and far-reaching trading activities. Within the framework of their natural philosophy, the Greeks tried to understand the origins and activities of volcanoes and earthquakes, in as much as their settlements in southern Italy and Sicily were endangered by these natural phenomena. Furthermore, Empedokles and Aristotle defined the meaning of elements and discussed their potential transmutations (see Sect. 9.1). Aristotle also speculated that the surface of the earth is subject to slow but permanent changes. Theophrastus (371–287 B.C.), his successor at the Lyceum, published a work *On Stones*, which presented for the first time a description and classification of ores and minerals. In their huge empire, the Romans systematically explored and exploited all kinds of earth treasures, and they considerably improved mining technology. Their activities included, for instance, mining and production of iron, copper, tin, lead, gold, and silver. Furthermore, Pliny the Elder published an extensive description and discussion of many more metals and minerals then used for practical purposes.

In the course of the migration of nations that destroyed the Roman Empire and under the influence of the Catholic Church, almost all the knowledge and techniques known before in central and western Europe were lost. A new upward trend in scientific thinking and knowledge began with the Renaissance after the thirteenth century. Inquisitive, avid researchers appeared on the scene who were not willing to confine their thinking and curiosity by religious dogmas. At the same time, mining and mining technology experienced an enormous impetus when copious deposits of silver were detected in the Valley of Inn (south of Munich in the Alps) and in the Erzgebirge (mountains along the border of Germany and the Czech Republic). The mining activities in those areas also entailed production of copper, lead, tin, and zinc. The improved access to new and known metals, ores, and minerals provided the alchemists with more and new chemicals. New experiments improved, in turn, the knowledge and understanding of rocks and minerals. Among the most widely known and famous naturalists of that time were the following:

Theophrastus Bombastus von Hohenheim (called Paracelsus, 1493–1541), who had a high reputation as physician and became famous for his book *Astronomia Magna* and for the first and still-valid definition of poison (*dosis sola facit venenum*: the quantity alone is decisive for the poisonous effect).

Georgius Agricola (1494–1555), who published influential books such as *De Ortu et Causis Subterraneorum*, *De Natura Fossilium*, and *De Veteris et Novis Metallis*.

Johann Joachim Becher (1635–1682), who wrote the famous book *Physica Subterranea*.

In 1669, the Danish naturalist Nicolaus Steno (1638–1686) invented the method of stratigraphy. This method is based on the law of nature that the spatial sequence of sediments or rock layers is a consequence of the timely sequence of events responsible for their formation. Almost at the same time, the British scientist Robert Hooke (1635–1703) postulated that the fossils embedded in sediments are indicative of the age of the sediments.

This hypothesis suggested for the first time an evolution of the earth's crust. However, this scientific view was not accepted by religious people. They accepted the existence of fossils in sediments, but they considered these fossils as evidence for the Deluge, as described in the Bible. In 1696, William Whiston (1667–1752) published a widely accepted book *A New Theory of Earth*, in which he attempted to prove that the Great Flood had indeed occurred. In summary, the seventeenth century witnessed the birth of new ideas, concepts, and experimental facts that formed the basis of geology as a modern science.

In the eighteenth century, the experimental basis of geology was broadened by mining engineers who developed the methods of mapping mines and of drawing stratigraphic profiles. Mining and, thus, accurate knowledge about ores and their natural distribution became of increasing economic interest for most European nations. A representative book covering this topic was published in 1774 by Abraham G. Werner (1749–1817) under the title *Von den äusserlichen Kennzeichen der Fossilien* (On the External Characters of Fossils). Werner's work also contained an indirect defense of the Deluge theory, because he believed that all rock layers and sediments, including granite and basalt, had precipitated from a large ocean. His work was influential and founded the theory of "Neptunism."

However, an alternative theory called "Plutonism" slowly gained increasing acceptance. This theory originated from the work *Histoire Naturelle* published by the French naturalist George-L. Leclerc, Comte de Buffon (1707–1780) in 1749 and it was expanded by the Scottish naturalist James Hutton (1726–1797) in the 1780s.

The Plutonists believed that volcanic processes were the driving force behind rock formation and that the earth's crust at that time was the result of a slow cooling process that still continued. According to this theory, the earth was at least 75,000 years old and not about 5000 years as inferred from the Bible. A fierce discussion about the age of earth continued until the end of the nineteenth century. The dispute between Neptunists and Plutonists was later nicknamed the first debate in the history of geology.

In the beginning of the nineteenth century, stratigraphy was turned into a highly informative scientific method. It helped to elucidate the history of the earth's crust and to establish the theory of evolution. The anatomists Georges Cuvier (1769–1832) and Alexandre Brongniart (1770–1847), working at the Ecole des Mines de Paris, found that the relative ages of fossils can be determined from the sequence of rock layers and the distance these rock layers have from the surface of the earth. They also realized that different strata could be identified by their characteristic fossil content.

Their book *Description Geologiques des Environ de Paris* published in 1811, made stratigraphy a very popular analytical method. At the same time, the mining surveyor William Smith (1769–1839) reached almost the same conclusions. He traveled Britain from north to south and from east to west working on a canal system. He elaborated the first geographical map of Britain (published in 1815), and published in 1816 a book entitled *Strata Identified by Organized Fossils*. He had observed that fossils were an effective means of distinguishing between otherwise similar formations of the landscape, and concluded that the distinct fossils in a stratum can be used to identify analogous strata (resulting from the same geological event) across regions. In 1831, the President of the Geological Society, Adam Sedgwick awarded Smith the first Wollaston Medal and called him the "Father of English geology."The years after 1830 witnessed the second debate in the history of geology. Cuvier, strongly influenced by the Plutonism of Comte de Buffon, developed a theory labeled "catastrophism." This theory held that the earth's surface was in the past strongly influenced by short-lived violent events, such as the rapid formation of mountain chains (including volcanic eruptions) and great floods. This theory was at least partially in agreement with the religious belief in Noah's flood, but Cuvier avoided any religious speculation in his treatises.

The opposite theory called "uniformitarianism" (or "gradualism") was formulated and advocated by the British geologist Charles Lyell (1797–1850). In his multivolume work *Principles of Geology*, Lyell proposed that all modifications of the earth's surface were consequences of slow incremental changes such as erosion. Uniformitarianism also held that the present is the key to the past and that all recent events and changes proceeded as in the distant past. Research in the twentieth century has clarified that the evolution of the earth's crust combines elements of both theories.

Lyell's work became very popular among geologists and his concept of slow, gradual changes was decisive for the formulation of the theory of evolution by Charles Darwin and Alfred R. Wallace (see Sect. 8.3). Darwin was also involved in a fierce discussion with William Thomson, First Baron Kelvin (1824–1907), about the age of the earth. Baron Kelvin was physicist and famous for his work on thermodynamics. This work included investigation of heat exchange between different states of matter. He stated: "This earth, certainly a moderate number of million years ago, was a red hot globe." From the cooling rate of the earth he finally estimated it to be approximately 20–30 million years old. Darwin never published an exact figure, but his understanding of sediments and evolution of animals required an age at least a ten times higher. In the early twentieth century, measurements of the radioactivity of various minerals and rocks shifted the earth's age to a figure of 2 billion years. At the end of the twentieth century, an age of around 4.5 billion years was widely accepted, a million times the age calculated by Bishop James Ussher.

In the twentieth century, geology was also confronted by another paradigm change. This paradigm change was caused by the German meteorologist Alfred Wegener (see Biography), who had studied astronomy, meteorology, and physics. In 1905 and 1912/1913 he accompanied two expeditions of Danish scientists to Greenland and died in 1930 during a third expedition headed by himself.

Wegener not only analyzed the shores of Greenland with regard to rock type, geological structure, and fossils, he also included the shores of western Europe, Africa, and North and South America in his studies. He found significant similarities between matching sites of the continents and developed the theory of "continental drift." This concept assumed that perhaps one billion years ago all continents were combined in one huge landmass named "Pangaea." Thereafter, they began to drift over the ocean floor, finally reaching their present positions. He speculated on the spreading of the sea floor: "The Mid-Atlantic Ridge should be regarded as a zone in which the floor of the Atlantic, as it keeps spreading, is continuously tearing open and making space for fresh, relatively fluid and hot sima [rising] from depth." This speculation later proved correct, but Wegener never repeated it in later editions of his book.

Wegener published a treatise with a first version of his concept in January 1912. Three improved and expanded editions appeared later as books under the title *Die Entstehung der Kontinente und Ozeane* (The Origin of Continents and Oceans). His first article and the first edition of his book (1915) were largely ignored because World War I had just broken out.

The second edition of his book, which appeared in 1922, initiated international discussion of his theory. A large number of skeptical or critical comments followed. A poor translation of the second edition of his book appeared in 1925 in the USA and had such a negative response that the American Organization of Petroleum Geologists organized a conference dedicated to criticism of the continental drift theory. The paradigm valid at that time can be summarized in three points:

1. Continents may slowly rise or sink a little, but the earth's crust is too compact to allow migration of continents
2. The shrinking of mountains upon rapid cooling is responsible for their folding

3. Almost identical fossils in neighboring continents separated by an ocean (e.g., west coast of Africa and east coast of South America) are the consequence of an ancient temporary land bridge

Wegener argued against point 1, stating that granite and gneiss, the main constituents of the continents, had a lower specific weight than the basalt of the sea floor, so that sinking was impossible. A rapid cooling of the earth's crust (point 2) was unlikely, because the radioactive decay of certain elements in the earth's crust produced heat (these processes were detected after World War I). The existence of land bridges (point 3) was disproven by the German research ship *Meteor* in the years 1924–1927 on the basis of echo depth measurements (unexpectedly, the Mid-Atlantic Ridge was detected then).

Wegener had several supporters, such as the paleontologist Edgar Dacqué (1878–1945), the astronomer Milutin Milankovic (1879–1958), and the geologist Alexander du Toit (1878–1948). Nonetheless, Wegener's theory was never fully accepted during his lifetime. For instance, his scientific enemy, the German geologist Franz Kosmat (1871–1939), stated that the earth's crust was too firm for the continents to "simply plough through." The American paleontologist Charles Schuchert (1858–1942) criticized that the coastlines were largely modified over millions of years by erosion and by the mouthing of rivers. However, he, like other critics, ignored the fact that Wegener's comparison of coastlines was based on the 200-m isobath (including the shelves) and not on their present form. An ironic compliment was contributed by the French geologist Pierre-Marie Termier (1859–1930) who said: "His theory is a wonderful dream of beauty and charm, the dream of a great poet."

In fact, Wegener's concept had a weak point from the very beginning, because a convincing explanation for the driving force behind the continental drift was lacking. At first (1912 and 1915), Wegener proposed that centrifugal forces resulting from the Earth's rotation or the astronomical precession caused the drift. However, the British geologist Harold Jeffreys (1891–1989) demonstrated in 1924 that all these forces were too weak. In the third edition of his book (1930), Wegener picked up a hypothesis of an Austrian colleague, Robert Schwinner (1878–1953), who postulated that a mobile phase of molten rocks existed beneath the earth's crust. Wegener speculated that a steady flow of this liquid layer caused the continental drift and this speculation came indeed close to reality, as evidenced by various scientists after World War II. The American geologists Bruce C. Heezen (1924–1977), Robert S. Dietz (1914–1995), and Harry H. Hess (1906–1969) found that the sea floor neighboring mid-ocean ridges was younger than the continents as a consequence of a permanent spreading of the seafloor away from the mid-ocean ridges. These findings led directly to the theory of plate tectonics, which confirmed Wegener's concept.

Alfred Wegener

Wegener was born on 1 November 1880 in Berlin as the youngest of five children of Richard Wegener and his wife. His father was priest and teacher of classical languages at the Berlin Gymnasium *Zum Grauen Kloster* (high school "Grey

Monastery"). In 1886, his family purchased a former manor house in the tiny village of Zechlinerhütte near the city of Reinsberg, where he spent most of his early years. He attended the Köllnisches Gymnasium in Berlin where he graduated as the best of his class. He began to study astronomy, meteorology, and physics in Berlin, attended for several months lectures at the Universities of Heidelberg and Innsbruck (Austria), and then returned to Berlin. In parallel to his studies, Wegener was assistant at the Urania astronomical observatory in the years 1902 and 1903. He obtained the doctoral degree in 1905 from the Friedrich Wilhelm University (today Humboldt University) with a dissertation in the field of astronomy. From this event on, Wegener concentrated his interest for the rest of his life on meteorology and physics.

Still in 1905, Wegener was appointed assistant at the Aeronautisches Observatorium Lindenberg near Beeskow. Here he joined his older brother Kurt, who was likewise working in the field of meteorology and polar research. The two tested the usefulness of balloon flights for retrieving meteorological data and for testing a new celestial navigation method based on a special quadrant. On 5–7 April 1906, the Wegener brothers set a new record for balloon flights with 52.5 hours remaining aloft. Later that year, A. Wegener participated in his first Greenland expedition, with the purpose of exploring the last unknown part of the northeastern coast. During this time in Greenland, Wegener constructed the first meteorological station near Danmarkshavn, where he launched kites and balloons to collect meteorological data. Unfortunately, the expedition came to a premature end when the leader, the Danish meteorologist Ludvig Mylius-Erichsen, and two colleagues died on an exploratory trip undertaken with sled dogs.

Between 1809 and the outbreak of World War I, Wegener was lecturer (Privatdozent) in astronomy, meteorology, and cosmic physics at the University of Marburg. His lectures became the basis of a textbook in meteorology, which included many of the results of the first Greenland expedition. This textbook was first published in 1910 under the title *Thermodynamik der Atmosphäre* (Thermodynamics of the Atmosphere). In 1912, he presented for the first time his concept of continental drift in a lecture at a session of the *Geologische Vereinigung* at the Senckenberg Museum (Frankfurt am Main), and published his thoughts in three articles in the journal *Petermanns Geographische Mitteilungen.*

Between the fall of 1912 and summer of 1913, Wegener participated in a second Greenland expedition. At the first stopover in Iceland, where he tested and eventually purchased ponies as pack animals, the expedition reached Danmarkshavn. Shortly before the start of exploration of the inland ice, the expedition was hit by a calving glacier, but fortunately nobody was killed. However, during the trip to the inland glaciers, the leader, the Dane Johan P. Koch, broke a leg and had to spend some months in bed. Therefore, Koch and Wegener were the first to spend a winter on the inland ice in northeast Greenland. In the summer of 1913, the complete team of four meteorologists started to cross the inland ice, but only a few kilometers from Kangersuatsiaq, a settlement on the west coast, the group ran out of food. They were struggling to find a route across a difficult glacier terrain and had eaten the last pony and the last dog. By chance they were picked up at a fjord by the clergyman of

Upernavik, who was on a trip to a remote congregation. After his return, Wegener resumed his lectureship in Marburg and, still in 1913, married Else Köppen, daughter of his former supervisor Wladimir Köppen.

When World War I began in August 1914, Wegener was immediately called up as an infantry reserve officer. He was involved in fierce fighting on the front in Belgium and after a few months was wounded twice. He was declared unfit for active service and eventually served in the army weather service. This position required traveling between numerous weather stations in Germany, along the Western Front, in the Baltic region, and on the Balkan. Fortunately, he had enough spare time to complete the first version of his major work *Die Entstehung der Kontinente und Ozeane* (The Origin of Continents and Oceans).

By the end of the war, Wegener had published about 20 more meteorological and physical papers, but due to the war his publications attracted little interest. After the war, Wegener's first position was that of meteorologist at the Deutsche Seewarte (German Naval Observatory) in Hamburg. Therefore, he moved with his wife and two daughters to Hamburg, where he was appointed associate professor at the new university in 1921. In 1922, Wegener published the third, fully revised, version of *The Origin of Continents and Oceans*. Discussion began on his concept of continental drift, at first in German-speaking countries and later internationally. In the years 1919–1923, he also worked on a book entitled *Die Klimate der Geologischen Vorzeit* (The Climates of the Geological Past), which was eventually published together with his father in law, W. Köppen.

In 1924, Wegener was appointed full professor in meteorology and geophysics at the University of Graz (Austria). He concentrated on studies of the atmosphere, including studies of tornados, and continued assessment of data from his second Greenland expedition. In 1926, Wegener published the fourth and final edition of *The Origin of Continents*. In the same year he made a short trip to Greenland, which included the testing of a new propeller-driven snowmobile, to lay the basis for a later expedition.

This third expedition started in 1930 with 14 participants under the leadership of A. Wegener. The main purpose of this expedition was the building of three permanent stations for year-round measurement of weather and of Greenland's ice sheet. The success of the expedition, which was financed by the German government, depended on the activities of two men who had to winter in the station *Eismitte* (Mid-Ice). Early in September, the crew of *Eismitte* sent a message to the West camp that they were running out of fuel. On 24 September, Wegener set out with the meteorologist Fritz Loewe and thirteen Greenlanders to supply the camp by means of dog sleds. The temperatures reached −60 °C and Loewe's feet became so frozen that his toes had to be amputated without anesthetic. He and 12 Greenlanders returned to West camp. Wegener and two other members arrived at *Eismitte*. Due to a shortage of food, Wegener and Rasmus Villumsen decided to return immediately with two dog sleds, but without any food for the dogs. They killed them one by one. However, they never returned.

On 12 May 1931, another expedition found Wegener's body halfway between *Eismitte* and West camp. Apparently he had died of a heart attack caused by overexertion. He had been carefully buried by Villumsen and the grave site marked with his skis. Wegener was reburied and the grave marked with a large cross. The body of Villumsen and Wegener's diary were never found.

Wegener belonged to the small group of famous scientists who never received any honor or award during their lifetime, but his recognition and fame grew steadily after his death.

In Bremerhaven, Germany, the Alfred Wegener Institute was established on his centenary in 1980. It awards the A. Wegener Medal in his name. Furthermore, the European Geosciences Union awards an A. Wegener Medal and Honorary Membership "for scientists who have achieved exceptional international standing in atmospheric, hydrological or ocean sciences, defined in their widest sense, for their merit and their scientific achievements." A crater on the Moon, a crater on Mars, the asteroid 29227, and the peninsula where he died (near Uummannaq, Greenland) were named after him.

Bibliography

Bowler PJ (2000) The earth encompassed: A history of environmental sciences, Norton, New York

Dansgaard W (2004) Frozen annals: Greenland ice sheet research. Narayana Press, Odder, Denmark

Erickson J (1992) Plate tectonics. Facts on File, New York

Frankel H (1987) The continental drift debate. In: Engelhardt Jr HT, Caplan AL (eds) Scientific controversies: Case solutions in the resolution and closure of disputes in science and technology. Cambridge University Press

Hullam A (1975) Alfred Wegener and the hypothesis of continental drift. Scientific American (2)

Jacoby WR (1981) Modern concepts of earth dynamics anticipated by Alfred Wegener in 1912. Geology 9:25

Palmer T (1994) Catastrophism, neocatastrophism and evolution. Society for Interdisciplinary Studies in Association with Nottingham Trent University

Rudwick MJS (1972) The meaning of fossils. The University of Chicago Press

Simon W (2001) The map that changed the world: William Smith and the birth of modern geology. Harper Collins, New York.

Spaulding NE, Namowitz SN (2005) Earth science. McDougall Littell, Boston, MA

Chapter 11
Concluding Remarks

From Albert Einstein came following insight: "I think that only daring speculation can bring us further and not accumulation of facts." However, the work of C. Darwin, A.R. Wallace, A.L. Lavoisier, H. Staudinger, W.H. Carothers, E. Hubble, and A. Wegener are examples illustrating that Einstein's insight is extremely one-sided, not to say wrong. New hypotheses, the "daring speculation" in Einstein's terminology, start out from accumulated facts and need confirmation by new experimental facts to become a reliable stage in the progress of science. Einstein's work is no exception. His new ideas did not emerge from nowhere. He had studied physics (including astronomy) and mathematics and thereby learned the experimental and theoretical facts elaborated by other scientists and mathematicians before him. Furthermore, it was the experimental confirmation of his theory of general gravity by the astronomer A. Eddington in 1919 that turned Einstein´s daring speculation into a corner stone of science.

Successful scientific research is based on an interplay between elaboration of experimental facts and intuitive invention of new concepts. A slightly modified version of an aphorism of the French mathematician and physicist Jules H. Poincaré (1854–1912) describes the reality of scientific research much better than Einstein's words: "It is through experiments that we prove, but through intuition that we discover." A complementary insight is reported from the American physician Jonas Salk (1914–1995): "Intuition will tell the thinking mind where to look next." Where the scientists have to look next are experiments that confirm intuitive speculations and revise previous mistakes and fallacies.

H.R. Kricheldorf, *Getting It Right in Science and Medicine*,
DOI 10.1007/978-3-319-30388-8_11

Index

H.R. Kricheldorf, *Getting It Right in Science and Medicine*,
DOI 10.1007/978-3-319-30388-8

Printed in the United States
By Bookmasters